T0335654

Information Systems for the Fashion and Apparel Industry

The Textile Institute and Woodhead Publishing

The Textile Institute is a unique organisation in textiles, clothing and footwear. Incorporated in England by a Royal Charter granted in 1925, the Institute has individual and corporate members in over 90 countries. The aim of the Institute is to facilitate learning, recognise achievement, reward excellence and disseminate information within the global textiles, clothing and footwear industries.

Historically, The Textile Institute has published books of interest to its members and the textile industry. To maintain this policy, the Institute has entered into partnership with Woodhead Publishing Limited to ensure that Institute members and the textile industry continue to have access to high calibre titles on textile science and technology.

Most Woodhead titles on textiles are now published in collaboration with The Textile Institute. Through this arrangement, the Institute provides an Editorial Board which advises Woodhead on appropriate titles for future publication and suggests possible editors and authors for these books. Each book published under this arrangement carries the Institute's logo.

Woodhead books published in collaboration with The Textile Institute are offered to Textile Institute members at a substantial discount. These books, together with those published by The Textile Institute that are still in print, are offered on the Elsevier website at: http://store.elsevier.com/. Textile Institute books still in print are also available directly from the Institute's web site at: www.textileinstitutebooks.com.

A list of Woodhead books on textiles science and technology, most of which have been published in collaboration with the Textile Institute, can be found towards the end of the contents pages.

Related titles

Fashion Supply Chain Management Using Radio Frequency Identification Technologies
(ISBN: 978-0-85709-805-4)

Design of Clothing Manufacturing Processes
(ISBN: 978-0-85709-778-1)

Optimizing Decision Making in the Apparel Supply Chain Using Artificial Intelligence (AI)
(ISBN: 978-0-85709-779-8)

Woodhead Publishing Series in Textiles: Number 179

Information Systems for the Fashion and Apparel Industry

Edited by

Tsan-Ming Choi

The Textile Institute

AMSTERDAM • BOSTON • CAMBRIDGE • HEIDELBERG
LONDON • NEW YORK • OXFORD • PARIS • SAN DIEGO
SAN FRANCISCO • SINGAPORE • SYDNEY • TOKYO
Woodhead Publishing is an imprint of Elsevier

Published by Woodhead Publishing in association with The Textile Institute
Woodhead Publishing is an imprint of Elsevier
The Officers' Mess Business Centre, Royston Road, Duxford, CB22 4QH, UK
50 Hampshire Street, 5th Floor, Cambridge, MA 02139, USA
The Boulevard, Langford Lane, Kidlington, OX5 1GB, UK

British Library Cataloguing-in-Publication Data
A catalogue record for this book is available from the British Library

Library of Congress Cataloging-in-Publication Data
A catalog record for this book is available from the Library of Congress

ISBN: 978-0-08-100571-2 (print)
ISBN: 978-0-08-100580-4 (online)

Publisher: Matthew Deans
Acquisition Editor: David Jackson
Editorial Project Manager: Edward Payne
Production Project Manager: Omer Mukthar
Designer: Greg Harris

Typeset by TNQ Books and Journals

Contents

List of contributors

L. Banica University of Pitesti, Pitesti, Romania

Brahmadeep University Lille Nord of France, ENSAIT-GEMTEX, 2 allée Louise et Victor Champier, Roubaix, France

H.-L. Chan The Hong Kong Polytechnic University, Kowloon, Hong Kong

H. Chaudhry Lahore University of Management Sciences, Lahore, Pakistan

Y. Chen Soochow University, Suzhou, People's Republic of China

T.-M. Choi The Hong Kong Polytechnic University, Kowloon, Hong Kong

A. Hagiu University of Pitesti, Pitesti, Romania

G. Hodge North Carolina State University, Raleigh, NC, United States

P.C.L. Hui The Hong Kong Polytechnic University, Kowloon, Hong Kong

L. Koehl Ecole Nationale Supérieure des Arts et Industries Textiles, Roubaix, France

X. Li Nankai University, Tianjin, China

Y. Li Nankai University, Tianjin, China

P.Y. Mok The Hong Kong Polytechnic University, Hung Hom, Hong Kong

B. Pan Seamsystemic Design Research, London, UK

S. Ren The Hong Kong Polytechnic University, Kowloon, Hong Kong

D.A. Serel Ipek University, Ankara, Turkey

S. Thomassey University Lille Nord of France, ENSAIT-GEMTEX, 2 allée Louise et Victor Champier, Roubaix, France

L. Wang Soochow University, Suzhou, People's Republic of China

Y.Y. Wu The Hong Kong Polytechnic University, Hung Hom, Hong Kong; Cornell University, Ithaca, NY, United States

F. Xu Nankai University, Tianjin, China

J. Xu The Hong Kong Polytechnic University, Hung Hom, Hong Kong; Wuhan Textile University, Wuhan, China

X. Zeng Ecole Nationale Supérieure des Arts et Industries Textiles, Roubaix, France

Woodhead Publishing Series in Textiles

1. **Watson's textile design and colour Seventh edition**
   Edited by Z. Grosicki
2. **Watson's advanced textile design**
   Edited by Z. Grosicki
3. **Weaving Second edition**
   P. R. Lord and M. H. Mohamed
4. **Handbook of textile fibres Volume 1: Natural fibres**
   J. Gordon Cook
5. **Handbook of textile fibres Volume 2: Man-made fibres**
   J. Gordon Cook
6. **Recycling textile and plastic waste**
   Edited by A. R. Horrocks
7. **New fibers Second edition**
   T. Hongu and G. O. Phillips
8. **Atlas of fibre fracture and damage to textiles Second edition**
   J. W. S. Hearle, B. Lomas and W. D. Cooke
9. **Ecotextile '98**
   Edited by A. R. Horrocks
10. **Physical testing of textiles**
    B. P. Saville
11. **Geometric symmetry in patterns and tilings**
    C. E. Horne
12. **Handbook of technical textiles**
    Edited by A. R. Horrocks and S. C. Anand
13. **Textiles in automotive engineering**
    W. Fung and J. M. Hardcastle
14. **Handbook of textile design**
    J. Wilson
15. **High-performance fibres**
    Edited by J. W. S. Hearle
16. **Knitting technology Third edition**
    D. J. Spencer
17. **Medical textiles**
    Edited by S. C. Anand
18. **Regenerated cellulose fibres**
    Edited by C. Woodings
19. **Silk, mohair, cashmere and other luxury fibres**
    Edited by R. R. Franck

20 **Smart fibres, fabrics and clothing**
Edited by X. M. Tao
21 **Yarn texturing technology**
J. W. S. Hearle, L. Hollick and D. K. Wilson
22 **Encyclopedia of textile finishing**
H.-K. Rouette
23 **Coated and laminated textiles**
W. Fung
24 **Fancy yarns**
R. H. Gong and R. M. Wright
25 **Wool: Science and technology**
Edited by W. S. Simpson and G. Crawshaw
26 **Dictionary of textile finishing**
H.-K. Rouette
27 **Environmental impact of textiles**
K. Slater
28 **Handbook of yarn production**
P. R. Lord
29 **Textile processing with enzymes**
Edited by A. Cavaco-Paulo and G. Gübitz
30 **The China and Hong Kong denim industry**
Y. Li, L. Yao and K. W. Yeung
31 **The World Trade Organization and international denim trading**
Y. Li, Y. Shen, L. Yao and E. Newton
32 **Chemical finishing of textiles**
W. D. Schindler and P. J. Hauser
33 **Clothing appearance and fit**
J. Fan, W. Yu and L. Hunter
34 **Handbook of fibre rope technology**
H. A. McKenna, J. W. S. Hearle and N. O'Hear
35 **Structure and mechanics of woven fabrics**
J. Hu
36 **Synthetic fibres: Nylon, polyester, acrylic, polyolefin**
Edited by J. E. McIntyre
37 **Woollen and worsted woven fabric design**
E. G. Gilligan
38 **Analytical electrochemistry in textiles**
P. Westbroek, G. Priniotakis and P. Kiekens
39 **Bast and other plant fibres**
R. R. Franck
40 **Chemical testing of textiles**
Edited by Q. Fan
41 **Design and manufacture of textile composites**
Edited by A. C. Long
42 **Effect of mechanical and physical properties on fabric hand**
Edited by H. M. Behery
43 **New millennium fibers**
T. Hongu, M. Takigami and G. O. Phillips

44 **Textiles for protection**
Edited by R. A. Scott
45 **Textiles in sport**
Edited by R. Shishoo
46 **Wearable electronics and photonics**
Edited by X. M. Tao
47 **Biodegradable and sustainable fibres**
Edited by R. S. Blackburn
48 **Medical textiles and biomaterials for healthcare**
Edited by S. C. Anand, M. Miraftab, S. Rajendran and J. F. Kennedy
49 **Total colour management in textiles**
Edited by J. Xin
50 **Recycling in textiles**
Edited by Y. Wang
51 **Clothing biosensory engineering**
Y. Li and A. S. W. Wong
52 **Biomechanical engineering of textiles and clothing**
Edited by Y. Li and D. X.-Q. Dai
53 **Digital printing of textiles**
Edited by H. Ujiie
54 **Intelligent textiles and clothing**
Edited by H. R. Mattila
55 **Innovation and technology of women's intimate apparel**
W. Yu, J. Fan, S. C. Harlock and S. P. Ng
56 **Thermal and moisture transport in fibrous materials**
Edited by N. Pan and P. Gibson
57 **Geosynthetics in civil engineering**
Edited by R. W. Sarsby
58 **Handbook of nonwovens**
Edited by S. Russell
59 **Cotton: Science and technology**
Edited by S. Gordon and Y.-L. Hsieh
60 **Ecotextiles**
Edited by M. Miraftab and A. R. Horrocks
61 **Composite forming technologies**
Edited by A. C. Long
62 **Plasma technology for textiles**
Edited by R. Shishoo
63 **Smart textiles for medicine and healthcare**
Edited by L. Van Langenhove
64 **Sizing in clothing**
Edited by S. Ashdown
65 **Shape memory polymers and textiles**
J. Hu
66 **Environmental aspects of textile dyeing**
Edited by R. Christie
67 **Nanofibers and nanotechnology in textiles**
Edited by P. Brown and K. Stevens

68 **Physical properties of textile fibres Fourth edition**
W. E. Morton and J. W. S. Hearle
69 **Advances in apparel production**
Edited by C. Fairhurst
70 **Advances in fire retardant materials**
Edited by A. R. Horrocks and D. Price
71 **Polyesters and polyamides**
Edited by B. L. Deopura, R. Alagirusamy, M. Joshi and B. S. Gupta
72 **Advances in wool technology**
Edited by N. A. G. Johnson and I. Russell
73 **Military textiles**
Edited by E. Wilusz
74 **3D fibrous assemblies: Properties, applications and modelling of three-dimensional textile structures**
J. Hu
75 **Medical and healthcare textiles**
Edited by S. C. Anand, J. F. Kennedy, M. Miraftab and S. Rajendran
76 **Fabric testing**
Edited by J. Hu
77 **Biologically inspired textiles**
Edited by A. Abbott and M. Ellison
78 **Friction in textile materials**
Edited by B. S. Gupta
79 **Textile advances in the automotive industry**
Edited by R. Shishoo
80 **Structure and mechanics of textile fibre assemblies**
Edited by P. Schwartz
81 **Engineering textiles: Integrating the design and manufacture of textile products**
Edited by Y. E. El-Mogahzy
82 **Polyolefin fibres: Industrial and medical applications**
Edited by S. C. O. Ugbolue
83 **Smart clothes and wearable technology**
Edited by J. McCann and D. Bryson
84 **Identification of textile fibres**
Edited by M. Houck
85 **Advanced textiles for wound care**
Edited by S. Rajendran
86 **Fatigue failure of textile fibres**
Edited by M. Miraftab
87 **Advances in carpet technology**
Edited by K. Goswami
88 **Handbook of textile fibre structure Volume 1 and Volume 2**
Edited by S. J. Eichhorn, J. W. S. Hearle, M. Jaffe and T. Kikutani
89 **Advances in knitting technology**
Edited by K.-F. Au
90 **Smart textile coatings and laminates**
Edited by W. C. Smith
91 **Handbook of tensile properties of textile and technical fibres**
Edited by A. R. Bunsell

92 **Interior textiles: Design and developments**
Edited by T. Rowe
93 **Textiles for cold weather apparel**
Edited by J. T. Williams
94 **Modelling and predicting textile behaviour**
Edited by X. Chen
95 **Textiles, polymers and composites for buildings**
Edited by G. Pohl
96 **Engineering apparel fabrics and garments**
J. Fan and L. Hunter
97 **Surface modification of textiles**
Edited by Q. Wei
98 **Sustainable textiles**
Edited by R. S. Blackburn
99 **Advances in yarn spinning technology**
Edited by C. A. Lawrence
100 **Handbook of medical textiles**
Edited by V. T. Bartels
101 **Technical textile yarns**
Edited by R. Alagirusamy and A. Das
102 **Applications of nonwovens in technical textiles**
Edited by R. A. Chapman
103 **Colour measurement: Principles, advances and industrial applications**
Edited by M. L. Gulrajani
104 **Fibrous and composite materials for civil engineering applications**
Edited by R. Fangueiro
105 **New product development in textiles: Innovation and production**
Edited by L. Horne
106 **Improving comfort in clothing**
Edited by G. Song
107 **Advances in textile biotechnology**
Edited by V. A. Nierstrasz and A. Cavaco-Paulo
108 **Textiles for hygiene and infection control**
Edited by B. McCarthy
109 **Nanofunctional textiles**
Edited by Y. Li
110 **Joining textiles: Principles and applications**
Edited by I. Jones and G. Stylios
111 **Soft computing in textile engineering**
Edited by A. Majumdar
112 **Textile design**
Edited by A. Briggs-Goode and K. Townsend
113 **Biotextiles as medical implants**
Edited by M. W. King, B. S. Gupta and R. Guidoin
114 **Textile thermal bioengineering**
Edited by Y. Li
115 **Woven textile structure**
B. K. Behera and P. K. Hari

116 **Handbook of textile and industrial dyeing Volume 1: Principles, processes and types of dyes**
Edited by M. Clark
117 **Handbook of textile and industrial dyeing Volume 2: Applications of dyes**
Edited by M. Clark
118 **Handbook of natural fibres Volume 1: Types, properties and factors affecting breeding and cultivation**
Edited by R. Kozłowski
119 **Handbook of natural fibres Volume 2: Processing and applications**
Edited by R. Kozłowski
120 **Functional textiles for improved performance, protection and health**
Edited by N. Pan and G. Sun
121 **Computer technology for textiles and apparel**
Edited by J. Hu
122 **Advances in military textiles and personal equipment**
Edited by E. Sparks
123 **Specialist yarn and fabric structures**
Edited by R. H. Gong
124 **Handbook of sustainable textile production**
M. I. Tobler-Rohr
125 **Woven textiles: Principles, developments and applications**
Edited by K. Gandhi
126 **Textiles and fashion: Materials design and technology**
Edited by R. Sinclair
127 **Industrial cutting of textile materials**
I. Viļumsone-Nemes
128 **Colour design: Theories and applications**
Edited by J. Best
129 **False twist textured yarns**
C. Atkinson
130 **Modelling, simulation and control of the dyeing process**
R. Shamey and X. Zhao
131 **Process control in textile manufacturing**
Edited by A. Majumdar, A. Das, R. Alagirusamy and V. K. Kothari
132 **Understanding and improving the durability of textiles**
Edited by P. A. Annis
133 **Smart textiles for protection**
Edited by R. A. Chapman
134 **Functional nanofibers and applications**
Edited by Q. Wei
135 **The global textile and clothing industry: Technological advances and future challenges**
Edited by R. Shishoo
136 **Simulation in textile technology: Theory and applications**
Edited by D. Veit
137 **Pattern cutting for clothing using CAD: How to use Lectra Modaris pattern cutting software**
M. Stott

138 **Advances in the dyeing and finishing of technical textiles**
M. L. Gulrajani
139 **Multidisciplinary know-how for smart textiles developers**
Edited by T. Kirstein
140 **Handbook of fire resistant textiles**
Edited by F. Selcen Kilinc
141 **Handbook of footwear design and manufacture**
Edited by A. Luximon
142 **Textile-led design for the active ageing population**
Edited by J. McCann and D. Bryson
143 **Optimizing decision making in the apparel supply chain using artificial intelligence (AI): From production to retail**
Edited by W. K. Wong, Z. X. Guo and S. Y. S. Leung
144 **Mechanisms of flat weaving technology**
V. V. Choogin, P. Bandara and E. V. Chepelyuk
145 **Innovative jacquard textile design using digital technologies**
F. Ng and J. Zhou
146 **Advances in shape memory polymers**
J. Hu
147 **Design of clothing manufacturing processes: A systematic approach to planning, scheduling and control**
J. Gersak
148 **Anthropometry, apparel sizing and design**
D. Gupta and N. Zakaria
149 **Silk: Processing, properties and applications**
Edited by K. Murugesh Babu
150 **Advances in filament yarn spinning of textiles and polymers**
Edited by D. Zhang
151 **Designing apparel for consumers: The impact of body shape and size**
Edited by M.-E. Faust and S. Carrier
152 **Fashion supply chain management using radio frequency identification (RFID) technologies**
Edited by W. K. Wong and Z. X. Guo
153 **High performance textiles and their applications**
Edited by C. A. Lawrence
154 **Protective clothing: Managing thermal stress**
Edited by F. Wang and C. Gao
155 **Composite nonwoven materials**
Edited by D. Das and B. Pourdeyhimi
156 **Functional finishes for textiles: Improving comfort, performance and protection**
Edited by R. Paul
157 **Assessing the environmental impact of textiles and the clothing supply chain**
S. S. Muthu
158 **Braiding technology for textiles**
Y. Kyosev
159 **Principles of colour appearance and measurement**
Volume 1: Object appearance, colour perception and instrumental measurement
A. K. R. Choudhury

160 **Principles of colour appearance and measurement**
Volume 2: Visual measurement of colour, colour comparison and management
A. K. R. Choudhury
161 **Ink jet textile printing**
C. Cie
162 **Textiles for sportswear**
Edited by R. Shishoo
163 **Advances in silk science and technology**
Edited by A. Basu
164 **Denim: Manufacture, finishing and applications**
Edited by R. Paul
165 **Fabric structures in architecture**
Edited by J. Ignasi de Llorens
166 **Electronic textiles: Smart fabrics and wearable technology**
Edited by T. Dias
167 **Advances in 3D textiles**
Edited by X. Chen
168 **Garment manufacturing technology**
Edited by R. Nayak and R. Padhye
169 **Handbook of technical textiles Second edition Volume 1: Technical textile processes**
Edited by A. R. Horrocks and S. C. Anand
170 **Handbook of technical textiles Second edition Volume 2: Technical applications**
Edited by A. R. Horrocks and S. C. Anand
171 **Sustainable apparel**
Edited by R. S. Blackburn
172 **Handbook of life cycle assessment (LCA) of textiles and clothing**
Edited by S. S. Muthu
173 **Advances in smart medical textiles: Treatments and health monitoring**
Edited by L. van Langenhove
174 **Medical textile materials**
Y. Qin
175 **Geotextiles**
Edited by R. M. Koerner
176 **Active coatings for smart textiles**
Edited by J. Hu
177 **Advances in braiding technology: Specialized techniques and applications**
Edited by Y. Kyosev
178 **Smart textiles and their applications**
Edited by V. Koncar
179 **Information systems for the fashion and apparel industry**
Edited by T. M. J. Choi
180 **Antimicrobial textiles**
G. Y. Sun
181 **Advances in technical nonwovens**
G. Kellie

Preface

The fashion apparel industry is one of the most influential industries in the world. Nowadays, to support fashion business operations, computerized information systems are essential. Recent developments in social media, big data analytics, RFID technology, artificial intelligent methods, and the commonly used enterprise resource planning (ERP) systems are all driving innovative business measures in the fashion apparel industry. For example, the RFID-supported system provides a completely new supply chain environment that streamlines the existing business processes. The widely implemented social media–based fashion retailing models have dramatically expanded multichannel retailing to the omni-channel domain. The use of ERP systems enhances supply chain integration between retailers and manufacturers and supports the implementation of various strategic partnership programs such as quick response and vendor-managed-inventory (VMI). There is no doubt that information systems play a crucial role in the fashion apparel industry.

In view of these points, I have edited this important Elsevier research handbook. This handbook includes two sections, namely Section 1: Principles of fashion information systems, and Section 2: Applications of information systems in the fashion and apparel industry. The featured papers are contributed by researchers from all around the world (Asia, Europe, and the United States). The specific topics covered include the following:

1. key decision points and information requirements in fast fashion supply chains
2. the use of artificial neural networks to improve decision-making in apparel supply chains
3. using evolutionary optimization techniques in apparel supply chains
4. applying fuzzy logic techniques in apparel supply chains
5. using RFID systems in apparel supply chains
6. apparel ERP systems
7. smart systems for improved customer choice in fashion retail outlets
8. intelligent demand forecasting systems for fast fashion
9. intelligent procurement systems to support fast fashion supply chains in the apparel industry
10. VMI systems in the apparel industry
11. intelligent systems for managing returns in apparel supply chains
12. fashion design using evolutionary algorithms and fuzzy set theory
13. intelligent risk-management systems

It is very encouraging to see that this handbook is very well balanced with analytical models, case studies, theoretical reviews, and new applications. A lot of promising future research areas are also outlined and proposed. I strongly believe that the findings reported in this handbook have laid the foundation for further research on information systems management for the fashion and apparel industry.

I would like to take this opportunity to sincerely thank Ms. Christina Cameron and Ms. Sarah Lynch for their helpful advice and support during the development of this important handbook. I am indebted to my Ph.D. students Hau-Ling Chan, Shu Guo, and Shuyun Ren for their helpful assistance. I am grateful to all the authors who have diligently revised their papers and contributed their research to this handbook. I also thank a few anonymous reviewers who reviewed the submitted papers and provided useful comments. I acknowledge the support of The Hong Kong Polytechnic University. Last but not least, I thank my family, colleagues, and students, who have been supporting me during the development of this important handbook.

Tsan-Ming (Jason) Choi
The Hong Kong Polytechnic University
October 2015

Introduction: key decision points and information requirements in fast fashion supply chains

1

T.-M. Choi
The Hong Kong Polytechnic University, Kowloon, Hong Kong

1.1 Introduction

The apparel industry is a very important industry. It is reported that the global market size of the apparel industry reached 1.7 trillion US dollars in 2012 (Caro and Martinez-de-Albeniz, 2015). Among all the apparel brands, the fast fashion brands have emerged as the major players in the top league. In fact, as reported by Interbrand.com, among the top 100 brands across all industries in the world in 2015, 11 of them are fashion apparel brands (see Table 1.1) and the fast fashion brands H&M and Zara are ranked top, just behind the sportswear giant Nike and the luxury brand leader Louis Vuitton.

In fact, in Table 1.1, the other top brands are all noble brands which are either the well-established giants in the sportswear market or the luxury brands which are known to be high-spenders in branding. Euromonitor (2015) comments that the fast fashion

Table 1.1 Top apparel brands

Brand	Category	Ranking (among top 100)
Nike	Sportswear	17
Louis Vuitton	Luxury	20
H&M	**Fast fashion**	**21**
Zara	**Fast fashion**	**30**
Hermes	Luxury	41
Gucci	Luxury	50
Adidas	Sportswear	62
Prada	Luxury	69
Burberry	Luxury	73
Ralph Lauren	Luxury	91
Hugo Boss	Luxury	96

Information Systems for the Fashion and Apparel Industry. http://dx.doi.org/10.1016/B978-0-08-100571-2.00001-4

brands are ambitious and expanding. They aim to have a prominent presence in the US mega stores and the emerging markets. Undoubtedly, the whole apparel industry is now driven by the fast fashion trend.

To a certain extent, the fast fashion business model was created because consumers demand newer and more trendy apparel products (Cachon and Swinney, 2011; Choi, 2014). By adopting fast fashion, the retailer can better deal with the strategic consumer behaviors (Cachon and Swinney, 2011; Lee et al., 2015). From the seller side (ie, the fashion supply chain), it is difficult to satisfy the market in a cost-effective manner if the lead time is long. Thus, driven by concepts such as quick response (Chow et al., 2012), fast delivery, and nearby sourcing strategies, Zara pioneered the magic model of requiring only two weeks to offer ready-to-sell merchandise to the market (from conceptual design). The other fashion retailers, especially those in the mass markets, are hence pressurized and have to react. Thus, there is an industrial trend of having shortened lead time in fashion apparel for the mass market brands over past decades. In the following, we will discuss some key decision points in fast fashion supply chains. We will then explore how information systems may help to enhance the decision quality.

1.2 Key decision points

It is well known that fashion design provides the soul to apparel products. It is probably the most fundamental yet critical process in the fashion supply chain. In the fast fashion era, fashion brands have to come up with the fashion design in the most time-efficient manner. Usually, the fashion brands either have their own in-house design teams (including chief fashion designers, middle-level fashion designers to junior fashion designers) or they outsource designs from other designer houses. However, in order to shorten lead time, technologies play a crucial role. In particular, it is known that it takes a long time for fashion designers (of different levels) to discuss and come up with a satisfactory final design because there is always a need to revise and improve the design so that it becomes as perfect as possible. Motivated by this, Yu et al. (2011) propose a novel method so that the collaborative fashion designs among design team members can be completed as soon as possible. The information technologies involved include fuzzy logic and artificial neural network (ANN) models.

After the designs have been created, important decisions have to be made on demand forecasting, sourcing and production, inventory and distribution management, and retailing. For demand forecasting, the supply chain agents (including fashion retailers, distributors, manufacturers, and suppliers) are all concerned as it will affect all of them. As a matter of fact, if the members are willing to contribute and share their vital data, the performance of the supply chain will be improved. One typical example is the sharing of market demand data (eg, the retailers' point-of-sales data). It is well known that such kind of information sharing and communication can significantly dampen the notorious bullwhip effect (Lee et al., 1997) and hence reduce the harms brought about by information distortion, which would naturally appear in the fashion supply chain system. On the other hand, it is also helpful if information regarding supply situations is known to the downstream members. For example, if a national distributor of an apparel brand knows that its products are in short supply, the respective information will be helpful for the fashion retailers to better plan and price the remaining inventory. To have proper demand forecasting, statistical tools and advanced artificial intelligence methods are both applicable (Yesil et al., 2012; Liu et al., 2013).

In sourcing and production, critical decisions include the selection of suppliers, proper coordination among multiple parties involved in the sourcing and production process, and the corresponding production scheduling. These are technical decisions which usually can be calculated as "sourcing and production" involves the "visible cost" which can be estimated. Usually, the "price tag" plays the most crucial role for a fashion brand to identify the right supplier. Of course, there are requirements on quality, and relationship matters, too. Nowadays, with the concern on environmental sustainability and corporate social responsibility (CSR) (Shen, 2014), the issue of carbon footprint cannot be ignored and hence local sourcing and production have become more and more important. As a remark, local sourcing and production would actually help to shorten lead time and can enhance the fast fashion model's performance (even though the production cost may be higher).

For inventory and distribution management, the key is to ensure supply is reliable and the allocation is sound, so that the right quantity is available in the right place at the right time. Some key decisions including the optimal stocking quantity, the optimal replenishment schedule, and the optimal inventory allocation scheme (including trans-shipment) should all be considered. Ideally, the consideration is not just for a single company's perspective (eg, the distributor), but it should include the supply network so that the decision is globally optimal for the respective system. Technologies definitely play a critical role in making the right decisions.

For retailing, in addition to getting the right products to the right market segments, it is important to price wisely and achieve an excellent service by combining the multiple channels together and achieve the holistic "omni-channel retailing" strategy (Brynjolfsson et al., 2013). The key decisions involved are: the optimal revenue management scheme which includes both the regular pricing and markdown pricing, the optimal inventory management scheme, the optimal promotion scheme (eg, product bundling), the customer service system (eg, VIP membership schemes), etc.

1.3 Information requirements

In Section 1.2, we have explored some key decision points in fast fashion supply chains. In this section, we proceed to examine the information requirements and the roles played by information systems.

First, in fashion design, to enhance the speed of creating a novel design, it is critical for the related companies to have the right computerized computer-aided design/computer-aided manufacturing (CAD/CAM) tools so that previous designs, proposed designs, and other references can all be consolidated and compared using computers.

The resulting fashion design process is not only more systematic, but also less time-consuming. In addition, to enhance communication among the involved parties, electronic data interchange (EDI) (Iacovou et al., 1995) platforms should be established so that the related draft designs can be exchanged and commented on in a timely manner. This also improves communication and encourages deeper collaboration. Finally, the endorsement and approval steps associated with the fashion design should also be supported by the information systems so that they can be done faster. The overall goal is to shorten lead time and to have a high-quality and creative fashion design which captures the important market elements and fashion trends.

Second, in demand forecasting, the most critical information is historical data. Datasets of both the same product and related products are useful. Then, there must be an efficient tool to help provide reliable and useful forecasts. Artificial intelligence (AI) methods, in the form of decision support systems, are known to be the right technological tools to help. Next, the forecast result can be useful for all supply chain agents. So, there should be a platform to share forecast. If the fast fashion supply chain is adopting a strategic alliance scheme, such as quick response (QR) (Iyer and Bergen, 1997), collaborative planning forecasting and replenishment (CPFR) (Panahifar et al., 2015), and vendor-managed inventory (VMI) (Waller et al., 1999), then there must be an EDI platform for the supplier and the buyer to share forecasting as well as other vital information. Finally, if there is a demand forecasting system which can also incorporate external market information (eg, the stock market performance, the retail market performance, and the performance of the major competitors) into the demand prediction process, the final forecasting result will be even more solid and comprehensive.

Third, in sourcing and production, as well as in inventory and distribution management, decision support systems (DSSs) are critically important. There should be analytical tools-based DSSs to help determine the related optimal decisions (see Section 1.2). As a remark, many of these tools are computational in nature and would require some computerized simulators to conduct analysis. Thus, it is important to provide high-quality data to support them because the recommendations are highly data-driven. Radio frequency identification (RFID) technology (Gaukler, 2011) provides the proper tool to help fashion companies collect data and process them in a timely manner. The appropriate use of RFID systems can improve data quality. In addition, in the supply chain context, coordination among the related decisions has to be achieved. This again requires the right platforms and systems. Usually, the enterprise resource planning (ERP) systems can serve this purpose (Choi et al., 2013).

Finally, in retailing, many marketing decisions (eg, pricing, promotion, inventory) have to be made dynamically with respect to the market situation (Caro and Gallien, 2007). Smart and intelligent systems are hence critical. In fact, it is important to adopt the "sense-and-respond" strategy (Haeckel, 1999) in which the retailer should have the right information system to detect the market changes (ie, "sense") and then the system should have built-in decision rules to provide the right plans to react and cope with the changes (ie, "respond"). In many cases, to facilitate the supply chain operations and to improve the whole supply chain's performance, the sense-and-respond function is not just for the retailer, but can be extended to help other supply chain agents.

1.4 Concluding remarks

In this introductory chapter, we have first discussed the importance of the fast fashion business model in the apparel supply chains, and the role played by information systems. Then, we have examined some key decision points, from fashion design, demand forecasting, sourcing and production, inventory and distribution management, to retailing, in fast fashion supply chain systems. We have then proposed the information requirements as well as the information technologies required to support these key decisions. Table 1.2 shows how the featured papers in this handbook provide support to decision making in fast fashion supply chains.

As a remark, the above discussions only provide a concise view of the highlighted issues. There are a lot more challenges and issues to be revealed and addressed. In particular, with the availability of a massive amount of data, we are moving toward

Table 1.2 **Supporting decision making in fast fashion supply chains**

Key decisions	Technologies involved	Related references (from this handbook)
Fashion design	ANN	Chapter "The use of fuzzy logic techniques to improve decision-making in apparel supply chains"
	Fuzzy logic	Chapters "The use of evolutionary optimization techniques to improve decision-making in apparel supply chains" and "Enterprise resource planning (ERP) systems for use in apparel supply chains"
	Smart retail systems	Chapter "Intelligent procurement systems to support fast fashion supply chains in the apparel industry"
	Evolutionary algorithms	Chapter "Enterprise resource planning (ERP) systems for use in apparel supply chains"
Demand forecasting	ANN	Chapter "The Use of fuzzy logic techniques to improve decision-making in apparel supply chains"
	Big data, cloud computing, and ENN	Chapter "Using radio frequency identification (RFID) technologies to improve decision-making in apparel supply chains"
	Extreme learning machine (ELM)	Chapter "Intelligent demand forecasting systems for fast fashion"

Continued

Table 1.2 **Continued**

Key decisions	Technologies involved	Related references (from this handbook)
Sourcing and production	ANN	Chapter "The use of fuzzy logic techniques to improve decision-making in apparel supply Chains"
	Big data, cloud computing, and ENN	Chapter "Using radio frequency identification (RFID) technologies to improve decision-making in apparel supply chains"
	RFID	Chapter "Artificial neural networks (ANN) to improve decision-making in apparel supply chains"
	ERP systems	Chapter "Smart systems for improved customer choice in fashion retail outlets"
Inventory and transportation management	ANN	Chapter "The use of fuzzy logic techniques to improve decision-making in apparel supply chains"
	Big data, cloud computing and ENN	Chapter "Using radio frequency identification (RFID) technologies to improve decision-making in apparel supply chains"
	RFID	Chapter "Artificial neural networks (ANN) to improve decision-making in apparel supply chains"
	ERP systems	Chapter "Smart systems for improved customer choice in fashion retail outlets"
	QR	Chapter "Fashion design using evolutionary algorithms and fuzzy set theory – a case to realize skirt design customisations"
	VMI	Chapter "Intelligent systems for managing returns in apparel supply chains"
	Returns management intelligent systems	Chapter "Vendor-managed inventory (VMI) systems in the apparel industry"
	Risk management stochastic optimization systems	Chapter "Intelligent demand forecasting supported risk management systems for fast fashion inventory management"

Table 1.2 **Continued**

Key decisions	Technologies involved	Related references (from this handbook)
Retailing	ANN	Chapter "The use of fuzzy logic techniques to improve decision-making in apparel supply chains"
	Big data, cloud computing, and ENN	Chapter "Using radio frequency identification (RFID) technologies to improve decision-making in apparel supply chains"
	RFID	Chapter "Artificial neural networks (ANN) to improve decision-making in apparel supply chains"
	Smart retail systems	Chapter "Intelligent procurement systems to support fast fashion supply chains in the apparel industry"
	VMI	Chapter "Intelligent systems for managing returns in apparel supply chains"
	Returns management intelligent systems	Chapter "Vendor-managed inventory (VMI) systems in the apparel industry"

the big data era. There will be all kinds of fascinating opportunities for the fashion industry, and information systems will continue to play a critical role in supporting its development.

References

Brynjolfsson, E., Hu, Y.J., Rahman, M.S., 2013. Competing in the age of omnichannel retailing. MIT Sloan Management Review 54, 23–29.

Cachon, G., Swinney, R., 2011. The value of fast fashion: quick response, enhanced design, and strategic consumer behavior. Management Science 57, 778–795.

Caro, F., Gallien, J., 2007. Dynamic assortment with demand learning for seasonal consumer goods. Management Science 53, 276–292.

Caro, F., Martınez-de-Albeniz, V., 2015. Fast fashion: business model overview and research opportunities. In: Agrawal, Smith (Eds.), Retail Supply Chain Management, pp. 237–264.

Choi, T.M. (Ed.), 2014. Fast Fashion Systems: Theories and Applications. CRC Press.

Choi, T.M., Chow, P.S., Liu, S.C., 2013. Implementation of fashion ERP systems in China: case study of a fashion brand, review and future challenges. International Journal of Production Economics 146, 70–81.

Chow, P.S., Choi, T.M., Cheng, T.C.E., 2012. Impacts of minimum order quantity (MOQ) constraint on a quick response supply chain. IEEE Transactions on Systems, Man, and Cybernetics – Part A 42, 868–879.

Euromonitor, 2015 (retrieved 19.10.15.). http://www.euromonitor.com/top-10-apparel-and-footwear-markets-growth-strategies-when-the-good-times-stall/report.

Gaukler, G.M., 2011. Item-level RFID in a retail supply chain with stock-out-based substitution. IEEE Transactions on Industrial Informatics 7, 362–370.

Haeckel, S.H., 1999. Adaptive Enterprise: Creating and Leading Sense-and-Respond Organizations. Harvard Business School Press.

Iacovou, C.L., Benbasat, I., Dexter, A.S., 1995. Electronic data interchange and small organizations: adoption and impact of technology. MIS Quarterly 19, 465–485.

Interbrand, 2015. http://interbrand.com/best-brands/best-global-brands/2015/ranking/ (retrieved 19.10.15.).

Iyer, A.V., Bergen, M.E., 1997. Quick response in manufacturer-retailer channels. Management Science 43, 559–570.

Lee, C.H., Choi, T.M., Cheng, T.C.E., 2015. Selling to strategic and loss-averse consumers: stocking, procurement, and product design policies. Naval Research Logistics 62, 435–453.

Lee, H.L., Padmanabhan, V., Whang, S., 1997. Information distortion in a supply chain: the bullwhip effect. Management Science 43, 546–558.

Liu, N., Ren, S., Choi, T.M., Hui, C.L., Ng, S.F., 2013. Sales forecasting for fashion retailing service industry: a review. Mathematical Problems in Engineering. http://dx.doi.org/10.1155/2013/738675. Article ID 738675.

Panahifar, F., Heavey, C., Byrne, P.J., Fazlollahtabar, H., 2015. A framework for collaborative planning, forecasting and replenishment (CPFR): state of the art. Journal of Enterprise Information Management 28, 838–871.

Shen, B., 2014. Sustainable fashion supply chain: lessons from H&M. Sustainability 6, 6239–6249.

Waller, M., Johnson, M.E., Davis, T., 1999. Vendor managed inventory in the retail supply chain. Journal of Business Logistics 20, 183–203.

Yesil, E., Kaya, M., Siradag, S., 2012. Fuzzy forecast combiner design for fast fashion demand forecasting. In: International Symposium on Innovations in Intelligent Systems and Applications, pp. 1–5.

Yu, Y., Choi, T.M., Hui, C.L., Ho, T.K., 2011. A new and efficient intelligent collaboration scheme for fashion design. IEEE Transactions on Systems, Man and Cybernetics – Part A 41, 463–475.

The use of fuzzy logic techniques to improve decision making in apparel supply chains

2

L. Wang[1], *X. Zeng*[2], *Y. Chen*[1], *L. Koehl*[2]
[1]Soochow University, Suzhou, People's Republic of China; [2]Ecole Nationale Supérieure des Arts et Industries Textiles, Roubaix, France

2.1 Introduction and background

In the apparel supply chain, fashion marketing is an application of a range of techniques and a business philosophy, centered upon the customer and potential customer of garments and related services in order to meet the long-term goals of the organization [1]. The very nature of fashion, where change is intrinsic, gives different emphasis to marketing activities. Furthermore, the role of fashion product development in both leading and reflecting consumer demand results in a variety of approaches for fashion marketing.

As the most important work from product development to target market, fashion marketing can help to provide the knowledge and skills needed to ensure that the creative component is used in the best way, allowing the business to succeed and further develop. However, the process of this work is always influenced by the human perception of fashion products, which is a dominant uncertain problem existing in the total apparel supply chain from manufacturers to consumers. Furthermore, with the development of human-centered design for fashion mass customization, the impact of human perception has been emphasized for supporting decision making in target market management.

2.1.1 *Fashion mass customization*

The concept of mass customization in the fashion industry is to design and produce various new fashion products with short life cycles and low costs, meeting personalized requirements of consumers at the level of comfort and fashion style. According to this concept, the key varieties are personalization, fit, and design [2]. Furthermore, the trends of advanced mass customization are quicker, more economical and professional.

Mass customization has been applied in the fashion mass market for more than 20 years. However, the related work mainly focuses on application of CAD tools, such as body shape modeling and garment modeling.

Information Systems for the Fashion and Apparel Industry. http://dx.doi.org/10.1016/B978-0-08-100571-2.00002-6

Fashion design and fashion marketing have not been involved systematically. In fact, when developing mass customized products, the study of human perception on products is necessary, including consumer's and design expert's perceptions, and it should be integrated into the whole personalized process from design to market.

2.1.2 Human perception for human-centered design

The designer's and consumer's perceptual process can be considered as a sequence of steps that work together to determine our experience and reaction to stimuli in the environment. The whole process is divided into four categories: Stimulus, Electricity, Experience and Action, and Knowledge. In the procedures of human perception, selection is the first stage, permitting to extract the important information. Next, raw sensory data are assembled in a meaningful way before they are useful. So, sensory data are organized in terms of form, constancy, depth, and color. In the final stage, interpretation, the brain uses this information to explain and make judgments about the external world [3].

For textile and apparel industrial applications, human perception of any physical object is often evaluated using a set of descriptive keywords. These keywords can be organized at two levels [4,5]:

1. Basic and concrete sensory descriptors, which only describe the basic nature of the object to be evaluated and are independent of its sociocultural background;
2. Abstract and complex concepts, which are strongly related to the sociocultural background of the object.

2.1.3 Sensory evaluation for acquiring human perception

Sensory evaluation is defined as a scientific discipline used to evoke, measure, analyze, and interpret reactions to the characteristics of products as they are perceived by the senses of sight, smell, taste, touch, and hearing [6]. It can be described as follows: under predefined conditions, a group of organized individuals evaluates attributes of a group of products with respect to a given target. Sensory evaluation has been widely applied in different industrial fields, especially for quality inspection, product design, and marketing. Classically, factorial multivariate methods are the only tool for analyzing and modeling perceptual data provided by experts, panelists, or consumers [7].

In practical sensory evaluations, companies work with different panels in order to obtain relevant data according to their specific needs, these panels can be generally classified into the following categories [8]:

1. Field panels: non-trained consumers randomly selected at shopping centers are invited to answer questions in a predefined questionnaire.
2. Consumer panels: selected non-trained consumers are invited to do evaluation in a laboratory under controlled conditions.
3. Free choice profiling panels: each trained panelist evaluates products with his linguistic terms selected freely.
4. Quantitative description analysis panels: trained panelists evaluate products with standard linguistic terms.
5. Expert panels: experts specialized in a specific technology evaluate typical products and define criteria of evaluation.

In practice, for the application of textile and apparel industry, sensory evaluation can be performed at two levels [9]:

1. Design-oriented sensory evaluation, which is done by a trained panel composed of experienced experts or technicians inside the enterprise for judging industrial products using a number of neutral linguistic descriptors. It aims at obtaining basic sensory attributes of products to improve the quality of product design and development.
2. Market-oriented sensory evaluation, which is given by untrained consumers using hedonic descriptors according to their preference on the products to be evaluated. It aims at identifying consumers' preferences in order to forecast market behavior.

2.1.4 Decision making with the uncertainty related to human perception in the fashion design

With respect to new product development in the fashion industry, the relation of fashion design and target marketing can be described by the segmentation in Fig. 2.1, where the vertical axis represents the effects of marketing, which is influenced by the consumers' perception, and the horizontal axis, the effects of design which is influenced by the experts' perception. There are four concepts in this relationship: failure, design-centered, marketing-centered, and compromise between design and marketing. Obviously, if a new developed product is relevant to neither design nor marketing, it can be regarded as a failure. In most cases, this is due to the fact that design ability is overestimated and customers' preferences and the need for profit are neglected.

From the point of view of establishing a compromise between design and marketing, it has more advantages than the design-centered and marketing-centered concepts for the following reasons. The business success of the design-centered concept only depends on good promotion of fashion style. Therefore, this concept is applicable to a very limited number of transactions only involving expensive garments for an elite market. Considering fashion design as a function of marketing research, the

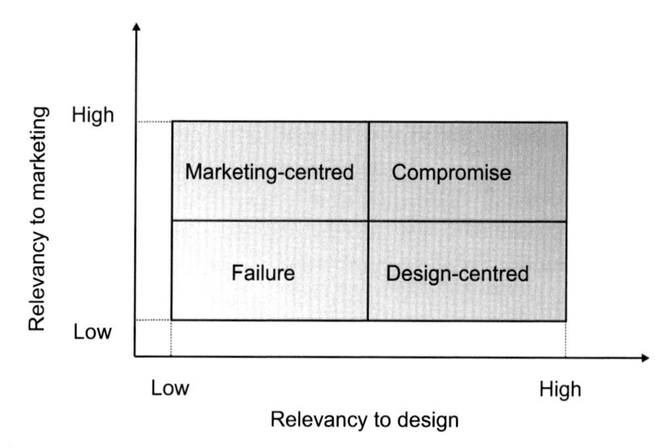

Figure 2.1 Segmentation of the relation between marketing and design.

marketing-centered concept fails due to the two following facts: (1) many people do not know their preference exactly until products are presented; (2) their preferences will change over time. Under the concept of compromise between design and marketing, a good fashion design should be accepted by both consumers and experts (designers, retailers, and marketers). As it has high relevancy in both aspects, this concept will lead to the highest profit. This concept attempts to embrace the positive aspects of high concern for design, customers, and profit by recognizing the interdependence of marketing and design. If designers understand how marketing can enhance the creation process and marketing personnel masters how fashion design can lead to an appropriate response to customer requirements, great business progress can be made. In this context, perceptions from consumers and experts are both important for the process of new product development [1]. Therefore, in practice, the compromise solution is recommended for supporting a decision from product development to target market in the apparel supply chain. But this solution has a high uncertainty, because it is mainly based on the decision rules mined from consumers' and experts' perceptions.

2.2 Fuzzy logic techniques

Human perception, including expert knowledge and consumer cognition, is often conceptual and ambiguous, which is difficult to characterize using classic computational tools such as statistics. In this context, fuzzy logic techniques will be more efficient for formalizing perceptual data, relations between concepts, and other uncertain problems in the design process. Three computational tools (fuzzy sets theory, decision tree, fuzzy cognitive map) are generally applied in the uncertainty study of human perception in apparel industry.

2.2.1 Fuzzy sets theory

Fuzzy sets appeared for the first time in 1965 in the paper by Lotfi A. Zadeh, entitled "Fuzzy Sets" [10]. Since then, theoretical results on fuzzy theory have quickly and massively been developed and applied in different industrial fields.

In the theory of crisp set (classic set), an element belongs to a set entirely or not. It does not support an element belonging to a set partially. However, the concept of fuzzy set is designed to allow the gradations of membership of a set for an element. That is to say that an element is allowed to partially belong to a set.

In Fig. 2.2, the element x belongs entirely to A. The element y does not belong to A. The element z belongs to B totally and the element σ belongs to B partially. The concept of fuzzy set looks like a relaxation of the subset of a universal set. Consequently, fuzzy set is a "vague boundary set" compared with a crisp set [11].

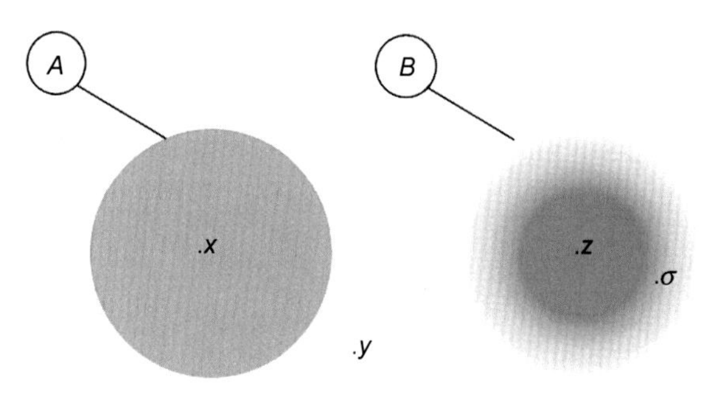

Figure 2.2 The difference between a crisp set and a fuzzy set.

Formally, the membership function μ_A of a crisp set maps all the members of the universal set X to the set $\{0,1\}$.

$$\mu_A : X \rightarrow \{0, 1\}.$$

For a fuzzy set, each element is mapped to [0,1] by its membership function:

$$\mu_A : X \rightarrow [0, 1]$$

The standard operations of fuzzy sets are given as follows.

1. The complement set of fuzzy set A is denoted as $\overline{A}$. The membership degree of $\overline{A}$ can be calculated as

$$\mu_{\overline{A}}(x) = 1 - \mu_A(x), \forall x \in X$$

2. Membership value of member x in the union takes the greater value of membership between fuzzy sets A and B

$$\mu_{A \cup B}(x) = \text{Max}[\mu_A(x), \mu_B(x)], \forall x \in X$$

Of course, A and B are subsets of $A \cup B$.

3. Intersection of fuzzy sets A and B takes a smaller value of membership function between A and B

$$\mu_{A \cap B}(x) = \text{Min}[\mu_A(x), \mu_B(x)], \forall x \in X$$

Intersection $A \cap B$ is a subset of A or B.

Fuzzy relations are developed by allowing the relationship between elements of two or more sets to take on an infinite number of degrees of relationship between the extremes of "completely related" and "not related," which are the only degrees of relationship possible in crisp relations. In this sense, fuzzy relations are to crisp relations as fuzzy sets are to crisp sets; crisp sets and relations are more constrained realizations of fuzzy sets and relations [12].

Fuzzy relations map elements of one universe, say, X, to those of another universe, say, Y, through the Cartesian product of the two universes. However, the "strength" of the relation between ordered pairs of the two universes is not measured with the indicator function (as in the case of crisp relations), but rather with a membership function expressing various "degrees" of strength of the relation on the unit interval [0,1].

Hence, a fuzzy relation R is a mapping from the Cartesian space $X \times Y$ to the interval [0,1], where the strength of the mapping is expressed by the membership function of the relation for ordered pairs from the two universes, or $\mu\_R(X,Y)$.

All fuzzy set operations can also be applied to fuzzy relations. For the two fuzzy relations $R \subseteq A \times B$ and $S \subseteq A \times B$, the operations are defined as follows:

1. Union relation
   Union of two relations R and S is defined as,

$$\forall (x, y) \in A \times B$$

$$\begin{aligned}\mu_{R \cup S}(x, y) &= \text{Max}[\mu_R(x, y), \mu_S(x, y)] \\ &= \mu_R(x, y) \vee \mu_S(x, y)\end{aligned}$$

In practice, "$\vee$" is generally realized using a Max operation. For n relations, it can be extended to the following form:

$$\mu_{R_1 \cup R_2 \cup R_3 \cup \ldots \cup R_n}(x, y) = \underset{R_i}{\vee} \mu_{R_i}(x, y)$$

2. Intersection relation
   The intersection relation $R \cap S$ of two fuzzy sets A and B is defined by the following membership function:

$$\forall (x, y) \in A \times B$$

$$\begin{aligned}\mu_{R \cap S}(x, y) &= \text{Min}[\mu_R(x, y), \mu_S(x, y)] \\ &= \mu_R(x, y) \wedge \mu_S(x, y)\end{aligned}$$

"$\wedge$" is usually realized using Min operation. In the same way, the intersection relation for n fuzzy relations is defined by:

$$\mu_{R_1 \cap R_2 \cap R_3 \cap \ldots \cap R_n}(x, y) = \overset{\wedge}{R_i}\, \mu_{R_i}(x, y)$$

3. Complement relation
The complement relation $\overline{R}$ of the fuzzy relation R is defined by the following membership function:

$$\forall (x,y) \in A \times B$$

$$\mu_{\overline{R}}(x,y) = 1 - \mu_R(x,y)$$

For extension, the composition of two fuzzy relations can be calculated as follows:

Definition 1: Considering the fuzzy relation R is defined on the universal sets A and B, ie, $R \subseteq A \times B$ and S on the universal sets B and C, ie, $S \subseteq B \times C$. The composition $R \circ S$ of two fuzzy relations R and S is expressed by the relation from A to C, which is defined by:

$$\mu_{S \circ R}(x,z) = \underset{y}{\text{Max}}[\text{Min}(\mu_R(x,y), \mu_S(y,z))]$$

$$= \bigvee_{y} [\mu_R(x,y) \wedge \mu_S(y,z)]$$

$S \circ R$ from this elaboration is a subset of $A \times C$. That is, $S \circ R \subseteq A \times C$.

If the relations R and S are represented by matrices M_R and M_S, the matrix $M_{S \circ R}$ corresponding to $S \circ R$ can be expressed by:

$$M_{S \circ R} = M_R \circ M_S$$

A fuzzy proposition can have its truth value in the interval [0,1]. The fuzzy expression function is a mapping function from [0,1] to [0,1].

$$f: [0,1] \rightarrow [0,1]$$

In a domain of n dimensions, the fuzzy function becomes:

$$f: [0,1]^n \rightarrow [0,1]$$

Therefore, the fuzzy expression can be interpreted as an n-array relation from n fuzzy sets to [0,1]. In the fuzzy logic, the operations such as negation ($\sim$), conjunction ($\wedge$), and disjunction ($\vee$) are defined in the same way as those of the classical logic.

Definition 2: Fuzzy logic is a logic represented by the fuzzy expression which satisfies the following conditions:

1. Truth values, 0 and 1, and variable $x_i \in [0,1]$, $(i = 1, 2, \ldots, n)$ are fuzzy expressions.
2. If f is a fuzzy expression, $\tilde{f}$ is also a fuzzy expression.
3. If f and g are fuzzy expressions, $f \wedge g$ and $f \vee g$ are also fuzzy expressions.

The operators in the fuzzy expression are defined as follows for a, $b \in [0,1]$.

1. Negation $\widetilde{a} = 1 - a$
2. Conjunction $a \wedge b = \text{Min}(a,b)$
3. Disjunction $a \vee b = \text{Max}(a,b)$
4. Implication $a \rightarrow b = \text{Min}(1,1 + b - a)$

2.2.2 Fuzzy decision tree

Decision tree induction is a learning strategy from a set of class-labeled training tuples. A decision tree is a flowchart-like tree structure, where each internal node (non-leaf node) denotes a test on an attribute, each branch represents an outcome of the test, and each terminal node (or leaf node) holds a class label. The topmost node in a tree is the root node [13].

During the late 1970s and early 1980s, J. Ross Quinlan, a researcher in machine learning, developed the well-known learning algorithm of decision tree, called ID3 algorithm (Iterative Dichotomiser). Quinlan later presented C4.5 (a successor of ID3), which became a benchmark to which newer supervised learning algorithms are often compared [14]. In 1984, a group of statisticians published the book "Classification and Regression Trees" (CART), which described the generation of binary decision trees [15]. ID3 and CART were invented independently of one another at around the same time, yet follow a similar approach for learning decision trees from training tuples. These two cornerstone algorithms spawned a flurry of work on decision tree induction. ID3, C4.5, and CART adopt a greedy (ie, non-backtracking) approach in which decision trees are constructed in a top-down recursive divide-and-conquer manner. Most algorithms for decision tree induction also follow such a top-down approach, which starts with a training set of tuples and their associated class labels. The training set is recursively partitioned into smaller subsets as the tree is being built. Differences in decision tree algorithms include how the attributes are selected in creating the tree and the mechanisms used for pruning [13].

Fuzzy decision trees aim at high comprehensibility normally attributed to ID3, with the gradual and graceful behavior attributed to fuzzy systems. Thus, they extend ID3 procedure, using fuzzy sets and approximate reasoning both for the tree-building and the inference procedures [16].

The features and rules in the process of Fuzzy ID3 algorithm are as follows [16]:

Step 1 of Fuzzy ID3: Splits creation in form of multiple splits, ie, for every attribute a single split is created where attributes' categories are branches of the proposed split.

Step 2 of Fuzzy ID3: Evaluation of best split for tree branching based on entropy measure and information gain, which are defined by:

1. Entropy: in a node, the entropy measure can be expressed by

$$E(S) = \sum_{i=1}^{c} [P(i) \log_2 P(i)]$$

where C represents the number of classes as the output variable, and $P(i)$ is the probability of the i-th class. S denotes the learning dataset in a node.

2. Information gain: ID3 uses information gain $I(X, S)$, as a measure of split quality measure, which is calculated by:

$$E(A,S) = \sum_{j=1}^{K} \left(\frac{|S_j|}{|S|} \cdot E(S_j) \right)$$

$$I(A,S) = E(S) - E(A,S)$$

where $E(A, S)$ is the expected entropy of an input attribute A that has K categories, $E(S_j)$ is the entropy responding to a certain category j of A, and $\frac{|S_j|}{|S|}$ is the probability of the j-th category in the attribute A. $I(X, S)$, representing the information gain of attribute A represents the amount of information held by attribute A for the class disambiguation. Furthermore, for all the attributes A's in one node, the node is split by using the most discriminatory attribute, whose information gain is maximal.

Step 3 of Fuzzy ID3: Checking of the stop criteria, and recursively applying steps 1 and 2 for generating new branches.

In classical decision trees, created by the ID3 algorithm, attributes can have only symbolic or discrete numerical values. In the case of fuzzy decision trees, attributes can also have linguistic values represented by fuzzy sets. Fuzzy decision trees have been obtained as a generalization of classical decision trees through application of fuzzy sets and fuzzy logic. The Fuzzy-ID3 algorithm is an extension of the ID3 algorithm. The difference between these two algorithms is in the method of computing the probability count of learning examples in each node.

In the Fuzzy-ID3 algorithm [17], the total examples count P in the node N are expressed as:

$$P = \sum_{j=1}^{L} P_j$$

where 1,…, L are corresponding to linguistic values for the decision attribute A. Assume x_i and y_i are input vector and output value, which correspond to attributes and class, respectively, and the examples count P_j for the i-th class is determined as follows:

$$P_j = \sum_{j=1}^{C} f\left(\mu^\circ (x_i), \mu_j(c_i)\right)$$

where f is the function employed to compute the value of fuzzy relation (the Min operator is used). μ° is the membership function of Cartesian product of fuzzy sets that appear on the path from the root node to node N and μ_j is the membership function of fuzzy set that determines class c_j.

Different from the ID3 algorithm, the equations of entropy takes the following forms:

$$E(S) = -\sum_{j=1}^{L} \frac{P_j}{P} \log_2 \frac{P_j}{P}$$

$$E(A,S) = \frac{\sum_{k=1}^{R} P_k^A E(k,S)}{\sum_{k=1}^{R} P_k^A}$$

where P_k^A is the total count of examples in node N concerning the linguistic value k of the attribute A, which has R linguistic values. The stopping conditions are the same as in the ID3 algorithm.

Most traditional decision tree methods (such as CART, ID3), which are used for extracting knowledge in classification problems, cannot deal with cognitive uncertainties such as vagueness and ambiguity associated with the human thinking process and perception. However, fuzzy decision trees (such as Fuzzy-ID3) can represent classification knowledge more naturally to the way of human thinking. They are more robust in tolerating imprecise, conflicting, and missing information for treating the problems of cognitive uncertainties [18].

2.2.3 Fuzzy cognitive map

In the 1970s, political scientist Robert Axelrod introduced cognitive maps for representing social scientific knowledge [19]. Shown in Fig. 2.3, Axelrod's cognitive maps are signed digraphs. Nodes are variable concepts. A positive edge from node *A* to node *B* means *A* causally increases *B*. A negative edge from *A* to *B* means *A* causally decreases *B*.

Based on the Axelrod's cognitive map, Bart Kosko introduced the fuzzy cognitive map in 1986. This is a fuzzy-graph structure for representing causal reasoning [20].

Most knowledge is specification of classifications and causalities. In general, the classes and causalities are uncertain (usually fuzzy or random). The fuzziness passes into knowledge representations and into knowledge bases, where it leads to a

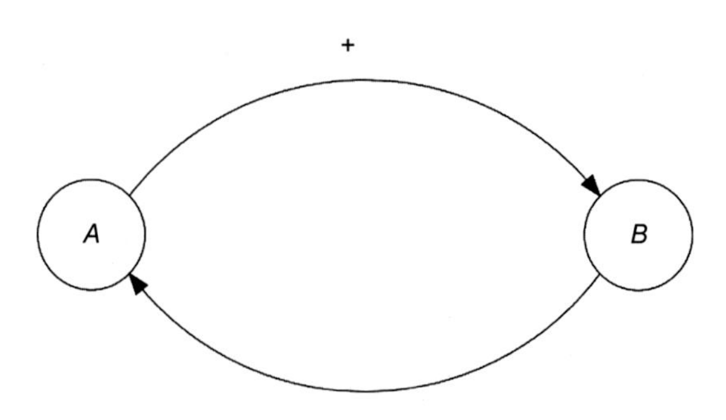

Figure 2.3 A simple example of Axelrod's cognitive map.

knowledge acquisition or processing tradeoff. If the knowledge representation is fuzzier, the knowledge acquisition is easier and the knowledge-source concurrence is greater. However, the knowledge processing is harder. The fuzziness of a fuzzy cognitive map allows hazy degrees of causality between hazy causal objects (concepts). Its graph structure allows systematic causal propagation, in particular forward and backward chaining, and it allows application in soft knowledge domains, where both the system concepts, relationships and the meta-system language, are fundamentally fuzzy [20].

The example of a typical fuzzy cognitive map in Fig. 2.4 has five concepts which are represented by C_1, C_2, C_3, C_4, C_5. Their causal relationships are expressed by the arcs with fuzzy linguistic values {"None," "Some," "Much," "A Lot"}.

According to the directions of the arcs, from C_1 to C_5, there are three causal paths:

$P_1 \rightarrow (\mathrm{arc}_{13}, \mathrm{arc}_{35})$
$P_2 \rightarrow (\mathrm{arc}_{13}, \mathrm{arc}_{34}, \mathrm{arc}_{45})$
$P_3 \rightarrow (\mathrm{arc}_{12}, \mathrm{arc}_{24}, \mathrm{arc}_{45})$

So the three effects of C_1 on C_5 by the paths are:

$I(P_1) = \min\{\text{Much, A Lot}\} = \text{Much}$
$I(P_2) = \min\{\text{Much, Some, Some}\} = \text{Some}$
$I(P_3) = \min\{\text{Some, A Lot, Some}\} = \text{Some}$

Finally, the total effect of C_1 on C_5 is:

$T(C_1, C_5) = \max\{I(P_1), I(P_2), I(P_3)\} = \max\{\text{Much, Some, Some}\} = \text{Much}$

In words, it means C_1 impacts much causality to C_5.

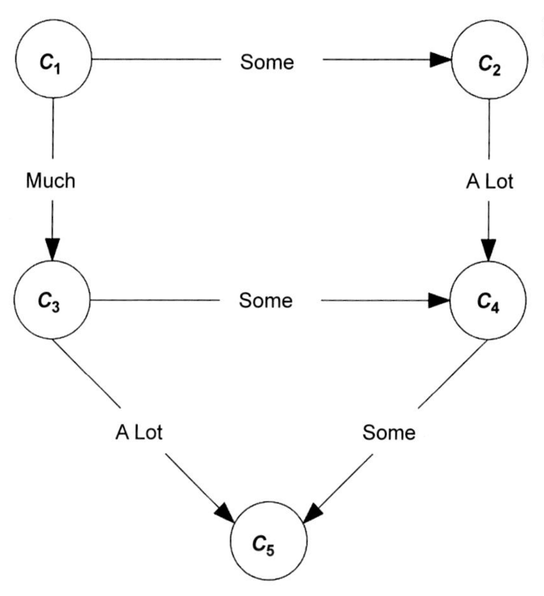

Figure 2.4 An example of a fuzzy cognitive map.

In order to be formalized more easily for calculation, a fuzzy cognitive map with n concepts can be generally represented by a $(n \times n)$-dimensional fuzzy relation matrix.

As the extension of a cognitive map with the additional capability of representing feedback through weighted causal links, fuzzy cognitive maps have been used for representing, analyzing, and aiding decision making in practices of many fields (such as finance, medicine, politics, and so on) [21,22].

The decision problems are usually characterized by numerous issues or concepts interrelated in complex ways. A fuzzy cognitive map can represent human knowledge and cognition in a form that lends itself to relatively easy integration into a collective knowledge base for a group involved in a decision process [23]. And, furthermore, it can be easy to aggregate the knowledge of various experts by using a fuzzy relation operation [22].

2.3 The target market selection in apparel supply chain using fuzzy decision making

In this application, we propose a new intelligent fashion recommender system based on human perception of fashion design to select the most relevant garment design scheme for a specific consumer in order to deliver new personalized garment products. This system integrates emotional fashion themes and human perception on personalized body shapes and professional designers' knowledge. The corresponding perceptual data are systematically collected from professionals using sensory evaluation techniques. The perceptual data of consumers and designers are formalized mathematically using fuzzy sets and fuzzy relations. The complex relationship between human body measurements and basic sensory descriptors, provided by designers, is modeled using fuzzy decision trees. The fuzzy decision trees constitute an empirical model based on learning data measured and evaluated on a set of representative samples. The complex relationship between basic sensory descriptors and fashion themes, given by consumers, is modeled using fuzzy cognitive maps. The combination of the two models can provide more complete information to the fashion recommender system, making it possible to evaluate whether a specific body shape is relevant to a desired emotional fashion theme and which garment design scheme can improve the image of the body shape. The proposed system has been carried out and validated in a case of target market selection through the evaluations of target consumers and fashion experts using a method frequently used in marketing studies.

This application utilizes the concepts and data formalized as follows:

Let $T = \{t_1, t_2,\ldots, t_n\}$ be a set of n fashion themes characterizing the sociocultural categories of body shapes.

Let $D = \{d_1, d_2,\ldots, d_m\}$ be a set of m basic sensory descriptors extracted by the fashion experts to describe various body shapes.

Let $W = \{w_1, w_2,\ldots, w_p\}$ be a set of p representative virtual body shapes generated from a garment CAD software.

Let BM = {bm_1, bm_2,...bm_h} be a set of h body measurement features characterizing body shapes. The body measurements are usually selected by fashion designers according to the garment type to be studied and the aesthetic effect of the concerned body part. In the case of designing men's overcoats, fashion designers usually consider the overall body shape and some details of the upper body part and then take six body measurements ($h = 6$) as follows: bm_1 = stature, bm_2 = total length of arm, bm_3 = chest circumference, bm_4 = neck circumference, bm_5 = waist circumference, bm_6 = hip circumference.

Let BR = {br_1, br_2,..., br_g} be a set of g body ratio indexes calculated from the body measurement features in BM. In each specific design scenario, the body ratio indexes are also determined by designers according to the garment type and the aesthetic effect of the concerned body part. In the case of designing men's overcoats, we have:
$br_1 = bm_2/bm_1$, $br_2 = bm_3/bm_1$, $br_3 = bm_4/bm_1$, $br_4 = bm_5/bm_1$, $br_5 = bm_6/bm_1$, $br_6 = bm_2/bm_3$, $br_7 = bm_4/bm_3$, $br_8 = bm_5/bm_3$, $br_9 = bm_6/bm_3$, $br_{10} = bm_5/bm_6$, $br_{11} = bm_2/bm_4$, $br_{12} = bm_2/bm_5$, $br_{13} = bm_2/bm_6$, $br_{14} = bm_4/bm_5$, $br_{15} = bm_4/bm_6$

For fashion experts, BR is more relevant than BM for quantitatively characterizing body shapes [24]. The body ratios of all the p virtual body shapes in W constitute a matrix, denoted as $(br_{ij})_{g \times p}$ with $i = 1,..., g$ and $j = 1,..., p$.

Let EX = {ex_1, ex_2,...ex_r} be a set of r fashion experts evaluating the relevancy of the sensory descriptors in D to naked body shapes in W by comparing them with CA170.

Let EC = {ec_1, ec_2,...ec_z} be a set of z consumers evaluating the relevancy of the sensory descriptors in D to the fashion themes in T.

Let $S = \{s_1, s_2,...s_\xi\}$ be a set of ξ existing reference garment styles used in the design process.

Let DE = {de_1, de_2,...de_λ} be a set of λ new garment styles generated for a specific body shape. These new styles are generated by making combinations of the reference styles.

2.3.1 Perception, evaluation, and formalization

The proposed recommender system is realized through the design of men's overcoats in different scenarios. In this context, three sensory experiments are carried out for collecting human perception data. *Experiments I* and *II* aim to extract, from their perceptions, fashion designers' knowledge and experience about body shapes and relations between body shapes and design styles. *Experiment III* enables identification of consumers' perceptions of fashion themes and their relationships with body shapes.

Experiment I: Sensory Evaluation of Naked Male Body Shapes

In this experiment, the fashion expert's perception of human body shapes was extracted through sensory evaluations on a set of basic and concrete sensory descriptors describing the basic nature of naked male body shapes. A classical normalized sensory evaluation procedure is adopted to perform this task [5].

First, 12 various virtual male body types ($p = 12$) are generated using the software MODARIS 3D Fit [25], covering the entire population of the South-East Region in China. These body shapes are expressed as:

W = {w_1: CY155, w_2: CY170, w_3: CY185, w_4: CA155, w_5: CA170, w_6: CA185, w_7: CB155, w_8: CB170, w_9: CB185, w_{10}: CC155, w_{11}: CC170, w_{12}: CC185}, in which *CY*, *CA*, *CB*, and *CC* represent four frequently used classes of upper body shapes, and *155*, *170*, and *185* are the corresponding heights in centimeters. In our study, we take *CA170* as the standard body shape.

Next, from each of these virtual naked body shapes, the fashion experts selected six key body measurements considered as the most relevant to the aesthetic effect of men's entire and upper body. These measurements include.

$BM = \{bm_1$*: stature,* bm_2*: total length of arm,* bm_3*: chest circumference,* bm_4*: neck circumference,* bm_5*: waist circumference,* bm_6*: hip circumference*$\}$.

In the sensory evaluation procedure, the trained fashion experts described the features of these body shapes using a set of normalized descriptors, selected from [26]. All evaluators used the same descriptors during their evaluations and were trained in order to understand the meaning of these descriptors. The method of semantic differential scale was used for expressing these descriptors. We obtained 22 normalized pairs of descriptors ($m = 22$) as follows:

$D = \{d_1$*: thin–fat,* d_2*: slim–bulgy,* d_3*: dented–swollen,* d_4*: atrophic–forceful,* d_5*: short–tall,* d_6*: narrow–wide,* d_7*: flat–thick,* d_8*: nonstreamline–streamline,* d_9*: knotty–smooth,* d_{10}*: shrive–fleshly,* d_{11}*: effeminate–lusty,* d_{12}*: fragility–strong,* d_{13}*: inelastic–elastic,* d_{14}*: deft–awkward,* d_{15}*: cabinet–huge,* d_{16}*: unbalanced–harmonious,* d_{17}*: unique–normal,* d_{18}*: unstylish–stylish,* d_{19}*: sebaceous–muscular,* d_{20}*: lazy–vivid,* d_{21}*: light–heavy,* d_{22}*: unsexy–sexy*$\}$

The evaluation was performed by comparing each virtual body shape with the standard reference *CA170* for each sensory descriptor. Each virtual body shape can be evaluated by multiple angles, like the example in Fig. 2.5. For each expert and each descriptor, the evaluation or comparison result takes a value from the set $\{C_1$ (very inferior), C_2 (inferior), C_3 (fairly inferior), C_4 (a little inferior), C_5 (identical), C_6 (a little superior), C_7 (fairly superior), C_8 (superior), C_9 (very superior)$\}$.

Experiment II: Sensory Evaluation of Male Body Shapes with Design Style

In this experiment, the objective was to extract the fashion expert's experience about how fashion style changes the perception of a human body shape. Five classical styles of male overcoats ($\xi = 5$), ie, $S = \{s_1$: "Chester", s_2: "Ulster", s_3: "Balmacaan", s_4: "Trench", s_5: "Duffle"$\}$, were selected as garment references.

The same fashion experts involved in *Experiment I* participated. The evaluators compared the naked standard body shape CA170 with each reference style for the same sensory descriptors defined in *Experiment I.* This treatment is due to the fact that designers of the mass market usually create styles on a standard mannequin. For each descriptor, each fashion expert also provided an evaluation score from the set $\{C_1, C_2, \ldots, C_9\}$.

Experiment III: Cognition of Fashion Themes and Their Relations with Body Shape Perceptions

Three fashion themes, namely, t_1: "sporty," t_2: "nature," and t_3: "attractive," were considered ($n = 3$).

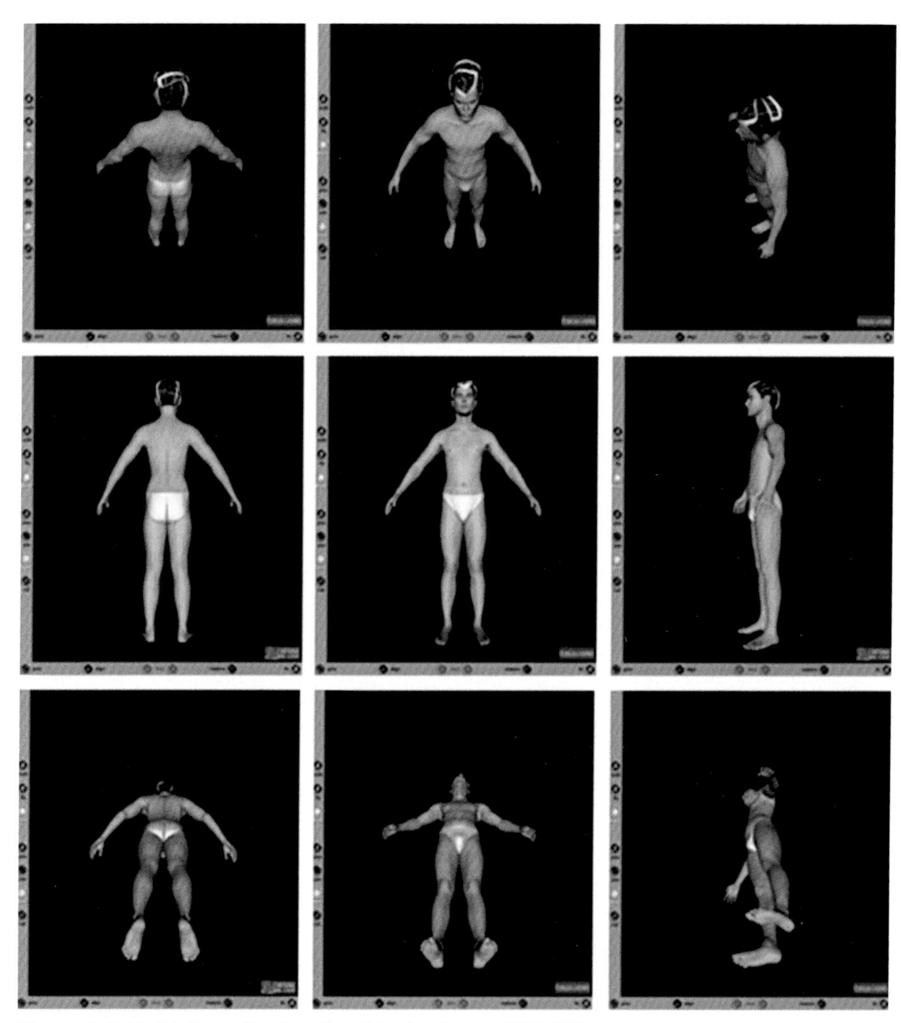

Figure 2.5 Multiple angles for virtual body shape of CA170.

During the evaluation, each evaluator was asked to provide a linguistic evaluation score for characterizing the relevancy degree of each sensory descriptor describing body shapes, to each fashion theme. These evaluation scores took values from the set {R_1 *(very irrelevant)*, R_2 *(fairly irrelevant)*, R_3 *(neutral)*, R_4 *(fairly relevant)*, R_5 *(very relevant)*}.

2.3.2 Modeling

According to the general structure of the proposed fashion recommender system in Fig. 2.6. The system is composed of three functional blocks: *Inputs*, *Learning data unit*, and *Decision support unit*. The perceptual data of fashion experts and consumers,

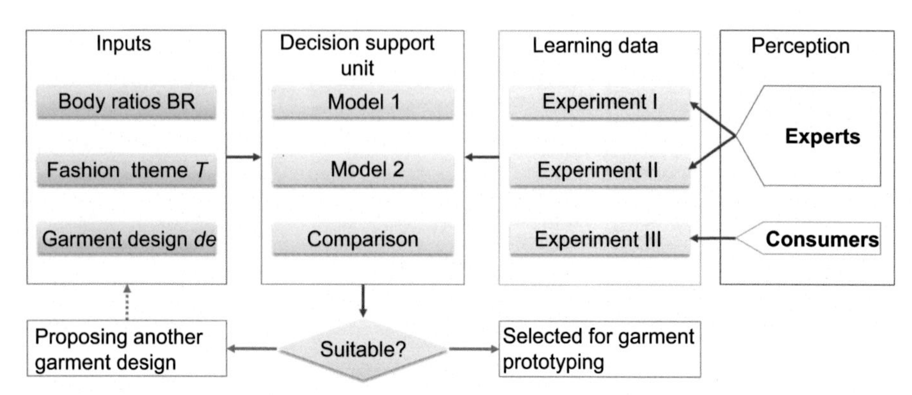

Figure 2.6 The proposed fashion recommender system.

obtained from the sensory experiments presented in the above section, will be memorized in the *Learning data unit*. For different garment collections, such as men's overcoats and women's professional skirts, the corresponding sensory data are stored in different cells and managed separately. When a new design case is introduced, a new cell is created in the *Learning data unit* for storing new experimental data. The *Decision support unit* will determine, through a series of computations, whether a specify body shape conforms to a given fashion theme. Its structure is stable and independent of the garment collections or design cases to be treated. *Inputs* is a user interface that allows the designer to input three parameters: a specific body shape, represented by its body ratios BR^Y, the desired fashion theme t_i, and a newly designed garment style de_ν, obtained by combining different existing reference styles.

The first stage of the *Decision support unit* is *Model I*, the aim of which is to evaluate the relevance of a specific naked body shape *Y*, expressed by its body ratios BR^Y, related to a desired fashion theme t_i. The second stage of this unit is *Model II*, the aim of which is to evaluate the relevancy of a specific body shape *Y* with a garment style de_ν related to t_i. In the last stage of this unit, the previous two relevancy results are compared by using a gravity center-based criterion to evaluate whether the newly designed style de_ν is suitable for the body shape *Y* in terms of image improvement toward the fashion theme t_i.

Model I is used to predict three relevancy degrees: the relevancy degree of the naked body ratios BR^Y to the basic sensory descriptors in *D*, denoted as $REL(D, BR^Y)$; the relevancy degree of the sensory descriptors in *D* to a given fashion theme t_i, denoted as $REL(t_i, D)$; and the relevancy degree of the naked body ratios BR^Y to the fashion theme t_i, denoted as $REL(t_i, BR^Y)$. Fig. 2.7 shows the functional structure of *Model I*.

$REL(D, BR^Y)$ characterizes the relation between *D* (human perception on body shapes) and BR^Y (specific body ratios derived from body measurements). It is modeled using a number of fuzzy decision trees by learning from the perceptual data obtained from *Experiment I* presented in Section 2.3.1. The modeling procedure is composed of two steps: (1) building a fuzzy decision tree for each evaluator (fashion expert) and each sensory descriptor, extracting all the fuzzy rules for the body shape *Y* and computing the relevancy degrees of *Y* to the sensory descriptors d_j's for these fuzzy

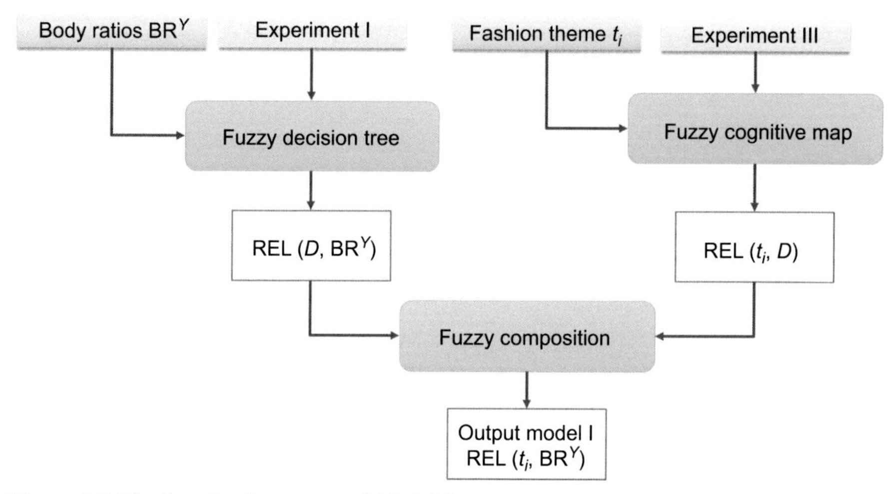

Figure 2.7 The functional structure of *Model I*.

rules; and (2) aggregating all the fuzzy decision trees for all the evaluators. REL(t_i, D) characterizes the relation between t_i and D. It is modeled using a fuzzy cognitive map, built from the consumer perceptions extracted from *Experiment III* presented in the above section.

Next, the two models, one characterizing a fashion designer's perception of body shapes and the other a consumer's perception of relations between body shapes and fashion themes, are combined using a fuzzy relation composition for computing the general relevancy degree of a specific body shape to a given fashion theme.

Model II is used to predict two fuzzy relations, namely, the relevancy degree of a number of reference styles in S to a newly designed garment style de_ν, denoted as REL(de_ν, S), and the relevancy degree of the standard body shape CA170 with a reference style s_λ in S to the sensory descriptors in D, denoted as REL(D, s_λ).

These two relevancy degrees are calculated from the sensory data of *Experiment II* presented in the above section.

By successively combining the two previous fuzzy relations and REL(t_i, D), calculated from *Model I*, we obtain the relevancy degree of the standard body shape CA170 with the newly designed style de_ν to the desired fashion theme t_i, denoted as REL(t_i, $BR^{CA170} \vee de_\nu$). Next, by applying the union operation between REL(t_i, $BR^{CA170} \vee de_\nu$) and REL(t_i, Y), we obtain the relevancy of any body shape Y with a newly designed style de_ν to the designed fashion t_i, denoted as REL(t_i, $BR^Y \vee de_\nu$). Fig. 2.8 shows the functional structure of *Model II*.

For a given body shape Y, the system compares its naked relevancy degree and that with the design style de_ν related to the fashion theme t_i in order to determine whether this new style can improve its image toward the direction of t_i. The style de_ν is considered a feasible design style if this improvement can be validated by the comparison of these two relevancy degrees.

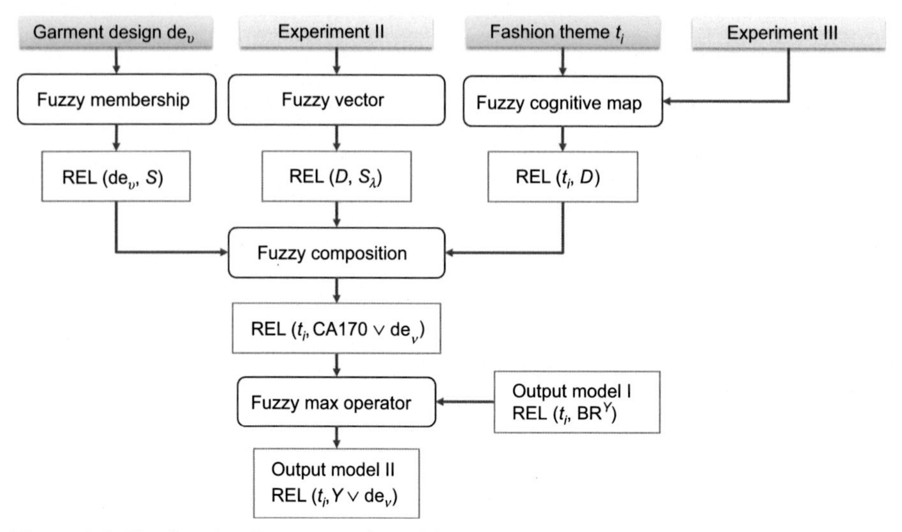

Figure 2.8 The functional structure of *Model II*.

2.3.3 Computation of the relevancy degrees for the perception of naked body shapes

2.3.3.1 Computation of the relevancy degree REL(D, Y)

Having acquired the body measurements and the dissimilarity degrees related to one sensory descriptor d_j for all the p virtual body shapes, obtained in *Experiment I*, we use the fuzzy decision trees for deducing or predicting the relations between d_j and the body ratios of the body shape Y. We build one fuzzy decision tree TR_{ij} for each sensory descriptor d_j and each evaluator ex_i by learning from the evaluated data of all the p representative virtual body shapes.

Fuzzy decision trees are more efficient for treating learning data of mixed type, including both numerical and categorical data [18]. They are more robust in tolerating imprecise, conflicting, and missing information. In our approach, the Fuzzy ID-3 algorithm [27] is used to build the decision trees. For any specific body ratio BR^Y, we obtain from this model a relevancy degree for each sensory descriptor in D.

When processing numerical learning data like body ratios with a Fuzzy-ID3 decision tree, the corresponding descriptors should be fuzzified. In our approach, the corresponding fuzzification procedure is given below.

Step 1: Normalization of Body Ratios

We normalize the numeric values of all the body ratios br_k's ($k = 1,\ldots, g$) into the range [0,1] by using $\text{br}_k := \text{bk}_i/\delta_k$. The coefficient δ_k is defined from the learning data of the virtual body shapes $(\text{br}_{kl})_{g \times p}$ according to [13] so that all learning data can be more uniformly distributed in [0,1]. We have:

$$\delta_k = \sqrt{\sum_{l=1}^{p} \text{br}_{kl}^2}$$

Step 2: Defining Linguistic Values and Membership Functions
Five linguistic values, VS (very small), S (small), M (medium), L (large), and VL (very large), are used for the fuzzification of all the body ratios. For each normalized body ratio, the membership functions for its five linguistic values are defined as follows.

For any normalized body ratio br_k, its triangular membership functions are characterized by $LC_{1k},\ldots, LC_{5k}$, obtained from the distribution of the learning data of the p virtual body shapes, ie:

$$\mathrm{LC}_{1k} = \min\{\mathrm{br}_{kl} | l = 1, \ldots, p\}$$

$$\mathrm{LC}_{5k} = \max\{\mathrm{br}_{kl} | l = 1, \ldots, p\}$$

$$\mathrm{LC}_{3k} = \mathrm{br}_{kv} | w_v \text{ is the standard virtual body shape CA170}$$

$$\mathrm{LC}_{2k} = (\mathrm{LC}_{1k} + \mathrm{LC}_{3k})/2$$

$$\mathrm{LC}_{4k} = (\mathrm{LC}_{3k} + \mathrm{LC}_{5k})/2$$

Step 3: Computing the Fuzzy Value from a Numerical Value
Given a specific numerical value x measured on a body ratio br_k (input value), its corresponding membership degree is computed according to Fig. 2.9. The corresponding fuzzy value of x can be expressed by:

$$x \to \left(\mu_k^{\mathrm{VS}}(x)\mu_k^{\mathrm{S}}(x)\mu_k^{\mathrm{M}}(x)\mu_k^{\mathrm{L}}(x)\mu_k^{\mathrm{VL}}(x)\right)^T$$

Next, this fuzzy value will be used in the implication of the Fuzzy-ID3 algorithm. The Fuzzy-ID3 algorithm is an extension of the classical ID3 procedure. By splitting at each node one attribute or variable into several branches, each corresponding to one linguistic value, we obtain a generalization of classical decision trees through the application of fuzzy sets and fuzzy logic. In a Fuzzy-ID3 decision tree, the terminal node of each path corresponds to a combination of all the output classes with different membership degrees. In our study, a terminal node, denoted as $C = (C_1, C_2,\ldots, C_9)$, is

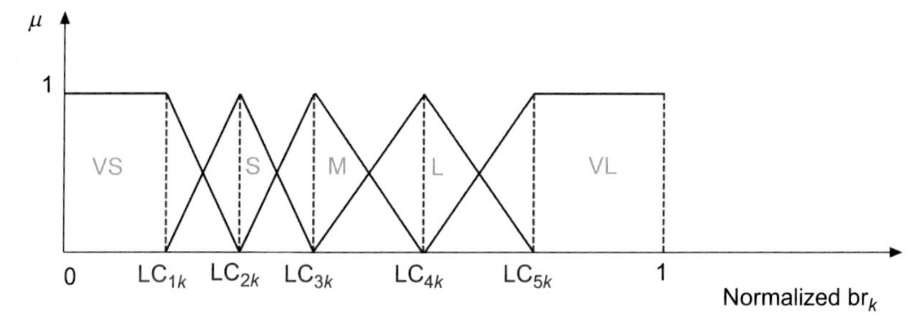

Figure 2.9 Membership functions of normalized body ratios.

considered a prediction of the evaluation score corresponding to x for the sensory descriptor d_j and the fashion expert ex_i.

For an example, one extracted fuzzy rule is as follows:

IF br_8 (Ratio Waist Circumference/Chest Circumference) = Small (S) AND br_3 (Ratio Neck Circumference/Stature) = Small (S) AND br_{15} (Ratio Neck Circumference/Hip Circumference) = Medium (M), THEN the corresponding Muscular level is (0 0 0.16 0.22 0 0 0 0 0)

The consequence of this rule means that *Muscular* level belongs to C_3 with a membership degree 0.16, to C_4 with a membership degree 0.22, and 0 for the other classes. Based on the fuzzy decision trees obtained for all the fashion experts ex_i's ($i = 1,\ldots, r$) and the sensory descriptor d_j, we aggregate them as follows.

Assuming that there exist t paths in the fuzzy decision tree T_{ij}, we can extract t following fuzzy rules:

RL_{ij}^{τ}: IF $\{(\text{br}_1 \text{ IS } v_1) \text{ OR } u_1\}$ AND... AND $\{(\text{br}_g \text{ IS } v_g) \text{ OR } u_g\}$, THEN the output of this terminal node IS $\rho_{ij}^{\tau}(C)$ with $\tau \in \{1, 2,\ldots, \text{t}\}$

In this rule, v_k ($k \in \{1,\ldots, g\}$) is a linguistic value separating one node of T_{ij} into multiple branches. They take values from the set {VS (very small), S (small), M (medium), L (large), VL (very large)}. The Boolean value u_k shows if there exists a body ratio variable br_k in this rule or not. We have $u_k = 0$ if br_k appears in the rule and $u_k = 1$ otherwise. The combination of all the nine output classes with different membership degrees is represented by:

$$\rho_{ij}^{\tau}(C) = \left(\rho_{ij}^{\tau}(C_1)\ldots\ldots\rho_{ij}^{\tau}(C_9)\right)$$

where $\rho_{ij}^{\tau}(C_1),\ldots, \rho_{ij}^{\tau}(C_9)$ are the membership degrees of $C_1,\ldots, C_9$ for the rule RL_{ij}^{τ}, respectively. They are calculated according to the Fuzzy-ID3 algorithm [16,27]. The corresponding membership degrees can be considered as weights of these linguistic values.

For each fuzzy decision tree T_{ij} and each output class C_k ($k = 1,\ldots, 9$), we aggregate all the t fuzzy rules by using the Sugeno method [28] and introducing the body ratios measured on Y to compute the related weights. By combining the results of all the output classes, the general aggregated result can also be considered as a fuzzy set C_{ij}, denoted by:

$$C_{ij} = \left(\mu_{ij}(C_1), \ldots, \mu_{ij}(C_9)\right) \text{ with } \sum_{k=1}^{9} \mu_{ij}(C_k) = 1$$

Next, we further aggregate the fuzzy sets C_{ij}'s given by all the r evaluators and obtain:

$$\text{RE}_j = \left(\mu_j(C_1), \ldots, \mu_j(C_9)\right) \quad \text{with}$$

$$\mu_j(C_k) = \sum_{i=1}^{r} \mu_{ij}(C_k) \Bigg/ \sum_{k=1}^{9} \sum_{i=1}^{r} \mu_{ij}(C_k) \quad \text{and } k = 1, \ldots, 9$$

By combining the results of all the sensory descriptors d_1, d_2,..., d_m, we obtain a fuzzy relation between these sensory descriptors and the body ratios of Y. It is a ($m \times 9$)-dimensional fuzzy matrix, denoted as:

$$\mathrm{REL}(D, \mathrm{BR}^Y) = \left(\mathrm{RE}_1^T, \ldots, \mathrm{RE}_m^T\right)^T$$

It represents the relevancy degree of a body shape Y related to a set of sensory descriptors describing human body shapes.

2.3.3.2 Computation of the relevancy degree *REL*(t_i, D)

From *Experiment III*, we have acquired the consumer data taken from $\{R_1, \ldots, R_5\}$, which characterize the relations between the desired fashion themes t_i's ($i = 1, \ldots, n$) and the sensory descriptors d_j's ($j = 1, \ldots, m$). In our study, this generalized relation is modeled using a fuzzy cognitive map. A fuzzy cognitive map [29] is a cognitive map within which the relations between the elements (eg, concepts, events, project resources) of a "mental landscape" can be used to compute the "strength of impact" of these elements. Fuzzy cognitive maps are signed fuzzy digraphs [20]. Any causal relation between two concepts is usually uncertain and varies between "positive and certain response" and "negative and certain response." Fuzzy cognitive maps offer a good approach to the stimulation and aggregation of generalized human perceptions provided by multiple evaluators [30]. A fuzzy cognitive map can effectively represent data provided by different consumers, which may be inconsistent.

Next, we aggregate the results of all the z consumer evaluators ec_k ($k = 1, \ldots, z$) and obtain a possibility distribution characterizing the relation between each fashion theme t_i and each sensory descriptor d_j. This relation aggregating all the z consumers can be expressed by a fuzzy value distributed on the set $\{R_1, \ldots, R_5\}$, ie:

$$\mathrm{REL}(t_i, d_j) = \left(\frac{\mathrm{NB}_{ij}(R_1)}{z} \ldots \frac{\mathrm{NB}_{ij}(R_5)}{z}\right) = \left(\mu_{ij}(R_1) \ldots \mu_{ij}(R_5)\right)$$

where $\mathrm{NB}_{ij}(R_k)$ ($k = 1, \ldots, 5$) is the number of evaluators selecting the linguistic value R_k when evaluating the relation between t_i and d_j. Evidently, the sum of all components of REL(t_i, d_j) is 1.

The ($m \times 5$)-dimensional fuzzy matrix REL(t_i, D) characterizes the fuzzy relation between a fashion theme t_i and all the sensory descriptors.

$$\mathrm{REL}(t_i, D) = \left(\mathrm{REL}(t_i, d_1)^T \mathrm{REL}(t_i, d_2)^T \mathrm{REL}(t_i, d_m)^T\right)^T$$
$$= \begin{bmatrix} \mu_{i1}(R_1) & \cdots & \mu_{i1}(R_5) \\ \vdots & \ddots & \vdots \\ \mu_{im}(R_1) & \cdots & \mu_{im}(R_5) \end{bmatrix}$$

2.3.3.3 Computation of the relevancy degree REL(t_i, BR^Y)

The relevancy of a specific naked body shape Y related to the desired fashion theme t_i, denoted as REL(t_i, BR^Y), is computed using the standard min–max composition "∘" of the previous fuzzy matrices REL(D, BR^Y) and REL(t_i, D), ie:

$$\mathrm{REL}\left(t_i,\mathrm{BR}^Y\right)=\mathrm{REL}(t_i,D)\circ\mathrm{REL}\left(D,\mathrm{BR}^Y\right)^T$$
$$=\begin{bmatrix}\mu_i^Y(R_1,C_1) & \cdots & \mu_i^Y(R_1,C_9)\\ \vdots & \ddots & \vdots\\ \mu_i^Y(R_5,C_1) & \cdots & \mu_i^Y(R_5,C_9)\end{bmatrix}$$

REL(t_i, BR^Y) is a (5 × 9)-dimensional fuzzy matrix.

Compared with the other aggregation operators, the composition "∘" can effectively combine individual fuzzy relations associated with the relation characterizing the importance of weights. It has the property of transitivity and is efficient in solving the problems of ordering and ranking for a set of alternatives [31].

2.3.4 Computation of the relevancy degrees for the perception of body shapes with a garment

2.3.4.1 Computation of the relevancy degree REL(de_v, S)

In a design process, fashion designers create a new fashion product using two approaches. In the first approach, designers consider the new style as a complete or partial revival of an existing reference style. In the second approach, a new style is generated from inspiration of several existing reference styles. In this context, we use the relevancy degrees to the reference styles to express a new garment design. The reference styles controlling new garment designs are denoted as $S=\{s_1, s_2,\ldots, s_\xi\}$. In the design of men's overcoats used in the proposed recommender system, we have $\xi=5$ with $s_1=$ "Chester," $s_2=$ "Ulster," $s_3=$ "Balmacaan," $s_4=$ "Trench," and $s_5=$ "Duffle."

The relevancy of de_v ($v\in\{1,\ldots,\lambda\}$) to the reference styles can be represented by:

$$\mathrm{REL}(\mathrm{de}_v,\ S)=\left(\mathrm{eq}(L(\mathrm{de}_v,s_1))\ldots\mathrm{eq}\left(L\left(\mathrm{de}_v,\ s_\xi\right)\right)\right)$$

where $L(\mathrm{de}_v, s_k)$ ($k\in\{1,\ldots, \xi\}$) is a linguistic relevancy degree of the garment design de_v related to the previous reference styles and evaluated by the fashion experts ex_i's using the following five linguistic scores: $L_1=$ "not belong to," $L_2=$ "a little belong to," $L_3=$ "belong to," $L_4=$ "quite belong to," $L_5=$ "totally belong to." The corresponding equivalence values are represented by eq(.), with $\mathrm{eq}(L_1)=0$, $\mathrm{eq}(L_2)=0.25$, $\mathrm{eq}(L_3)=0.5$, $\mathrm{eq}(L_4)=0.75$, and $\mathrm{eq}(L_5)=1$.

2.3.4.2 Computation of the relevancy degree REL(D, s_k)

From the sensory evaluation of the virtual male body shapes with design style (*Experiment II*), the relevancy of a reference style s_k to the sensory descriptors describing body shapes should be extracted for determining how each reference style can change the image of the standard body shape (CA170) related to these sensory descriptors. For example, a duffle-style overcoat can make the male wearer of standard body shape more vivid. In *Experiment II*, the fashion experts ex_i's give their evaluations according to their experience.

Similar to the sensory evaluation of the virtual body shapes, for an expert ex_i, the evaluation score of a standard body shape with a specific reference style s_k related to a sensory descriptor d_j also takes on value from the linguistic set $\{C_1, \ldots, C_9\}$. Furthermore, the evaluation results from r expert are aggregated in a similar way to that presented in Section 2.3.4.1, ie:

$$\mathrm{REL}(s_k, d_j) = \left(\frac{\mathrm{NB}_{kj}(C_1)}{r} \ldots \frac{\mathrm{NB}_{kj}(C_9)}{r}\right) = (\mu_{kj}(C_1) \ldots \mu_{kj}(C_9))$$

where $\mathrm{NB}_{kj}(C_k)$ ($k = 1, \ldots, 9$) is the number of evaluators selecting the linguistic value C_k when evaluating the relation between s_k and d_j. Evidently, the sum of all components of $\mathrm{REL}(s_k, d_j)$ is 1.

The $(m \times 9)$-dimensional fuzzy matrix $\mathrm{REL}(s_k, D)$ characterizes the fuzzy relation between a reference style s_k and all the m sensory descriptors of D.

$$\mathrm{REL}(s_\lambda, D) = \left(\mathrm{REL}(s_\lambda, d_1)^T \mathrm{REL}(s_\lambda, d_2)^T \ldots \mathrm{REL}(s_\lambda, d_m)^T\right)^T$$

2.3.4.3 Computation of the relevancy degree $REL(t_i, BR^{CA170} \vee de_v)$

The relation between a fashion theme t_i and the standard body shape (CA170) with a new garment design de_v represents how this garment design changes the image of this body shape for the fashion theme t_i. Considering that a garment design de_v can be decomposed into a combination of reference styles, we first evaluate the relations between t_i and the standard body shape with all the reference styles and then aggregate these relations using the fuzzy composition operation.

The fuzzy relation between the fashion theme t_i and the standard body shape with the style s_λ is calculated by:

$$\mathrm{REL}(t_i, s_\lambda) = \mathrm{REL}(t_i, D) \circ \mathrm{REL}(D, s_\lambda)^T$$

This is a (5×9)-dimensional fuzzy matrix.

Taking $\mathrm{eq}(L(de_v, s_\lambda))$ as the weight of $\mathrm{REL}(t_i, s_\lambda)$, we use the scalar multiplication of fuzzy matrices [32] to compute the weighted relevancy degree of t_i to the reference style s_λ, ie:

$$\mathrm{WREL}(t_i, de_v, s_\lambda) = \mathrm{eq}(L(de_v, s_\lambda)) \cdot \mathrm{REL}(t_i, s_\lambda)$$
$$= \begin{pmatrix} \mathrm{wrel}_{11}(t_i, de_v, s_\lambda) & \cdots & \mathrm{wrel}_{19}(t_i, de_v, s_\lambda) \\ \cdots & \ddots & \cdots \\ \mathrm{wrel}_{51}(t_i, de_v, s_\lambda) & \cdots & \mathrm{wrel}_{59}(t_i, de_v, s_\lambda) \end{pmatrix}$$

Finally, the fuzzy relation between the fashion theme t_i and the standard body shape with the new garment design de_v is calculated by:

$$\mathrm{REL}\left(t_i, \mathrm{BR}^{\mathrm{CA170}} \vee \mathrm{de}_v\right) = \bigcup_{\lambda=1}^{\xi} \mathrm{WREL}(t_i, \mathrm{de}_v, s_\lambda)$$

$$= \begin{pmatrix} \max\limits_{\lambda}\{\mathrm{wrel}_{11}(t_i, \mathrm{de}_v, s_\lambda)\} & \cdots & \max\limits_{\lambda}\{\mathrm{wrel}_{19}(t_i, \mathrm{de}_v, s_\lambda)\} \\ \cdots & \ddots & \cdots \\ \max\limits_{\lambda}\{\mathrm{wrel}_{51}(t_i, \mathrm{de}_v, s_\lambda)\} & \cdots & \max\limits_{\lambda}\{\mathrm{wrel}_{59}(t_i, \mathrm{de}_v, s_\lambda)\} \end{pmatrix}$$

This is also a (5 × 9)-dimensional fuzzy matrix.

2.3.4.4 Computation of the relevancy degree REL(t_i, $BR^Y \vee de_v$)

The fuzzy matrix REL(t_i, BR^Y), can be used to characterize the relation of the fashion theme t_i and a specific body shape Y. In the same way, the fuzzy matrix REL(t_i, $\mathrm{BR}^{\mathrm{CA170}} \vee \mathrm{de}_v$) can be used to characterize the relation between t_i and the standard body shape with the garment design de_v.

The relation between the fashion theme t_i and the specific body shape Y with the garment design de_v is calculated by using the fuzzy union operation (maximum operator) of the above two fuzzy relation matrices.

2.3.5 Comparison of the two relevancy degrees

In this section, for a specific body shape Y, we compare the two (5 × 9)-fuzzy relation matrices or relevancy degrees, REL(t_i, BR^Y) and REL(t_i, $\mathrm{BR}^Y \vee \mathrm{de}_v$), obtained in Section 2.3.4.4, to determine if the new style de_v can really improve the image of the body shape toward the direction of t_i. The style de_v is considered a feasible design style if this improvement can be validated by the comparison of these two relevancy degrees.

In this approach, the gravity centers of fuzzy relation matrices are used to determine the comparison criteria. To compute the gravity center of REL(t_i, BR^Y), we first transform the linguistic evaluation values $\{R_1, R_2, R_3, R_4, R_5\}$ into their equivalence numerical values {0, 0.25, 0.5, 0.75, 1}, and $\{C_1, C_2, C_3, C_4, C_5, C_6, C_7, C_8, C_9\}$ into {−1, −0.75, −0.5, −0.25, 0, 0.25, 0.5, 0.75, 1}.

The gravity center of the fuzzy relation matrix

$$\mathrm{REL}\left(t_i, \mathrm{BR}^Y\right) = \begin{pmatrix} \mu_i^Y(R_1, C_1) & \ldots & \mu_i^Y(R_1, C_9) \\ \vdots & \ddots & \vdots \\ \mu_i^Y(R_5, C_1) & \cdots & \mu_i^Y(R_5, C_9) \end{pmatrix}$$

is calculated by:

$$G_{\mathrm{REL}(t_i,\mathrm{BR}^Y)} = \sum_{j=1}^{5}\sum_{k=1}^{9} \mu_i^Y(R_j, C_k)\mathrm{eq}(R_j)\mathrm{eq}(C_k)$$

where $\mathrm{eq}(R_j)$ and $\mathrm{eq}(C_k)$ are the equivalence values of R_j and C_k, respectively.

The projection of the gravity center on the axis Y is calculated by:

$$\alpha^i = G_{\mathrm{REL}(t_i,\mathrm{BR}^Y)} \Bigg/ \sum_{j=1}^{5}\sum_{k=1}^{9} \mu_i^Y(R_j, C_k)\mathrm{eq}(R_j)$$

Similarly, the projection of the gravity center on the axis t_i is calculated by:

$$\beta^i = G_{\mathrm{REL}(t_i,\mathrm{BR}^Y)} \Bigg/ \sum_{j=1}^{5}\sum_{k=1}^{9} \mu_i^Y(R_j, C_k)\mathrm{eq}(C_k)$$

The combination of these two projections is:

$$\mathrm{GP}_{\mathrm{REL}(t_i,\mathrm{BR}^Y)}^{Y\times t_i} = \left(\alpha^i, \beta^i\right)$$

The product of the above two projections is:

$$\omega^i = \alpha^i \cdot \beta^i$$

For the fuzzy matrix $\mathrm{REL}(t_i,\ \mathrm{BR}^Y \vee \mathrm{de}_v)$, its gravity center and projections, denoted as $\overline{\overline{\alpha_v^i}}$ and $\overline{\overline{\beta_v^i}}$, respectively, are calculated in the same way. The product of these two projections is denoted as $\overline{\overline{\omega_v^i}}$.

The image difference between the naked body shape Y and Y with the newly designed garment style de_v related to the fashion theme t_i, can be characterized by calculating the corresponding projections, namely, $\Delta\alpha_v^i = \overline{\overline{\alpha_v^i}} - \alpha^i$ and $\Delta\beta_v^i = \overline{\overline{\beta_v^i}} - \beta^i$.

And the corresponding difference of the projection products is $\Delta\omega_v^i = \overline{\overline{\omega_v^i}} - \omega^i$.

To compare the relevancy degrees obtained from *Model I* (for the naked body shape) and *Model II* (for the body shape with a new garment style), we define the following rules.

Rule 1: For the value of $\alpha\left(\alpha^i \text{ or } \overline{\overline{\alpha_v^i}}\right)$:

The bigger α is, the higher is the relevancy degree of the body shape Y (naked or with garment design style) to the fashion theme t_i for design expert perception. Otherwise, the relevancy degree is lower.

Rule 2: For the value of $\beta\left(\beta^i \text{ or } \overline{\overline{\beta^i_v}}\right)$:
The bigger β is, the higher is the relevancy degree of the body shape Y (naked or with garment design style) to the fashion theme t_i for consumer perception. Otherwise, the relevancy degree is lower.
Rule 3: For the value of $\omega\left(\omega^i \text{ or } \overline{\overline{\omega^i_v}}\right)$:
The bigger ω is, the higher is the relevancy degree of the body shape Y (naked or with garment design) to the fashion theme t_i for the compromise between design expert perception and consumer perception. Otherwise, the relevancy degree is lower.

Therefore, if $\Delta\omega^i_v$ is a positive value, the relevancy degree of the body shape Y with the garment design de_v to the fashion theme t_i is higher than that of the naked body shape Y for the compromise between the expert perception and the consumer perception.

If $\Delta\omega^i_v$ is 0, there is no difference.

If $\Delta\omega^i_v$ is a negative value, the relevancy degree of the body shape Y with the garment design de_v to the fashion theme t_i is smaller than that of the naked body shape Y for the compromise between the expert perception and the consumer perception.

According to these rules of comparison and analysis, we define the evaluation criterion of new garment design styles as follows.

If the value of $\Delta\omega^i_v$ is positive, the garment design style de_v should be selected as a feasible design style for improving the relevancy of the body shape Y to the fashion theme t_i. Otherwise, the design style de_v should be rejected.

2.3.6 Application and validation

The proposed fashion recommender system had been applied to target market selection scenario to validate its effectiveness. The aim of this scenario is to select a set of feasible design styles meeting the needs of one specific population instead of a personalized body shape. For this purpose, we select 60 male customers with different body shapes. The objectives of this design can handle each several fashion themes. In this application, it proposes three objectives, as follows:

Objective 1: "Sporty" and "Nature"
Objective 2: "Sporty" and "Attractive"
Objective 3: "Nature" and "Attractive"

Six new styles are described using their linguistic membership degrees related to the five reference styles:

de_1: totally belonging to "Chester"
de_2: totally belonging to "Duffle"
de_3: fairly belonging to "Chester" and a little to "Ulster"
de_4: a little belonging to "Chester" and fairly to "Ulster"
de_5: fairly belonging to "Balmacaan" and a little to "Trench"
de_6: a little belonging to "Balmacaan" and totally to "Trench"

Table 2.1 gives the rates of feasible responses of all 60 body shapes for each new design style and each objective. As each objective is composed of two fashion themes,

Table 2.1 Statistical results of feasible responses for all the body shapes

	de_1 (%)	de_2 (%)	de_3 (%)	de_4 (%)	de_5 (%)	de_6 (%)
Objective 1	53	17	53	43	40	65
Objective 2	68	17	63	43	32	60
Objective 3	53	22	53	55	32	58

a response is feasible if and only if the evaluation criteria $\Delta\omega_v^i$ are positive for both fashion themes. In any case, the higher the evaluation criterion, the more feasible the corresponding design style is to the design objective.

In this application, it defines the following rules for selecting relevant design styles for the whole population:

Rule 1: If the rate of feasible responses is ≥50% (the image of body shape is improved for more than 50% of the population), the corresponding style will be selected for the population.
Rule 2: The design style having the highest rate of feasible responses will be regarded as the best solution.

Table 2.2 shows the final recommendations for three different design objectives according to these rules. The table shows that de_6 is considered the best design style for Objective 1 and Objective 3, whereas de_1 is the best for Objective 2.

These results are also validated at the consumer level and the fashion expert level. Table 2.3 gives the results of the consumers' validation.

From Table 2.3, we find that, apart from de_6 for Objective 2 (60%), de_3 for Objective 3 (47%), and de_4 for Objective 3 (75%), the rates of positive responses (absolutely acceptable, fairly acceptable, and acceptable) for the other seven feasible recommended styles, given by 60 randomly selected consumers, are higher than 80%. This means that the recommender system can generally be accepted by the target market.

Table 2.4 shows the validation results given by the fashion experts. We can find that at least five of seven fashion experts accept the feasible recommended styles, and at least six experts accept the best recommended styles. At least five fashion experts consider that the proposed recommender system can be applied to the fashion market.

Table 2.2 Recommendation results for the whole population

	Feasible selection	Best selection
Objective 1	de_1, de_3, de_6	de_6
Objective 2	de_1, de_3, de_6	de_1
Objective 3	de_1, de_3, de_4, de_6	de_6

Table 2.3 **Validation of the recommended styles by the consumers**

	Feasible recommended styles	Statistics of the consumers' evaluation scores			Consumers' most favorable style
		Absolutely acceptable (%)	Fairly acceptable (%)	Acceptable (%)	
Objective 1	de_1	11.67	23.33	45.00	de_6
	de_3	6.67	26.67	53.33	
	de_6	21.67	26.67	41.67	
Objective 2	de_1	18.33	30.00	41.67	de_1
	de_3	11.67	11.67	65.00	
	de_6	1.67	23.33	33.33	
Objective 3	de_1	11.67	21.67	41.67	de_4
	de_3	6.67	18.33	21.67	
	de_4	11.67	21.67	43.33	
	de_6	11.67	18.33	50.00	

Table 2.4 **Validation of the recommended styles by the fashion experts**

Rates of positive responses	Effectiveness	Acceptability	Realizability
Objective 1	5/7	7/7	6/7
Objective 2	6/7	7/7	7/7
Objective 3	5/7	6/7	5/7

The proposed recommender system has been validated by the previous target mass market-oriented design selection with high acceptability for consumers and fashion experts. However, for fashion designers, there exist some hesitations in applying it directly to the fashion market (realizability is relatively low), because their classical design process will also be modified with any fashion recommender system. This problem can be solved with more and more successful stories in application of the proposed system.

2.4 Conclusion

In practice within the apparel industry, not only the professional fashion designer's knowledge and human perceptions on fashion products, but also the consumer's perception and emotion play an important role in the decision to classify the new fashion product toward the specific target market. Concerning fashion design, the human perception, including expert knowledge and consumer cognition, is often conceptual and ambiguous, which is difficult to be characterized using classic computational tools such as statistics. In this context, fuzzy logic techniques will be more efficient for formalizing perceptual data, relations between concepts, and other uncertain problems from product development to target market.

In this chapter, as the application of decision making based on fuzzy logic in apparel supply chain, a perception-based fashion design recommender system is presented to support fashion marketing in selecting the best personalized fashion design scheme and in designing new products. Unlike existing fashion recommender systems, which mainly deal with basic or generic garment styles by exploiting cognitive domain expertise and user interaction, this system originally integrates emotional fashion themes and personalized wearer's body shapes.

This system also acquires the designer's and the consumer's perceptual data in a different but more systematic way. The acquired fashion designer's perception is strongly related to the designer's professional experience and knowledge, whereas the consumer's perception characterizes the evolution of the target market. The techniques of sensory evaluation are applied to extract the designer's perception of human body shapes, expressed by a number of normalized descriptors, and the consumer's perception of relations between body shapes and emotional fashion themes. The consumer's perceptual data are strongly related to a sociocultural context. A fashion theme can be considered a social image desired by the consumer and an orientation of new design. The proposed fashion recommender system integrates the principle of garment design and permits the acquisition of more complete perceptual data on normalized keywords (sensory descriptors and fashion themes) from consumers and designers and is more appropriate for fashion products.

Because of the capacity of interpretation and robustness of the proposed fashion recommender system, fuzzy logic constitutes the main computational tool for the formalization and modeling of perceptual data and their relations. The perceptual data of both consumers and designers are formalized mathematically using fuzzy sets and fuzzy relations. The relationship between human body measurements and basic sensory descriptors, provided by designers, is modeled using fuzzy decision trees. Fuzzy decision trees are adopted due to their capacity to learn from data. They can also be interpreted by fashion designers. The fuzzy decision trees used here constitute an empirical model based on learning data measured and are evaluated on a set of representative samples. The complex relationship between basic sensory descriptors and fashion themes, given by consumers, is modeled using fuzzy cognitive maps, which are a robust and interpretable tool for modeling a consumer's cognition about the causal relations between fashion themes and body shapes. The combination of two models can provide more complete information to the fashion recommender system, making it possible to evaluate a specific body shape related to a desired emotional fashion theme and to obtain the design orientation in order to improve the image of the body shape. The standard min−max composition operation of fuzzy relations is used for aggregating the different fuzzy relevancy degrees obtained from the previous fuzzy decision trees and cognitive maps and for building an overall evaluation criterion for new personalized design schemes.

References

[1] Easey M. Fashion marketing. 3rd ed. Oxford: Wiley Blackwell Press; 2009. 278 pp.
[2] Ives B, Piccoli G. Custom made apparel and individualized service at lands' end. Commun Assoc Inf Syst 2003;11:79–93. Article 3.
[3] Carpenter S, Huffman K. Visualizing psychology. John Wiley & Sons Inc.; 2007. 544 pp.
[4] Brangier E, Barcenilla J. Concevoir un Produit Facile à Utiliser. Paris: Editions d'Organisation; 2003. p. 662.
[5] Zhu Y, et al. A general methodology for analyzing fashion oriented textile products using sensory evaluation. Food Qual Prefer 2010;21(8):1068–76.
[6] Stone H, Sidel JL. Sensory evaluation practices. 3rd ed. California: Elsevier Academic Press; 2004. 408 pp.
[7] Zeng XY, et al. Intelligent sensory evaluation: concepts, implementations and applications. Math Comput Simul 2008;77:443–52.
[8] Dijksterhuis GB. Multivariate data analysis in sensory and consumer science. Trumbull, Connecticut, USA: Food & Nutrition Press Inc; 1997.
[9] KOEHL L, et al. Sensory evaluation of industrial products for exploiting consumer's preferenc. Math Comput Simul 2008;77:443–52.
[10] Zadeh LA. Fuzzy sets. Inf Control 1965;8:338–53.
[11] Chevrie F, Guely F. La Logique Floue. Cah Tech 1998;191. Available in: http://www.schneider-electric.com.
[12] Ross TJ. Fuzzy logic with engineering applications. 2nd ed. UK: John Wiley & Sons Inc.; 2004. 628 pp.
[13] Han J, Kamber M. Data mining: concepts and techniques. San Francisco: Morgan Kaufmann; 2006. 800 pp.
[14] Quinlan JR. An empirical comparison of genetic and decision-tree classifiers. In: Proc. 1988 int. Conf. Machine Learning (ML'88), San Mateo, CA; 1988. p. 135–41.
[15] Breiman L, et al. Classification and regression trees. Monterey, CA: Wadsworth & Brooks/Cole Advanced Books & Software; 1984. 358 pp.
[16] Janikow CZ. Exemplar learning in fuzzy decision trees. In: Proceedings of FUZZ-IEEE; 1996. p. 1500–5.
[17] Bartczuk L, Rutkowska D. A new version of the fuzzy-ID3 algorithm. In: ICAISC 2006, LNAI 4029; 2006. p. 1060–70.
[18] Yuan YF, Shaw MJ. Introduction of fuzzy decision trees. Fuzzy Sets Syst 1995;69(2):125–39.
[19] Axelrod R. Structure of decision: the cognitive map of political elites. Princeton University Press; 1976. 421 pp.
[20] Kosko B. Fuzzy cognitive maps. Int J Man-Machine Stud 1986;(24):65–75.
[21] Vasantha Kandasamy WB, Smarandache F. Fuzzy cognitive maps and neutrosophic cognitive maps. Phoenix: Xiquan; 2003. 211 pp.
[22] Papageorigiou EI, et al. Fuzzy cognitive map based decision support system for thyroid diagnosis management. In: FUZZ-IEEE 2008, Fuzzy Systems; June 1–6, 2008. p. 1204–11.
[23] Khan MS, Quaddus M. Group decision support using fuzzy cognitive maps for causal reasoning. Group Decis Negot 2004;(13):463–80.
[24] Fan J, Yu W, Hunter I. Clothing appearance and fit: science and technology. Cambridge: Woodhead Publishing Limited and CRC Press LLC; 2004.

[25] Provot X. Deformation constraints in a mass-spring model to describe rigid cloth behavior. In: Proc. Graph. Interface; 1995. p. 147–54.
[26] Baek SY, et al. Kansei factor space classified by information for Kansei image modeling. Appl Math Comput 2008;205(2):874–82.
[27] Janikow CZ. Fuzzy decision trees: issues and methods. IEEE Trans Syst Man Cybern 1998;28(1):1–14.
[28] Takagi T, Sugeno M. Fuzzy identification of systems and its applications to modeling and control. IEEE Trans Syst Man Cybern 1985;15(1):116–32.
[29] Stylios CD, et al. Fuzzy cognitive map architectures for medical decision support systems. Appl Soft Comput 2008;8:1243–51.
[30] Khan MS, Quaddus M. Group decision using fuzzy cognitive maps for causal reasoning. Group Decis Negot 2004;13(5):463–80.
[31] Peneva V, Popchev I. Aggregation of fuzzy preference relations to multicriteria decision making, fuzzy optim. Decis Mak 2007;6:351–65.
[32] Lee KH. First course on fuzzy theory and applications. Berlin: Springer Press; 2005. 345 pp.

Using radiofrequency identification (RFID) technologies to improve decision-making in apparel supply chains

3

H.-L. Chan
The Hong Kong Polytechnic University, Kowloon, Hong Kong

3.1 Introduction

In the fashion and apparel industries, products are featured by short product life cycle, highly volatile and unpredictable demand, and long lead time (Sen, 2008). Ever-changing fashion trends, a comprehensive product mix, and quick production and delivery pressure make inventory management more arduous. According to Dehoratius and Raman (2008), inventory discrepancy between the inventory record and the actual physical level is commonly observed in retailing, where only 35% of the collected records are accurate. Besides inaccurate inventory records will lead to overstocking or a stockout scenario, and result in lost sales of about $1350 per week (DeHoratius and Raman, 2008). The revenue loss from stockout is considerably larger than the inventory loss value (Kang and Gershwin, 2004).

The fashion and apparel industry is one of the sectors that has the longest supply chain (Lowson et al., 1999) and the market is highly competitive (Bhardwaj and Fairhurst, 2010). In order to manage these short-life products (e.g., fashionable items) effectively, an efficient supply chain management is the key to success (Karkkainen, 2003). It is hence crucial to adopt an information technology to monitor the products and facilitate communication and information exchange among supply chain partners so as to satisfy customer needs, enhance operation efficiency, and improve decision-making in the supply chain context.

Barcoding systems have been widely adopted in different industries since the 1970s for product identification (Attaran, 2007). This is an optical technology that requires line-of-sight to read the data. Data are stored in the lines and spaces (i.e., linear barcode) or in the symbols (i.e., two-dimensional barcode). When a barcode reader scans the barcode, it gathers the encoded information and transmits it to a computer for storage (Sriram et al., 1996). In terms of inventory management, barcoding systems are used to identify the product type (White et al., 2007) and enhance ordering accuracy (IT Reseller, 2006). Even though a barcoding system has a relatively low implementation cost and high read-rate accuracy, it demands extensive human efforts to scan every single item manually and the users have to scan the barcode carefully so that the reader can read the data successfully. This practice is time consuming, especially when it is applied

Information Systems for the Fashion and Apparel Industry. http://dx.doi.org/10.1016/B978-0-08-100571-2.00003-8

for inventory management at the item level. In addition, the external environment condition also affects the performance of the barcoding system. For example, if the barcode is covered by dirt or used in a humid environment, it may not function properly.

In light of the deficiency of the barcoding system, radiofrequency identification (RFID) systems have been proposed to facilitate an efficient supply chain (Kärkkäinen, 2003). RFID is a technology in which the information can be transmitted in real time through radio waves (Sarac et al., 2010). A typical RFID system consists of three main components: an RFID tag, an RFID reader, and middleware. At present, two types of RFID tags—passive and active—are widely used (Zhu et al., 2012). The choice of RFID tag depends on many factors and the application. Inside the tag, there is a chip (which contains the product information) and an antenna (which sends the radio signals). The RFID reader is also connected to an antenna to emit and receive the signals from the RFID tag, and convert them to digital data. The data are then filtered and formatted in the middleware; finally, all the processed data are stored in the enterprise applications (Wong and Guo, 2014). Specifically, different kinds of RFID readers emit different frequency ranges and affect the corresponding read distance (see Chawla and Ha, 2007 for details). The nature of the RFID system can help to identify the product and collect the product information automatically without physical contact. Because both RFID and barcode systems can be used to identify an object, their characteristics have been compared extensively. Table 3.1 summarizes the key differences between RFID and barcode systems.

Even though RFID systems can improve product traceability and visibility along the supply chain (Sarac et al., 2010), they have not yet been used on a large scale in real-world applications. The system investment cost and its return, tag placement, and successful read-rate are driving fashion retailers to delay full RFID adoption. In practice, retailers have conducted various pilot test projects to evaluate the true performance of RFID systems. This chapter aims to illustrate how the RFID technology can be implemented in the fashion industry through a case study and discusses possible future research directions for RFID applications in an apparel supply chain. Based on the case study analysis, we demonstrate that RFID systems can be used for shop floor management, logistics and distribution management, and customer relationship management. Our findings generate managerial insights for academicians and practitioners to better understand the application scopes and potential benefits of RFID systems in apparel supply chains.

A literature review on RFID implementation in the fashion industry is presented in Section 3.2. In Section 3.3, case studies are conducted to demonstrate real-world applications of RFID in apparel supply chains. Finally, our conclusions and suggestions for future research directions are presented in Section 3.4.

3.2 Literature review

Numerous studies have examined RFID system adoption and discussed the potential benefits and concerns in the healthcare industry (Tseng et al., 2008; Yao et al., 2012),

Table 3.1 **Comparison of bar-coding and RFID systems**

	Bar-coding system	RFID system
Operations principle	Requires line-of-sight so that the information can be read successfully	Uses radio waves and generates electromagnetic field to transmit the data between the reader and the tag
Product	Also suitable for metal or liquid products	May not be suitable for metal or liquid products
Information	Cannot be updated	Can be updated for passive tags
System security	Data cannot be protected	Data can be protected
Efficiency	Lower efficiency: • Only one label can be scanned each time by one scanner • Requires human efforts to conduct product scanning	Higher efficiency: • Multiple tags can be read at the same time by one reader • Human effort is not necessary because it can work automatically
Functioning condition	Cannot function properly in humid environments or when the label is covered by dirt	Functions well with excessive dirt, moisture, and extreme temperatures
Implementation cost	Relatively cheaper	Much more expensive

Based on the findings by Aguilar et al. (2006), Michael and McCathie (2005), Wyld (2006), White et al. (2007).

aircraft industry (Ngai et al., 2007), and travel industry (Meingast et al., 2007) through case studies. The examination of RFID systems in the fashion industry can be divided into four aspects: a descriptive case of RFID applications, examination of RFID investment and return, exploration of the motivation and impact of RFID system adoption, and RFID-enabled systems development. Table 3.2 summarizes the existing literature on the use of RFID systems in the apparel supply chain.

3.2.1 Descriptive case of RFID applications

Inventory inaccuracy occurs when the recorded inventory level does not match the actual on-hand level perfectly (Raman et al., 2001). This problem can be caused by four reasons: transaction errors, shrinkage errors, misplacement, and supply errors. Among these causes, shrinkage results in 1.46% of turnover in western European retailing (Bamfield, 2004). Inventory inaccuracy (also referred to as *stock loss*) means that the ready-to-sell products are lost. Shoplifting and employee theft are the major contributors to this problem (Sarac et al., 2010). Existing studies propose using an RFID system to mitigate the inventory inaccuracy problem.

Table 3.2 Summary of the existing literature on the use of RFID systems in apparel supply chains

Literature	Research aspect	Objectives
Al-Kassab et al. (2009)	RFID-enabled system development	To develop a cost-benefit calculator to evaluate the performance of an RFID system
Azevedo and Ferreira (2009)	Descriptive case on RFID applications	To present how the fashion retailer Throttleman benefited from using RFID technology
Azevedo and Carvalho (2011)	Descriptive case of RFID applications	To discuss the reasons for RFID technology adoption
Azevedo et al. (2014a)	Descriptive case of RFID applications	To analyze the advantages, disadvantages, and barriers associated with RFID deployment from different supply chain members' perspectives
Azevedo et al. (2014b)	Descriptive case of RFID applications	To analyze the advantages, disadvantages, and barriers associated with RFID deployment from different supply chain members' perspectives
Bottani et al. (2009)	Examination of an RFID investment using cost-benefit analysis	To quantify the returns from the RFID execution in the fashion supply chain based on the 11 targeted sites in Italy (6 distribution centers and 5 retail stores)
Choi et al. (2014)	RFID-enabled system development	To study a track-and trace anticounterfeiting system with a discussion of e-pedigree formatting, data synchronization, and traceability control
Choy et al. (2009)	RFID-enabled system development	To develop an innovative system to track swatches and fabric status
De Marco et al. (2012)	Exploration of the impacts of RFID system using simulation analysis	To access the impacts of an RFID system
Garrido Azevedo and Carvalho (2012)	Descriptive case of RFID applications	To indicate the advantages, disadvantages, and barriers of RFID adoption in fashion supply chains, specifically in logistics activities based on a proposed conceptual framework
Guo et al. (2014)	RFID-enabled system development	To develop a decision support system for production control decision-making

Table 3.2 Continued

Literature	Research aspect	Objectives
Lee et al. (2013)	RFID-enabled system development	To develop a decision support system to better allocate resources for garment manufacturing
Loebbecke et al. (2006)	Descriptive case of RFID applications	To illustrate how an RFID system was applied in the Kaufhof department store and Gerry Weber
Lui and Lo (2014)	Examination of an RFID investment using statistical analysis	To conduct an empirical analysis of the financial performance of fashion companies with the use of an RFID system
Massimo et al. (2012)	Examination of an RFID investment using cost-benefit analysis	To conduct a pilot test to quantify the benefits of using an RFID system in an Italian fashion supply chain
Moon and Ngai (2008)	Exploration of the impacts of RFID system adoption using content analysis	To explore the impacts and challenges of an RFID system
Ngai et al. (2012)	Exploration of the motivation for RFID system adoption with a technology push and need pull framework	To examine the motivation for RFID system adoption from a manufacturer's perspective.
Quetti et al. (2012)	Exploration of the motivation and impact of RFID system adoption	To investigate the factors attributed to RFID adoption in a vertical silk supply chain
Tajima (2012)	Exploration of the motivation and impact of RFID system adoption using game theory and stability analysis	To explore the decision for RFID system adoption from a small manufacturer's perspective
Teucke and Scholz-Reiter (2014)	RFID-enabled system development	To discuss a prototypical software tool for making decisions on delivery schedules to retailers.
Thiesse et al. (2009)	Descriptive case of RFID applications	To compare the application area for an RFID pilot test and the full adoption in the Kaufhof case

Continued

Table 3.2 Continued

Literature	Research aspect	Objectives
Ustundag et al. (2013)	Examination of an RFID investment using cost-benefit analysis	To present a case study to determine the value of using an RFID system in a Turkish apparel retail company through a cost-benefit analysis
Wong et al. (2006)	RFID-enabled system development	To develop an infrastructure so that all supply chain members are able to access the data with a reduced operations cost
Wong et al. (2012)	RFID-enabled system development	To develop two intelligent RFID-based systems for smart dressing and cross-selling
Zhou and Piramuthu (2013)	RFID-enabled system development	To develop an authentication protocol to tackle a shrinkage problem

Loebbecke et al. (2006) conducted a descriptive case study to illustrate how an RFID system was adopted in a European fashion supply chain, including the Kaufhof department store and manufacturer Gerry Weber, to prevent inventory loss from stolen items. Thiesse et al. (2009) compared the RFID application areas in a pilot project and full adoption in the Kaufhof case. For the full adoption, the RFID system was incorporated with the point-of-sale (POS) system so as to update and visualize the physical in-store level. In addition, they adopted a conceptual model to analyze the company's performance under different RFID investments. Azevedo and Ferreira (2009) presented how the fashion retailer Throttleman benefited from using an RFID system.

There are various concerns with RFID deployment from the supply chain members' perspectives. Azevedo and Carvalho (2011) and Azevedo et al. (2014a) reviewed the existing literature to analyze the advantages, disadvantages, and barriers associated with RFID deployment. Azevedo and Carvalho (2011) focused on manufacturers (VF Corporation, Lawsgroup, Gerry Weber), distributers (Jobstl Warehousing and Fashion), and retailers (Charles Vogele Group, Kaufhof, Trottleman, American Apparel), while Azevedo et al. (2014a) integrated different manufacturers (The Basic House, Crystal Group, Griva, VF Corporation, Lawsgroup, Gerry Weber), distributers (Jobstl Warehousing and Fashion, LTC-Logistics, DHL Solutions fashion) and retailers (Charles Vogele Group, Kaufhof, Trottleman, American Apparel, S. Oliver Bernd Freier, Benetton). Later, Azevedo et al. (2014b) presented another cross-case analysis to scrutinize the applications, advantages, and reasons for RFID technology adoption, with the additional consideration of a textile manufacturer (Griva) and two manufacturers (Gaardeur AG, and Lemmi Fashion). Garrido Azevedo and Carvalho (2012) proposed a conceptual framework to indicate the advantages, disadvantages, and barriers of RFID adoption in the fashion supply chain, especially focusing on the impact of logistics activities.

3.2.2 Examination of RFID investments and returns

RFID implementation requires continuous commitment and significant investment, and numerous studies have evaluated the return of this technology using cost-benefit analysis. Cost-benefit analysis is used to determine the present value of balance (total benefit minus total cost) with given constraints (Prest and Turvey, 1965). Bottani et al. (2009) studied a two-echelon fashion supply chain, including a distribution center and retail store, and conducted a profitability evaluation for an RFID system. They quantified the returns from RFID execution in a fashion supply chain based on 11 targeted sites (six distribution centers and five retail stores) in Italy and discussed the business process reengineering to implement such technology in the retail store and distribution center successfully. To estimate the value of the RFID system, an economic analysis was conducted and a breakeven curve was derived in terms of the number of retail stores and tag cost.

Ustundag et al. (2013) presented a case study on RFID system adoption in a Turkish apparel supply chain, and then conducted a cost-benefit analysis of RFID investment. By estimating the total cost savings and total expenses for tagging 25,000 items in a retail store, it was found that the return on the RFID investment was about 3 years. The authors concluded that it is more desirable to apply a reusable RFID tag than a disposable one because this can decrease the total cost when used continuously.

Massimo et al. (2012) conducted an experiment to evaluate the potential benefits of RFID in the apparel supply chain. They applied RFID tags to track 20,000 items in an Italian fashion brand, with sales and turnover from replenishment increasing by 0.52% per hour and 4.91%, respectively. In addition, the RFID system was able to collect valuable data (e.g., counting the number of fittings on a particular item) to facilitate appropriate inventory planning (and sales) and improve customer satisfaction upon the requests.

Lui and Lo (2014) conducted a statistical analysis to examine and compare the financial performance of 18 clothing and textile companies and other firms based on public data. They found that the RFID system can significantly reduce the inventory days (i.e., the number of days that on-hand inventory can be sold) in the textile and clothing industry than that of the other manufacturing firms in the long run.

3.2.3 Exploration of the motivation and impact of RFID system adoption

Apart from illustrating RFID adoption cases from the retailer's side, Moon and Ngai (2008) conducted interviews with five different fashion retailers to explore the perception and challenges of RFID technology in fashion retailing using a content analysis approach. The respondents believed that an RFID system could facilitate a prompt reaction to highly volatile demand, get better knowledge about their customers' preferences to improve the customer relationship, generate mix-and-match suggestions for cross-selling, and enhance inventory and logistics management by allowing information access in real time. However, they also were concerned about the RFID adoption cost, compatibility of the RFID system with the existing information systems, reading accuracy, top management support, and consent from staff members.

To better understand the driving force of RFID adoption from an industrial perspective, Ngai et al. (2012) presented an RFID-enabled system for apparel manufacturing in a China factory and conducted an interview to investigate the motivation for such practice based on the "technology-push" and "need-pull" conceptual frameworks. They also discussed the successful factor for RFID implementation. Quetti et al. (2012) conducted a more comprehensive discussion on the factors attributed to RFID adoption in a vertical silk supply chain. Moreover, Tajima (2012) analyzed the decision-making of small apparel manufacturers regarding RFID system adoption when trading with large retailers via a 2×2 apparel game model. The authors first analyzed the outcome stability and then assessed the decision of the players. Finally, a simulation analysis was conducted. The results indicated that the retailer's pressure tactic and collaborative strategy do not effectively motivate a small manufacturer to adopt an RFID system. De Marco et al. (2012) modeled the situation of an Italian apparel retailer and used a simulation approach to access the impact brought by an RFID system. The authors found that an RFID system can help to increase sales and improve inventory and operations management.

3.2.4 RFID-enabled systems development

The existing literature comprehensively discusses the protocol and infrastructure associated with the use of RFID technology in the fashion and apparel industries. Wong et al. (2006) integrated RFID technology and electronic product code (EPC) standard to develop an infrastructure in which the information flow along the supply chain and all the supply chain members are able to access the data with a reduced operations cost. Al-Kassab et al. (2009) developed a cost-benefit calculator to evaluate the performance of RFID technology, specifically in apparel retailing. The tool considers the cost savings factors and other benefits of the RFID system. Zhou and Piramuthu (2013) considered a shrinkage problem in fashion retailing that occurs when the consumers switches the price tag of a particular apparel item with a cheaper one in order to pay less at the cashier. To tackle this problem, they developed an authentication protocol and tested its security level using GNY (Gong-Needham-Yahalom) logic. Choi et al. (2014) studied a track-and trace anticounterfeiting system with a discussion on e-pedigree formatting, data synchronization, and traceability control in the apparel industry.

Some studies incorporated RFID technology to develop an innovative system for supporting decision-making. Choy et al. (2009) developed an innovative system to track swatches and fabric status in a fabric sample storeroom. This system aims to support the choice of the appropriate fabric for producing new apparel products. Wong et al. (2012) developed a smart dressing system (SDS) and an intelligent product cross-selling system (IPCS) for fashion retailing. The SDS was used to collect relevant information such as product identification and customer preference of the product available at the shop floor; the IPCS provides customer with mix-and-match product recommendation and supports multiple-criteria decision-making for cross-selling.

Lee et al. (2013) adopted fuzzy logic to collect and manage the inaccurate information and incorporate the RFID technology to improve the knowledge and decision-making for resource allocation in garment manufacturing. Guo et al. (2014) developed a decision support system for production control decision-making that incorporates the RFID technology and a production control decision support (PCDS) model. The RFID technology is used to capture the data, while the PCDS models the optimization process. Teucke and Scholz-Reiter (2014) discussed a prototypical software tool that is used to provide decision-making for a delivery schedule to retailers.

3.3 Case studies

In this section, we illustrate RFID adoption in apparel supply chains through a case study. Case study research methodology addresses the exploration of real-life situations, together with data collection and context analysis (Yin, 2013). Over the past few years, many fashion retailers have started to expand RFID deployment in different international retail stores; however, RFID is still in the emerging status and the discussion on how to improve the decision-making is not well discussed. Therefore, we conducted a face-to-face interview with a staff member of Zara to better understand its daily operations, supply chain management, and RFID application scope. Apart from gathering the primary data, we also collected secondary data, including information from annual reports and white papers to supplement the findings. In addition, we gathered public secondary data to discuss RFID adoption in Marks and Spencer (M&S) and American Apparel, as they have not implemented RFID technology in Hong Kong retail stores.

3.3.1 *Zara*[1]

Established in 1975, Zara is a well-known Spanish fast fashion brand owned by Inditex. As a fast fashion retailer, Zara sells men's, women's, and children's wears that are characterized by having a short life cycle, high demand uncertainty, and wide product variety. It produces the fashionable products at its own factory while outsourcing basic items to third-party suppliers globally, such as in China, India, and Turkey. All of the merchandise products are then shipped to the warehouse in Spain before distributing to each individual international retail store. Specifically, the ready-to-sell apparel products are delivered from the distribution center to each individual market by air (except European retail stores), taking about 2–3 days to arrive in Hong Kong retail stores. Each retail store receives two lots of products separately every week; one is the new arrival product, while the other one is a replenishment order. The life cycle of the fashionable products is about 2 weeks on average, and the leftover items are either

[1] For the Zara case, information about the daily operations and RFID adoption was collected from a face-to-face interview as well as the report by Swedberg (2014).

consolidated at one store for continuous selling or sold at a salvage value at the end of the selling season. The demand for new arrival products is monitored and a replenishment order may be placed.

With the expansion of both retail stores and the online business, Zara was driven to adopt RFID technology. In 2014, Zara implemented the EPC standard UHF Passive RFID tag supplied by Tyco in 700 stores in 22 different regions worldwide for item-level tagging. The RFID tag is sent to Zara's factory, placed inside the security alarm, and attached to each apparel item. When the mechanizing products are shipped to Spain's warehouse, the product information (such as the size and color of that particular item) are written into the RFID tag by encoding a unique identification (ID) number, and then the tag is connected to inventory management software (Swedberg, 2014). When an apparel item is sold, the RFID tag is removed using the RFID detacher that transmitted the product ID to the software system to update the inventory record, and deactivated eventually. The tags are collected and sent to Tyco to remove all of the memory and then are shipped to Zara's factory for reuse. The tags are also used for tagging the returned products on the shop floor. For example, in Hong Kong, a shop assistant rewrites the product information into the RFID tag and attaches it to the returned product. On the shop floor, about 50,000 RFID tags are applied on item-level tagging and at least 10 readers are used in daily operations. This real-time transaction record through RFID deployment can help Zara plan the inventory replenishment.

In addition, Zara has applied the RFID system to enhance its operations efficiency and improve customer satisfaction in different ways. First, a Hong Kong Zara's shop assistant commented that a shrinkage problem was observed and there were significant amounts of apparel product being stolen. Therefore, Zara made use of the RFID system to tackle this problem. Zara implemented the RFID system to keep updating the inventory level through the daily inventory counting. Prior to the RFID adoption, the shop assistant had to check the product code and match it with the record manually. This method required a lot of human effort, and sometimes human error occurred, which would generate an inaccurate information record. However, after the RFID implementation, it was much easier and more convenient to get the inventory information. The shop assistant could carry the RFID reader, walk around the store, and scan every single item automatically. Compared with manual operations, the RFID system helped to reduce roughly 75% of time to conduct inventory checking, which dramatically improved the daily operation efficiency.

Second, Zara was alerted to missing products immediately with the use of the RFID system. With this information, the shop assistant could observe and analyze the current situation, such as what kinds of products suffered from serious shrinkage problems and which location of the store had the highest product missing rate; consequently, a constructive measure could be taken. Third, Zara also implemented the RFID system to satisfy customer requests in real-time basis. For instance, if a customer was looking for a particular size of a product, the shop assistant could identify the product model by scanning the barcode printed on the price label. Next, the availability of the product would be examined through the system. If the item was available in store, an internal message would be sent to the storeroom for pickup. Otherwise, it would facilitate a cross-store inventory check in real time and show the product availability in other Hong Kong retail stores.

Zara plans to further extend the use of the RFID system for product tracking in Hong Kong. The company has proposed to install an RFID reader at the front door of the retail store so that inventory can be scanned and counted automatically when arriving on the shop floor.

3.3.2 Marks and Spencer[2]

Marks and Spencer (M&S) was founded in 1884 in the United Kingdom as a market stall. Nowadays, it is a leading international retailer, selling clothing, food, beauty products, and home products, with more than 3000 suppliers globally. Its fundamental business value is to provide quality, value, service, innovation, and trust to the customers (M&S, 2015). In 2014, M&S had 1253 retail stores worldwide, with about 64% of them operating in the United Kingdom; the remainder are international stores spread over 50 regions in European, Middle East, and Asia (M&S annual report, 2014).

M&S started using an RFID system in 2001 and applied it to the food supply chain to trace and track the location of the fresh food from its suppliers to distribution centers (Violino, 2013). In the past, the company used a barcoding system in daily operations. However, this manual system was not error free, so several operational problems in the apparel supply chain were encountered (Retail Technology, 2014). The first problem was inaccurate stock level records. M&S did not conduct inventory counting with the use of the barcoding system because of the extensive human effort requirement and only 400–600 items could be scanned every hour; as a result, shrinkage due to employee theft or shoplifting emerged. The second problem was high stockout level. Because the barcoding system was not able to update the inventory level and sales performance in real time, decisions on the inventory planning were made based on inaccurate information. It was reported that customers preferred getting products immediately instead of waiting for the backorder; hence, M&S experienced lost sales. As a result, M&S launched an RFID pilot project using 868 MHz Gen2 Passive RFID for item-level tagging in one of the retail stores (High Wycombe Stores) in 2003 to improve apparel product availability (Collins, 2004; Retail Technology, 2014). This project mainly focused on tagging 10,000 menswear items, including suits, shirts, and ties because these three categories of apparel products shipped from distribution centers to retail stores with different packaging (McCue, 2003). For example, the suits were hanged on hangers and the shirts were wrapped in bags during the delivery along the supply chain. Ties were firstly packed in boxes by suppliers and then repacked with hangers in the distribution centers before distributing to the retail stores. Information about the apparel item written in the RFID tag in the form of a unique ID number was attached to the "intelligent label" alongside the price tag.

[2] For the M&S case, information about the daily operations and the RFID adoption was retrieved from Collins (2004), Hadfield (2006), McCue (2003), Retail Technology (2014), and Violino (2013).

M&S implemented the RFID system for apparel products in several ways. First, it was used to track the inventory level and visualize the inventory flow along the supply chain. When the shirts entered or left the distribution center, the portal would scan the products quickly. In addition, it supported the inventory counting on shop floor via the handheld RFID reader. When the frontline staff walked along the shop with the wireless handled reader, the information stored in the intelligent label would be scanned automatically. The collected unique identification numbers of the product's details would be transmitted from the reader to the database. Finally, all the data would be sent to the head office's application system. A replenishment order could be placed promptly when the on-hand inventory level at the shop floor did not match the legacy record perfectly (Hadfield, 2006). The intelligent label could be detached and taken away from the shirt when it was purchased.

The RFID system implementation was a great success. M&S found that customers purchased the peripheral items simultaneously (e.g., men's shirts and ties), and the sales increased considerably. In addition, M&S could better match the demand and supply and significantly reduced the stockout rate by 30–40% (Retail Technology, 2014), resulting in better customer satisfaction. In terms of the internal operations, the RFID system facilitated efficient inventory counting and faster real-time inventory tracking. Last but not least, the information accuracy was enhanced because of the increased information about the inventory position. M&S planned to implement 100% RFID item-level tagging for all apparel products as well as the cosmetics goods by 2015.

3.3.3 American Apparel[3]

American Apparel is a US-based fashion retailer and manufacturer founded in 1998. Initially, it supplied basic apparel items to wholesalers and the first retail store was launched in California in 2003. At the end of 2014, American Apparel offered wide ranges of products for men, women, and children, operating 136 retail stores in the United States and 106 international stores in 19 different countries (American Apparel annual report, 2015).

Apart from selling the apparel products to the customer, manufacturing and distribution are also done by American Apparel; specifically, this process is called vertical integration. American Apparel believes that this centralized operation strategy can help the company to react quickly to the ever-changing demand and fashion trends, as well as understand the status and flow of the apparel products (American Apparel, 2015). American Apparel has the largest manufacturing factory in North American, with five separate factories and one distribution center (all located in California).

Traditionally in retail operations, labor efforts on inventory replenishment and conducting physical inventory taking are substantial (Reik, 2009). This issue is also observed in the case of American Apparel. On American Apparel's shop floor,

[3] For the American Apparel case, information about the daily operations and the RFID adoption was retrieved from O'Connor (2011), Reik (2009), and Roberti (2011, 2014).

interestingly, only one piece of apparel garment of each item is displayed. When that particular piece of garment is sold, a sales assistant has to replenish the shelf on the shop floor from the storeroom instantly while the transaction is recorded by the POS system through the barcode scanner. On average, each retail store had 12,000 SKUs while the storeroom held 26,000 items. At the end of each day, a replenishment order would be placed to the factory based on the total demand. In addition, inventory counting was conducted on the sales floor twice a week because of an inventory inaccuracy problem. It was found that there were about 100–300 items missing for each inventory counting. A replenishment order would be placed when there was mismatch between the on-hand inventory level and the legacy record. Therefore, manual checking was an essential way to correct the discrepancy between the real physical inventory level and the record with the use of the barcode system. However, this operation was time consuming, requiring four staff members to check every item with 8 hours.

The severe shrinkage and lost sales problems motivated American Apparel to consider RFID deployment. In 2007, it commenced an RFID pilot test with the UHF Gen2 RFID passive tag and firstly tagged about 40,000 apparel items in one single store in New York. The items were tagged at the factory and scanned when they were delivered from the factory to the storeroom of the retail store. When the products arrived at the retail store, they were scanned and compared with the factory's shipment information to detect if any item was lost during delivery (O'Connor, 2011). On the other hand, if an item was sold, the POS system would generate a signal and the storeroom assistant had to pick up that item. The shelves on the shop floor were replenished from the storeroom when the quantity sold reached the predetermined level.

After implementing the RFID system, only two employees were needed to conduct the inventory counting and the whole checking process could be completed within 2 hours. American Appeal found that the RFID system improved its financial performance. It was also reported that sales increased by 14% in each store as the product availability improved and reached the 99% level. More importantly, the inventory accuracy increased to 99.8% without increasing the labor workload (Roberti, 2011). The shrinkage problem reduced by more than half (Roberti, 2014), and the RFID investment was estimated to have payback of about 4.5 months on average. As a result, American Apparel planned to expand the RFID deployment to 38 more stores and adopt it at the distribution center for inventory checking to reduce human errors in packing (Roberti, 2011).

3.3.4 Discussion

From the above case studies, it can be observed that fashion retailers have adopted UHF passive RFID tags for item-level tagging. Even though Zara, Marks and Spencer, and American Appeal use their RFID systems for different purposes, they can be categorized into three main application scopes according to Moon and Ngai (2008): shop floor management, logistics and distribution management, and customer relationship management. For shop floor management, the RFID system facilitates inventory counting efficiently on the shop floor to mitigate shrinkage problems without increasing the human effort needed to scan the products when compared with a barcoding system. Therefore, shop assistants can focus more on sales and customer service. Second, for logistics and inventory management, the RFID system can keep tracking the location and status of the inventory, which improves product visibility and identifies any inventory that is lost during delivery along the supply chain. Finally, for customer service management, fashion retailers can obtain more accurate on-hand inventory information and share this information with the headquarters to improve inventory planning decisions. With the use of an RFID system, fashion retailers can react quickly to demand changes and improve product availability. As a result, stockout rates are lower and customer service levels are increased, which is crucial under the keen competition in the business environment. Our findings on the business value of RFID technology (in terms of shop floor management, logistics and inventory management, and customer relationship management) in the apparel supply chain are summarized in Table 3.3.

Table 3.3 Summary of the business value of RFID system adoption at Zara, Marks and Spencer (M&S), and American Apparel

	Zara	M&S	American Apparel
Type of RFID adopted	UHF passive RFID tag for item-level tagging	UHF passive RFID tag for item-level tagging	UHF passive RFID tag for item-level tagging
Shop floor management	Conduct inventory counting and use this information to develop constructive measures to avoid shrinkage.	Conduct inventory counting	Conduct inventory counting and facilitate shelf replenishment from storeroom to shop floor
Inventory and distribution management	Update the demand information in real time and share it with headquarters for inventory planning	Track the location and status of products from distribution center to retail outlets	Identify any product lost from factory to retail store and reduce human errors when packing the products at the distribution center
Customer relationship management	Facilitate cross-store inventory checking in real time to satisfy customer needs.	Improve product availability and reduce the stockout rate by 30–40%	Improve product availability to a 99% level

3.4 Conclusions and future research directions

This chapter illustrates how RFID systems can be implemented in the fashion industry and identifies how RFID systems are commonly used in shop floor management, logistics and distribution management, and customer relationship management.

Apart from revealing the application scopes and corresponding benefits of using an RFID system, it is also interesting to explore how RFID systems can support decision-making in the supply chain context. In the fashion industry, renowned fashion retailers such as Benetton and Levi's have adopted a quick response program (QRP) (Forza and Vinelli, 1997) for inventory management. This program aims to reduce the lead time to have a quicker response to market changes and observe market signals to make better inventory planning decisions (Iyer and Bergen, 1997). With a QRP, a fashion retailer is able to match the supply and demand, reduce the chance of stockout, and facilitate express delivery (Choi and Sethi, 2010; Choi, 2013). It is believed that Zara and M&S can adopt this strategy to improve the initial forecast through the RFID technology:

- Zara integrates the sales assistants' feedback on the latest fashion trend/elements and prior sales performance of the apparel garments (e.g., best-selling items) to create the products and determine the production quantity. Zara can exploit the RFID system to capture all sales data and keep updating the inventory record automatically. This information provides important market signals about the demands for each fashion product at each shop floor; as a result, Zara can use it to adjust the demand forecasting, determine the production quantity, reallocate the products to different international stores, and react quickly to market changes.
- In the past, M&S outsourced product design and production parts to third-party suppliers. Nowadays, it revolutionizes the authority and plans to design the products by itself to cope with the demand changes and reduce the markdown. To facilitate this new strategy, an RFID system will play a critical role in capturing the demand information, allowing M&S to adjust its initial forecasting.

Because demands for fashionable products vary significantly among all retail stores, fashion retailers are driven to address a risk pooling strategy to reduce demand uncertainty. One measure is to aggregate the inventory at a warehouse so as to offset the demand fluctuations across different markets. This centralized system will result in a lower total inventory requirement than that of the decentralized system with two or more warehouses (Simchi-Levi et al., 2000).

- Inventory planning and replenishment decisions at each of Zara's international retail stores is centralized. Zara consolidates and aggregates all demand information of the merchandise products to determine the total production quantity of a particular apparel product. With the use of an RFID system, information collection becomes easier and more efficient. Even though the Hong Kong retail stores can place a replenishment order to the headquarters with a determined quantity, the final amount of backordered products arriving at the stores may be less than the requested level because the decision is made centrally from Spain's headquarters. Thus, it is important to share the demand information of each retail store with the headquarters in real time so that a better inventory planning decision can be made in a timely manner.

- American Apparel is vertically integrated from production to retailing, so the RFID system facilitates real-time information sharing from the sales floor to the production factories. Vertical integration helps to avoid the bullwhip effect as the production factory observes the final market demand to determine the production quantity. Therefore, factories can make use of such information to plan the production scheduling, driving American Apparel to make an optimal decision.

With reference to the existing literature presented in Section 3.2 and the RFID application case studies in Section 3.3, we identified the following research gaps that warrant more in-depth investigation and discussion.

First, the scope of RFID applications in the apparel supply chain in real-life practice are limited. Existing pilot studies and analytical studies (Atali et al., 2006; Dai and Tseng, 2012; Gaukler et al., 2007; Heese, 2007; Kök and Shang, 2014; Uçkun et al., 2008; Zhu et al., 2013) extensively illustrate how RFID technology can be used to tackle shrinkage and stockout problems. However, other applications in the apparel supply chain, such as reducing the bullwhip effect, are limited and should also be considered. Bottani et al. (2010) and Zhou (2012) provide good references on the subject. Bottani et al. (2010) analytically evaluated the quantity of bullwhip effect reduction for an Italian fast moving consumer goods (FMCG) supply chain based on data collected from manufacturers and distributors. Zhou (2012) conducted a similar study to quantify the performance (e.g., production lead time, inventory counting accuracy) of a Chinese FMCG supply chain. Moreover, employing a new technology into the supply chain may require business process reengineering, as addressed in Bottani et al. (2009). A detailed exploration and more comprehensive discussion, especially in the apparel industry, will generate important insights.

Second, it has been reported that the read-rate accuracy of an RFID system is 89–98% (Thilmany, 2007). In other words, transaction errors exist because RFID technology is not able to provide perfect information on the on-hand inventory level. Sahin and Dallery (2009) analytically analyze the impact of using barcoding and RFID systems and the efficiency loss due to the transaction error with the newsvendor framework. Chan et al. (2012) analytically derived the condition in healthcare organizations in which it is optimal to switch from the barcode to RFID systems with the consideration of the transaction errors using an (s, S) policy. On the other hand, RFID technology can reduce shrinkage errors when compared with a barcode system. This inventory inaccuracy problem was also explored by Cakici et al. (2011), who analytically compared the benefits under different information technologies for managing pharmaceutical products. The business value of both information technologies under a situation with the co-existence of the transaction errors and shrinkage errors in the fashion and apparel industries also needs investigation.

On a strategic level, Tajima (2012) found that a retailer's pressure tactic and collaborative strategy cannot help to motivate a small manufacturer to deploy an RFID system. Hence, it is interesting to first explore a real-world case study of such a situation and evaluate the relationship and strategy that is possible to implement the RFID technology along the supply chain. In addition, Balocco et al. (2011) stated that a large fashion manufacturer requests the retailer to share larger part of the RFID tag cost.

The issue of how to implement a contract to share the RFID cost between the supplier and retailer has not yet been well discussed in the apparel industry. On the contrary, the effect on how big fashion brand owners use RFID technology will affect the decision of the dealers is also a timely and interesting topic.

Performance evaluations of the RFID system in apparel supply chain based on the empirical and analytical modeling study are limited. In the existing literature, Lui and Lo (2014) is the only study that empirically tests how an RFID system is able to improve the financial performance of 18 fashion companies. They found that RFID technology can be a competitive advantage in the apparel supply chain performance, where it can significantly decrease the inventory days. However, this study evaluates the statistical results based on inventory days, account receivable days, and operating cycles. It is therefore also proposed to investigate whether an RFID system can improve the financial performance of a company even under financial downturn, and specifically whether an RFID system can reduce the risk level of the supply chain members. Other financial performance indicators, such as net profit and return on investment, also provide good references to explore this issue. Because RFID tags can be reused or disposed, it is also worthwhile to examine the corresponding return on the RFID investment.

Finally, an RFID system is able to capture an extensive amount of data. It is interesting to explore how to manage the data and make use of it to improve decision-making, such as in Wamba et al. (2015). Developing an algorithm using the RFID data in alignment with real-world situations for supply chain optimization will also provide important insights in academia. For example, Choy et al. (2009) employed RFID technology to propose a sample management system for both fashion designers and clothing merchandisers to determine the most suitable fabric for product development. This system supports knowledge sharing, improves communication efficiency between the partners, reduces fabric searching time, and enhances information accuracy. Kang and Lee (2013) proposed algorithms to enhance the traceability service in logistics and distribution management, which can efficiently facilitate information exchange with the EPC standard.

References

Aguilar, A., van der Putten, W., Mauire, G., 2006. Positive patient identification using RFID and wireless networks. In: 11th Annual Conference and Scientific Symposium (HISI), Dublin, Ireland.

Al-Kassab, J., Mahmoud, N., Thiesse, F.G., Fleisch, E., 2009. A cost benefit calculator for RFID implementations in the apparel retail industry. In: Proceedings of the Fifteenth Americas Conference on Information Systems, San Francisco, California, pp. 478–489.

American Apparel Annual Report, 2015. Retrieved from: http://files.shareholder.com/downloads/APP/314231439x0xS1336545%2D15%2D58/1336545/filing.pdf.

American Apparel, 2015. Retrieved from: http://www.americanapparel.net/.

Atali, A., Lee, H.L., Özer, Ö., 2006. If the Inventory Manager Knew: Value of RFID under Imperfect Inventory Information. Technical Report. Graduate School of Business, Stanford University.

Attaran, M., 2007. RFID: an enabler of supply chain operations. Supply Chain Management: An International Journal 12 (4), 249–257.

Azevedo, S.G., Carvalho, H., 2011. RFID technology in the fashion supply chain: an exploratory analysis. In: Choi, T.M. (Ed.), Fashion Supply Chain Management: Industry and Business Analysis: Industry and Business Analysis. IGI Gobal, pp. 303–326.

Azevedo, S.G., Ferreira, J., 2009. RFID technology in retailing: an exploratory study on fashion apparels. IUP Journal of Managerial Economics 7 (1), 7–22.

Azevedo, S.G., Carvalho, H., Cruz-Machado, V.A., 2014a. A cross-case analysis of RFID deployment in fast fashion supply chain. In: Proceedings of the Eighth International Conference on Management Science and Engineering Management. Springer Berlin Heidelberg, pp. 605–617.

Azevedo, S.G., Prata, P., Fazendeiro, P., 2014b. The role of radio frequency identification (RFID) technologies in improving process management and product tracking in the textiles and fashion supply chain. In: Wong, W.K., Guo, Z.X. (Eds.), Fashion Supply Chain Management Using Radio Frequency Identification (RFID) Technologies. Woodhead Publishing, Cambridge, pp. 42–69.

Balocco, R., Miragliotta, G., Perego, A., Tumino, A., 2011. RFID adoption in the FMCG supply chain: an interpretative framework. Supply Chain Management: An International Journal 16 (5), 299–315.

Bamfield, J., 2004. Shrinkage, shoplifting and the cost of retail crime in Europe: a cross-sectional analysis of major retailers in 16 European countries. International Journal of Retail & Distribution Management 32 (5), 235–241.

Bhardwaj, V., Fairhurst, A., 2010. Fast fashion: response to changes in the fashion industry. The International Review of Retail, Distribution and Consumer Research 20 (1), 165–173.

Bottani, E., Ferretti, G., Montanari, R., Rizzi, A., 2009. The impact of RFID technology on logistics processes of the fashion industry supply chain. International Journal of RF Technologies: Research and Applications 1 (4), 225–252.

Bottani, E., Montanari, R., Volpi, A., 2010. The impact of RFID and EPC network on the bullwhip effect in the Italian FMCG supply chain. International Journal of Production Economics 124 (2), 426–432.

Çakıcı, Ö.E., Groenevelt, H., Seidmann, A., 2011. Using RFID for the management of pharmaceutical inventory—system optimization and shrinkage control. Decision Support Systems 51 (4), 842–852.

Chan, H.L., Choi, T.M., Hui, C.L., 2012. RFID versus bar-coding systems: transactions errors in health care apparel inventory control. Decision Support Systems 54 (1), 803–811.

Chawla, V., Ha, D.S., 2007. An overview of passive RFID. IEEE Communications Magazine 45 (9), 11–17.

Choi, S.H., Cheung, H.H., Yang, B., Yang, X.Y., 2014. Implementation issues in RFID-based anti-counterfeiting for apparel supply chain. International Journal of Intelligent Computing Research (IJICR) 5 (1/2), 279–389.

Choi, T.M., 2013. Local sourcing and fashion quick response system: the impacts of carbon footprint tax. Transportation Research Part E: Logistics and Transportation Review 55, 43–54.

Choi, T.M., Sethi, S., 2010. Innovative quick response programs: a review. International Journal of Production Economics 127 (1), 1–12.

Choy, K.L., Chow, K.H., Moon, K.L., Zeng, X., Lau, H.C., Chan, F.T., Ho, G.T., 2009. A RFID-case-based sample management system for fashion product development. Engineering Applications of Artificial Intelligence 22 (6), 882–896.

Collins, J., 2004. Marks & Spencer expands RFID trial. RFID Journal. Retrieved from: http://www.nen.troots.com/TradeRoots_Advisors/TradeRoots_Library/TradeRoots_Library_-_News/RFID_Journal_-_Marks___Spencer_Expands_RFID_Trial.pdf.

Dai, H., Tseng, M.M., 2012. The impacts of RFID implementation on reducing inventory inaccuracy in a multi-stage supply chain. International Journal of Production Economics 139 (2), 634–641.

DeHoratius, N., Raman, A., 2008. Inventory record inaccuracy: an empirical analysis. Management Science 54 (4), 627–641.

Forza, C., Vinelli, A., 1997. Quick response in the textile-apparel industry and the support of information technologies. Integrated Manufacturing Systems 8 (3), 125–136.

Garrido Azevedo, S., Carvalho, H., 2012. Contribution of RFID technology to better management of fashion supply chains. International Journal of Retail & Distribution Management 40 (2), 128–156.

Gaukler, G.M., Seifert, R.W., Hausman, W.H., 2007. Item-level RFID in the retail supply chain. Production and Operations Management 16 (1), 65–76.

Guo, Z.X., Wong, W.K., Leung, S.Y.S., Fan, J.T., Chan, S., 2014. The role of radio frequency identification (RFID) technologies in improving garment assembly line operations. In: Wong, W.K., Guo, Z.X. (Eds.), Fashion Supply Chain Management Using Radio Frequency Identification (RFID) Technologies. Woodhead Publishing, Cambridge, pp. 99–125.

Hadfield, W., 2006. Marks & Spencer expands RFID trial as it moves closer to decision over full roll-out. Computer Weekly.com. Retrieved from: http://www.computerweekly.com/feature/Marks-Spencer-expands-RFID-trial-as-it-moves-closer-to-decision-over-full-roll-out.

Heese, H.S., 2007. Inventory record inaccuracy, double marginalization, and RFID adoption. Production and Operations Management 16 (5), 542–553.

Iyer, A.V., Bergen, M.E., 1997. Quick response in manufacturer-retailer channels. Management Science 43 (4), 559–570.

Kang, Y.S., Lee, Y.H., 2013. Development of generic RFID traceability services. Computers in Industry 64 (5), 609–623.

Kang, Y., Gershwin, S.B., 2004. Information inaccuracy in inventory systems—stock loss and stockout. IIE Transactions 37, 843–859.

Kärkkäinen, M., 2003. Increasing efficiency in the supply chain for short shelf life goods using RFID tagging. International Journal of Retail & Distribution Management 31 (10), 529–536.

Kök, A.G., Shang, K.H., 2014. Evaluation of cycle-count policies for supply chains with inventory inaccuracy and implications on RFID investments. European Journal of Operational Research 237 (1), 91–105.

Lee, C.K.H., Choy, K.L., Ho, G.T.S., Law, K.M.Y., 2013. A RFID-based resource allocation system for garment manufacturing. Expert Systems with Applications 40 (2), 784–799.

Loebbecke, C., Palmer, J., Huyskens, C., 2006. RFID's potential in the fashion industry: a case analysis. In: Proceedings of the 19th Bled eConference, Slovenia.

Lowson, B., King, R., Hunter, A., 1999. Quick Response: Managing the Supply Chain to Meet Consumer Demand. Wiley, Chichester, West Sussex, England.

Lui, K., Lo, C.K.Y., 2014. Measuring the impact of radio frequency identification (RFID) technologies in improving the efficiency of the textile supply chain. In: Wong, W.K., Guo, Z.X. (Eds.), Fashion Supply Chain Management Using Radio Frequency Identification (RFID) Technologies. Woodhead Publishing, Cambridge, pp. 187–202.

De Marco, A., Cagliano, A.C., Nervo, M.L., Rafele, C., 2012. Using system dynamics to assess the impact of RFID technology on retail operations. International Journal of Production Economics 135 (1), 333–344.

M&S Annual Report, 2014. Retrieved from: http://corporate.marksandspencer.com/investors/b73df1d3e4f54f429210f115ab11e2f6.

M&S, 2015. Retrieved from: http://www.marksandspencer.com.

Massimo, B., Eleonora, B., Gino, F., Antonio, R., Andrea, V., 2012. Experimental evaluation of business impacts of RFID in apparel and retail supply chain. International Journal of RF Technologies Research and Applications 3, 257–282.

McCue, A., 2003. The Retailer Has Begun a Trial Involving the Fixing of Radio Frequency Identification Tags to Men's Clothing. Retrieved from: http://www.zdnet.com/article/marks-spencer-tags-shirts-with-rfid/.

Meingast, M., King, J., Mulligan, D.K., 2007. Embedded RFID and everyday things: a case study of the security and privacy risks of the US e-passport. In: Proceedings of IEEE International Conference on RFID, pp. 7–14 (Grapevine, Texas).

Michael, K., McCathie, L., 2005. The pros and cons of RFID in supply chain management. In: Proceedings of International Conference Mobile Business, pp. 623–629.

Moon, K.L., Ngai, E.W.T., 2008. The adoption of RFID in fashion retailing: a business value-added framework. Industrial Management & Data Systems 108 (5), 596–612.

Ngai, E.W.T., Chau, D.C.K., Poon, J.K.L., Chan, A.Y.M., Chan, B.C.M., Wu, W.W.S., 2012. Implementing an RFID-based manufacturing process management system: lessons learned and success factors. Journal of Engineering and Technology Management 29 (1), 112–130.

Ngai, E.W.T., Cheng, T.C.E., Lai, K.H., Chai, P.Y.F., Choi, Y.S., Sin, R.K.Y., 2007. Development of an RFID-based traceability system: experiences and lessons learned from an aircraft engineering company. Production and Operations Management 16 (5), 554–568.

O'Connor, M.C., 2011. American Apparel adding 50 more stores in aggressive RFID rollout. RFID Journal. Retrieved from: http://www.rfidjournal.com/articles/view?8374.

Prest, A.R., Turvey, R., 1965. Cost-benefit analysis: a survey. The Economic Journal 75 (300), 683–735.

Quetti, C., Pigni, F., Clerici, A., 2012. Factors affecting RFID adoption in a vertical supply chain: the case of the silk industry in Italy. Production Planning & Control 23 (4), 315–331.

IT Reseller, 2006. Accuracy Tops UK Barcode Benefits League. Retrieved from: http://www.itrportal.com/articles/2006/12/12/3487-accuracy-tops-uk-barcode-benefits.

Raman, A., DeHoratius, N., Ton, Z., 2001. Execution: the missing link in retail operations. California Management Review 43 (3), 136–152.

Reik, R., 2009. American Apparel case study write-up. RFID Monthly. Retrieved from: http://rfidjournal.net/PDF_download/American_Apparel_Case_Study.pdf.

Retail Technology, 2014. Accurate Inventory Tracking Benefits M&S. Retrieved from: http://www.retailtechnology.co.uk/news/5115/accurate-inventory-tracking-benefits-ms/.

Roberti, M., 2011. RFID delivers unexpected benefits at American Apparel. RFID Journal. Retrieved from: http://www.rfidjournal.com/articles/view?8843/.

Roberti, M., 2014. How RFID could save retailers $42 billion annually. RFID Journal. Retrieved from: http://www.rfidjournal.com/articles/view?12410.

Sahin, E., Dallery, Y., 2009. Assessing the impact of inventory inaccuracies within a newsvendor framework. European Journal of Operational Research 197 (3), 1108–1118.

Sarac, A., Absi, N., Dauzère-Pérès, S., 2010. A literature review on the impact of RFID technologies on supply chain management. International Journal of Production Economics 128 (1), 77–95.

Şen, A., 2008. The US fashion industry: a supply chain review. International Journal of Production Economics 114 (2), 571–593.
Simchi-Levi, D., Kaminsky, P., Simchi-Levi, E., 2000. Designing and Managing the Supply Chain: Concepts, Strategies, and Case Studies. Irwin/McGraw-Hill, Burr Ridge, IL.
Sriram, T., Rao, K.V., Biswas, S., Ahmed, B.A., 1996. Applications of barcode technology in automated storage and retrieval systems. In: Proceedings of the 1996 IEEE IECON 22nd International Conference on Industrial Electronics, Control, and Instrumentation, pp. 641–646 (Taipei).
Swedberg, C., 2014. Tyco wins chain-wide contract from Inditex. RFID Journal. Retrieved from: http://www.rfidjournal.com/articles/view?12009/.
Tajima, M., 2012. Small manufacturers versus large retailers on RFID adoption in the apparel supply chain. In: Supply Chain Management: Concepts, Methodologies, Tools, and Applications: Concepts, Methodologies, Tools, and Applications. IGI Global, pp. 196–220.
Teucke, M., Scholz-Reiter, B., 2014. Improving order allocation in fashion supply chains using radio frequency identification (RFID) technologies. In: Wong, W.K., Guo, Z.X. (Eds.), Fashion Supply Chain Management Using Radio Frequency Identification (RFID) Technologies. Woodhead Publishing, Cambridge, pp. 126–157.
Thiesse, F., Al-Kassab, J., Fleisch, E., 2009. Understanding the value of integrated RFID systems: a case study from apparel retail. European Journal of Information Systems 18 (6), 592–614.
Thilmany, J., 2007. TAL at 60: faster than ever. Apparel Magazine. Retrieved from: http://wpc.32bf.edgecastcdn.net/0032BF/en/docs/200711ApparelMagazine2.pdf.
Tzeng, S.F., Chen, W.H., Pai, F.Y., 2008. Evaluating the business value of RFID: evidence from five case studies. International Journal of Production Economics 112 (2), 601–613.
Uçkun, C., Karaesmen, F., Savaş, S., 2008. Investment in improved inventory accuracy in a decentralized supply chain. International Journal of Production Economics 113, 546–566.
Ustundag, A., Ustundag, A., Bal, M., 2013. Economic potential of RFID use in apparel retail industry. In: Ustybdag, A. (Ed.), The Value of RFID. Springer, London, pp. 129–139.
Violino, B., 2013. Marks & Spencer rolls out RFID to all its stores. RFID Journal.
Wamba, S.F., Akter, S., Edwards, A., Chopin, G., Gnanzou, D., 2015. How 'big data' can make big impact: findings from a systematic review and a longitudinal case study. International Journal of Production Economics 165, 234–246.
White, G.R.T., Gardiner, G., Prabhakar, G., Razak, A.A., 2007. A comparison of barcode and RFID technologies in practice. Journal of Information, Information Technology, and Organizations 2, 119–132.
Wong, K.H.M., Chan, A.C.K., Hui, P.C.L., Patel, C.A., 2006. A framework for data flow in apparel supply chain using RFID technology. In: IEEE International Conference on Industrial Informatics, pp. 61–66 (Singapore).
Wong, W.K., Guo, Z.X., 2014. The role of radio frequency identification (RFID) technologies in the textiles and fashion supply chain: an overview. In: Wong, W.K., Guo, Z.X. (Eds.), Fashion Supply Chain Management Using Radio Frequency Identification (RFID) Technologies. Woodhead Publishing, Cambridge, pp. 42–69.
Wong, W.K., Leung, S.Y.S., Guo, Z.X., Zeng, X.H., Mok, P.Y., 2012. Intelligent product cross-selling system with radio frequency identification technology for retailing. International Journal of Production Economics 135 (1), 308–319.
Wyld, D.C., 2006. RFID 101: the next big thing for management. Management Research News 29 (4), 154–173.
Yao, W., Chu, C.H., Li, Z., 2012. The adoption and implementation of RFID technologies in healthcare: a literature review. Journal of Medical Systems 36 (6), 3507–3525.

Yin, R.K., 2013. Case Study Research: Design and Methods. Sage Publications.

Zhou, W., Piramuthu, S., 2013. Preventing ticket-switching of RFID-tagged items in apparel retail stores. Decision Support Systems 55 (3), 802–810.

Zhou, Z., 2012. Applying RFID to reduce bullwhip effect in a FMCG supply chain. In: Lee, G. (Ed.), Advances in Computational Environment Science. Springer Berlin Heidelberg, pp. 193–199.

Zhu, L., Hong, K.-S., Lee, C., 2013. Optimal ordering policy of a risk-averse retailer subject to inventory inaccuracy. Mathematical Problems in Engineering 2013. Article ID 951017.

Zhu, X., Mukhopadhyay, S.K., Kurata, H., 2012. A review of RFID technology and its managerial applications in different industries. Journal of Engineering and Technology Management 29 (1), 152–167.

Using big data analytics to improve decision-making in apparel supply chains

L. Banica, A. Hagiu
University of Pitesti, Pitesti, Romania

4.1 Introduction

Decision-making in the clothing and fashion industry is different from other industries due to the fast-paced changes occurring and the diversity of consumer preferences.

The management has at its disposal reports, charts, and statistics of sales, as well as processed data of some surveys and a Web portal of communication with customers. But this information either is obtained after months from the time of occurrence, or it is formal and does not give a complete and accurate image of the customer's view. The data must be completed with the information circulating on social media that could give the management sincere, just-in-time information. This could improve the creativity in design and establish the utility, quality, and comfort in use when developing new products.

Without giving up the advantages of an attractive and useful Web portal, the apparel management decisions leverage on two new information technology and communication (IT&C) facilities that offer the following:

- The knowledge of customer preferences (wishes). Though until now it has been possible to communicate through comments posted on a website, perform surveys, and study the profile of loyal customers who frequently access the site, more is needed. A good manager also needs to know what the clients really want in fashion matters for the future.
- The increasing capacity to select and process the reviews of the social networks (eg, Twitter and Facebook), perhaps even to extract information from messages received by mail, without violating the privacy rights of potential clients. The information technology (IT) industry has to meet the business environment needs through software packages that process unstructured data from social media through social network analysis (SNA) software to extract useful information from this huge volume of data, known as big data.

This chapter has three main objectives:

1. to present the actual stage of Romanian clothing and fashion by key indicators of performance
2. to specify the competitiveness of Romanian clothing products and find the ways to improve that by using new technologies in the area of IT&C

Information Systems for the Fashion and Apparel Industry. http://dx.doi.org/10.1016/B978-0-08-100571-2.00004-X

3. to study how social media information, retrieved and processed through big data and analyzed with analytic tools, can influence decision-making in supply chains

The chapter is divided in several sections, the next section being a literature review.

In Section 4.3, we will approach the Romanian case from the management perspective, analyzing the clothing and fashion industry face-to-face with the economic crisis, as well as its slow but steady recovery since 2013 (according to official data available on the Website of the Romanian Institute of Statistics). There are a number of indicators to be interpreted in light of the Romanian reality, meaning that many prestigious companies, at least nationally but also at the world level, are still working in the "lohn" system for big brands. But, we must welcome that some are trying to (re-) launch by promoting their own brand in the market place.

Also in this section, the authors emphasize the three directions by which companies' management from Romania can become competitive on the international market:

- originality: by encouraging creativity, supporting research and development in the area, promoting in fashion the stylized features of the Romanian outfit
- quality: by using the most modern manufacturing technologies, training, and specialization of the Romanian clothing and fashion staff
- online commerce and marketing: using the Internet not only as a means of product presentation but as a capitalization market (e-commerce), as a way to find out customer preferences, and to meet their requirements

In Section 4.4, we will focus on presenting the latest IT&C technologies that can be used in the clothing and fashion industry to select the real opinions of potential customers, moving from the rigidity and formalism of ordered polls to the processing of massive amounts of data from social networks in the cloud environment. Also, we propose a model for integrating big data with an existing enterprise resource planning (ERP) system, resulting in a four-layer architecture based on a cloud Hadoop implementation.

Exploring social media using SNA software already has become an effective marketing technique, as it gives knowledge about customer satisfaction, opinion leaders, and centers of influence. There are several examples of famous fashion houses, such as Burberry and Ralph Lauren, that have used big data analysis to forecast the latest trends and improve their presence on Internet. Another leading fashion brand increased its sales by 33% after using big data analysis services. Also, there are powerful companies that combine ERP and business intelligence (BI) to analyze data and predict consumer preferences (Gupta, 2015).

But, we intend to demonstrate the advantages of using big data analytics throughout the entire fashion supply chain, not just in certain activities.

The conclusions section will summarize the main ideas emerging from this article, emphasizing the benefits for decision-making in apparel supply chains obtained by exploring the social media via two IT technologies: big data and SNA software. Also, we will present some future research directions in this fascinating and ever changing industry of fashion.

4.2 Literature review

More and more researchers and businessmen come to think that big data is a concept that will affect organizations across industries, sectors, and economies. The frequent changes in fashion and the difficulty to make forecasts, even in the short term, will propel firms in the clothing and fashion industry, as well as retailers, to quickly understand and apply this concept that can change customer relationships. Data analysis will become a significant part of the business process, instead of an activity performed only by specialists.

According to the group of researchers from McKinsey Global Institute (Manika et al., 2011), the feedback obtained by analyzing social network conversations and blog comments is faster and more efficient than customer surveys and focus group impressions.

Gathering a huge volume of data involves a new approach in terms of management, analysis, and BI. In this context, it is necessary for a company's IT infrastructure, generally an ERP, to be extended to be capable of collecting the data, processing it, and extracting information about which trends are gaining momentum and which are losing ground, and to be able do this at any point in the product life cycle (Trites, 2013).

Important IT vendors, such as Microsoft, SAP, and Hewlett-Packard have been players in the big data domain since 2009, offering Hadoop-related technology or business analytical tools that can turn massive amounts of data into real business value.

4.2.1 A brief insight into big data in business

Starting with the definition of big data and some of its characteristics, we will emphasize its impact on the business world and ways of applying it in optimal conditions at low cost. Big data is a popular term used to describe the exponential growth and availability of data, both structured and unstructured. And big data may be as important to business—and society—as the Internet has become. Why? More data may lead to more accurate analyses (Davenport and Dyché, 2013).

The main characteristics of big data are described by experts through the five V's:

- **Volume**: scalability is the most important aspect for every domain of application. There is a lot of unstructured data coming from social networks, weblogs, click streams, sensors, and mobile networks in comparison with structured data, but the ratio of useful information is reversed in favor of the latter (Banica and Hagiu, 2015). The relevance of the volume of data collected may be obtained by filtering using analytic tools to identify important patterns and metrics that are found in the business field.
- **Velocity**: the increasing flow of data requires hardware and software solutions to process data streaming in as fast as possible; management needs answers to the questions posed from the virtual marketplace in real time (Dijcks, 2013).
- **Variety**: there are many types of data formats that big data platforms must recognize to cover any opportunity to know customer opinions.
- **Veracity**: the trustworthiness of the information must be evaluated as the data may not be significant or there could be disparities in the sample of data processed (Rouse, 2014).

- **Value:** this is the most important feature of big data because it turns the large amount of data into commercial value and, thus, reaches the final target of developing the business and obtaining a stronger competitive position (Banica and Hagiu, 2015).

The companies active in any domain could improve their business using big data, and we will give some significant examples (Rosenbush and Totty, 2013):

1. Great software companies, such as Google, Facebook, and Amazon, showed their interest in processing big data in the cloud environment many years ago. They play a dual role: providing such services for other firms and also collecting information from social media to analyze for improving their own IT tools.
2. Manufacturing companies, especially those selling online, have begun to monitor Facebook and Twitter, analyzing data from different angles, to forecast customer behavior and use the information in designing new products.
3. Retailers are able to analyze large volumes of data to obtain insights about individual consumer behavior and preferences and to offer personalized recommendations, tailored to merchandise selection and pricing.
4. More and more companies invest in big data on virtual clouds to monitor their marketing strategies. Big data is extremely useful in a marketing capacity, using information such as customer demographics, spending habits, and oscillation between simplicity, comfort, and glamour.

There are many IT technologies that can be used to manipulate, manage, and analyze big data, and researchers continue to develop new techniques and improve on existing ones in response to the requirements of enterprises (Manyika et al., 2011). Some of these are briefly described next:

1. **Association rule learning** refers to a set of techniques for discovering interesting relationships (association rules) among variables in large databases. One application is market basket analysis, in which a retailer can determine which products are frequently bought together and use this information for marketing (Manyika et al., 2011).
2. **Cluster analysis** is a statistical method for classifying objects by splitting a large group into smaller segments of similar objects, after defining the characteristics of similarity. An example of cluster analysis is segmenting consumers into groups that have targeted marketing (Mooi and Sarstedt, 2011).
3. **Data fusion and data integration** is a set of techniques that combine data from multiple sources and analyze it to develop new insights. For example, data from social media, analyzed by natural language processing, can be combined with real-time sales data to determine the effect of a marketing campaign on customer sentiment and purchasing behavior (Manyika et al., 2011).
4. **Sentiment analysis** refers to the processing of natural language from source text materials to identify and extract subjective information. For example, sentiment analyses have been launched in social media to determine the attitude of customers regarding some products and services; these are expressed by type (positive, negative, or neutral) and the degree of the sentiment (Socher et al., 2013).
5. **BI** is an advanced analytic tool designed to collect and analyze data and also to create reports, dashboards, and data visualizations. BI tools generally process structured data stored in data warehouses. For processing semi-structured and unstructured data (NoSQL databases), a newer class of BI applications has been launched that supports Hadoop and related tools such as MapReduce, Spark, and Hive (Rouse, 2014).

6. **Cassandra** is an open-source database management system designed to handle huge amounts of data on a distributed system. This system was originally developed at Facebook and is now managed as a project of the Apache Software foundation (Manyika et al., 2011).
7. **Hadoop** is an open-source, free software for processing large amounts of data, similar to Google's MapReduce and Google File System. It was originally developed at Yahoo! and is now managed by Apache Software foundation (Bhosale and Gadekar, 2014). The components of Hadoop and its functions will be detailed in Section 4.4 of this chapter.
8. **Stream processing** includes technologies that process continuous data flows in real-time and identify meaningful events or patterns from data sources that come from the business environment and governmental institutions (Banica et al., 2014). These events may be related to an organization and involve their business data, or they may include other types of information, such as text messages, social media, and traffic reports. Software companies such as Oracle, SAP, and Microsoft have launched a fusion between complex event processing and Hadoop technologies that integrates real-time processing and batch processing for big data.
9. **SNA** is a new and modern set of techniques used to characterize relationships among discrete nodes in a graph or a network. In SNA, connections between individuals in a community or organization are analyzed from different angles, such as the circuit of information, a node having more influence than another, and similarities or differences by gender, age, occupation, or educational achievement.

For many clothing manufacturers and fashion producers as well as retailers, their presence on the Web and social networks is increasingly important to their business. But, they must overcome this passive stage and try to listen to online impressions and emotions, summarized into a report. This would help to generate realistic forecasts.

4.2.2 Leveraging big data on cloud computing

Collecting and processing large volumes of data used in the business environment requires significant hardware and software resources at low cost. The IT world has received this new challenge and found the solution through two technologies that support each other and offer advantages to small- and medium-sized enterprises (SMEs) as well as great companies. The two evolving technologies are big data and cloud computing.

Because ordinary businesses cannot afford such advanced technological resources, we consider that the development of big data relies mostly on public cloud implementations.

There is a deep connection between these two technologies: cloud implementation includes powerful capabilities to process and store large volumes of data in a secure environment, and big data platforms precisely need these facilities.

Powerful IT corporations such as Google, IBM, Sun, Amazon, Cisco, Intel, and Oracle offer a wide range of cloud-based solutions, which include many aspects: the storage capacity, the services provided, the security of data, and the subscription cost (Banica et al., 2014).

From a global CEO survey, in the Romanian case, about 90% of Internet users store their data on clouds, using different services such as Microsoft Skydrive, Google, or

Dropbox. And, 33% of enterprises consider clouds to be a major infrastructure (Iuga and Simion, 2015).

But, according to European statistics (Eurostat, 2014), only 5% of Romanian enterprises used cloud computing solutions. The most frequently used software tools are salary management, data analysis, ERPs (finance and accountability), customer relationship managements (CRMs), HR management, and project management applications.

The low level of investment capital and the managers' distrust of the importance of such tools for increasing efficiency, customer satisfaction, and overall business competitiveness have led to Romania being close to the tail of the list of European countries. These systems are still not recognized as being necessary.

4.2.3 Big data–driven fashion supply chains

According to Mentzer, "a supply chain is defined as a set of three or more entities (organizations or individuals) directly involved in the upstream and downstream flows of products, services, finances, and/or information from a source to a customer" (Mentzer et al., 2001). Supply chain management (SCM) involves the coordination and integration of the three main flows of the supply chain: product flow, the information flow, and the finances flow (Naslund &Williamson, 2010).

The supply chain includes the activities of manufacturing, storage, and distribution, as well as a model integrating these flows for decision-making based on minimal cost and/or optimal time consumption (Nagurney and Yu, 2010).

Big data analytics (BDA) influences the SCM at all levels. The fashion companies that invested in implementing solutions for gathering, processing, and analyzing internal data, combined with external consumer and market data, have a competitive advantage in the market place. Consumer feedback, identified through social media comments, conversations, and surveys, can be the key to eliminating inefficient supply chain practices and therefore to coordinating decisions along the entire supply chain.

Some of the benefits of big data for SCM are the following: real-time delivery tracking, automatic product supply, dynamic sales and pricing, customized production, and service (Kotlik et al., 2015). Despite these advantages, many apparel companies ignore or have no confidence in this strategy, and some prefer to use BDA only in certain segments of the supply chain (eg, marketing) to keep pace with changing times and to make realistic forecasts.

All these new data sources lead to more informed buyers and sellers who can make decisions faster by leveraging the aggregated information from business transactions (Report from the Economist Intelligent Unit, 2014). From social media, managers could discern that a supplier may have problems (eg, cash flow), and they can also help to establish and implement more rapid and effective recovery plans should problems arise.

Particularly in fashion more than in other areas, it is difficult to stay ahead of the competition; therefore companies must anticipate the risks and trends in the market and adjust the processes of their supply chain to meet those challenges, thus improving their competitiveness.

4.3 Romanian clothing and fashion industry

4.3.1 An overview of the actual stage

Within the manufacturing industry in Romania, apparel manufacture is detailed in CAEN code number 14.

Amid the financial crisis, we have seen an involution of the clothing industry from 2008–2011 and a slight recovery from 2012–2013, according to data released by the Romanian National Institute of Statistics (http://www.insse.ro, 2015).

The number of active units and the total number of employees is presented in Table 4.1. We observe that during 2008–2011, the number of firms decreased, though there was a slight increase from 2012–2013. In 2013, there were about 17.1% fewer companies than in 2008.

The average number of employees are shown in Fig. 4.1, representing a decrease of approximately 32.5% in 2013 compared to 2008.

Table 4.1 The evolution of clothing companies by number of active firms

Year	2008	2009	2010	2011	2012	2013
Number of active firms	5867	5313	4480	4111	4231	4378
Total number of employees	211,241	176,070	138,147	144,261	145,374	142,540

Romanian National Institute of Statistics.

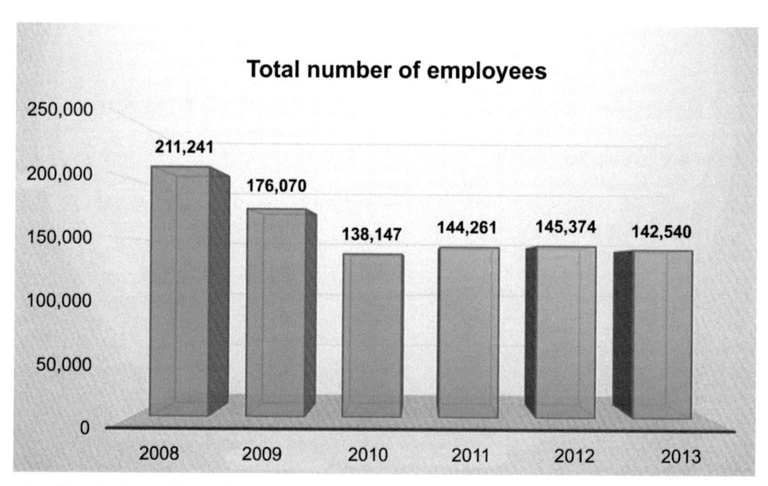

Figure 4.1 Evolution of clothing companies by number of employees, 2008–2013. Adapted from Romanian National Institute of Statistics.

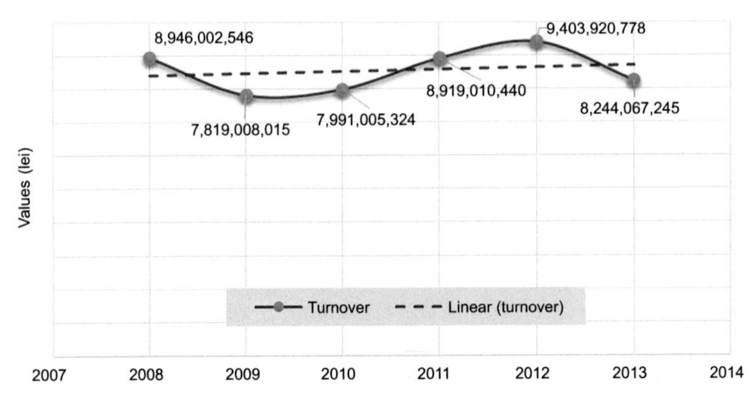

Figure 4.2 Evolution of clothing companies by turnover, 2008–2013.
Adapted from Ministry of Public Finances, 2014.

The turnover recorded a strong decline in 2009, after entering on an upward trend, so 2012 exceeded the value registered in 2008. Looking at Ministry of Public Finances data (http://www.mfinante.ro), we see the turnover recorded a decrease in 2013 (Fig. 4.2).

Appendix A shows the evolution of another indicator, the level of investments, which follows a similar trend as the turnover. In the 2010–2014 period, Romanian exports and imports increased continuously (Appendix B). Concerning the clothing industry, the annual value of exports increased by 29.7%, which can be seen as an expansion of markets and an increase in the quality of exported products (Appendix C).

In the structure of exports and imports, two groups prevail, namely, "machinery and transport equipment" (35.05% at export and 39.52% at import) and "manufactured goods" (20.23% at export and 16.91% at import), as is shown in Table 4.2 and Fig. 4.3.

The manufactured goods group contains clothes, accessories, and footwear.

Table 4.2 **The trade structure by product groups, first semester 2015**

Trade structure in August, 2015	Exports FOB (%)	Imports CIF (%)
Food, beverages, and tobacco	11.48	9.56
Crude materials, inedible, except fuels	4.60	2.9
Mineral fuels, lubricants, and related materials	5.53	7.27
Chemicals and related products	5.38	14.50
Manufactured goods	16.91	20.23
Machinery and transport equipment	39.52	35.05
Miscellaneous manufactured articles	15.59	10.11

Romanian National Institute of Statistics.

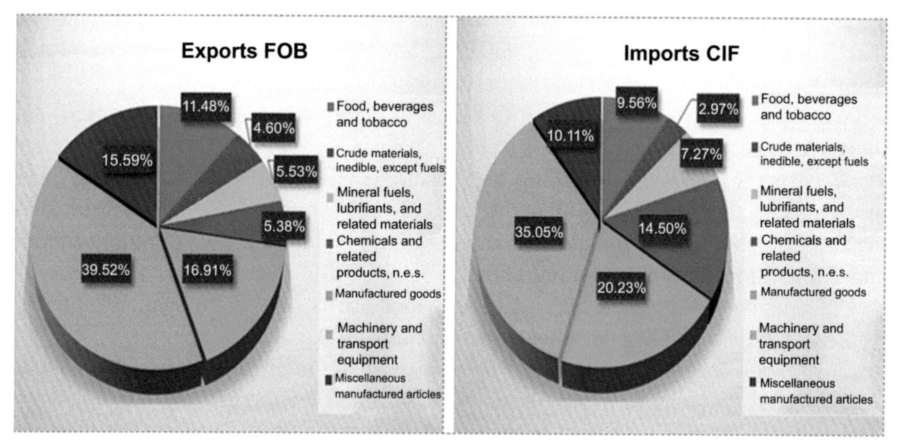

Figure 4.3 The exports and imports structure by product groups, first semester 2015. *n.e.s.*, not elsewhere specified.
Author's interpretation of statistical data from the Romanian National Institute of Statistics.

The Romanian clothing and fashion enterprises have serious problems in staying with the market due to the lack of capital to expand their businesses and to appropriately market their products.

If during 2004–2008 Romania had registered a large increase in the clothing manufacture field (Banica et al., 2014) in terms of turnover (from $2,284,209,202 in 2004 to $2,509,557,965 in 2008) and total exports (from $1,382,901,903 in 2004 to $1,664,186,290 in 2008), the economic crisis hit also this sector, which had a sharp decline in 2009 ($1,721,204,418 turnover and $1,338,257,445 total exports).

Among the causes that led to the decline, and even destruction, of some Romanian brands, we emphasize the application of the "lohn system," which also contributed to the low labor costs in this production model combined with more appealing employment opportunities in sectors such as retailing.

To withstand the economic crisis, many Romanian firms have been working for companies like H&M, C&A, and Zara with disadvantageous contracts, where the labor force is poorly paid and "Made in Romania" does not appear on garment labels. The Swedish company H&M is a multinational retail clothing company that was imposed on the Romanian market relatively quickly through 28 stores existing at the end of 2013. The turnover has increased since 2011, when it was the first store founded in Romania (37 million euros) until 2013 (101 million euros) by about 2.7 times (Shen, 2014). The C&A company is another important chain of fashion and retail clothing stores, having stores in many European countries, including Romania. During 2009–2014, C&A opened 30 large stores on the Romanian market (http://www.c-and-a.com/ro/ro/corporate/company/).

The company Inditex, which produces Zara clothes, also registered success on the Romanian market. The number of Zara stores in Romania in 2014 was 21, and they sell 40,000 models of clothes per year. The turnover reported by Inditex for Romanian stores was EUR 85.4 million and generated a EUR 12.3 million profit (Bazavan, 2014).

The effects of the crisis are still present in the clothing industry, despite the fact that Romanians are interested in fashion, and the sales value advanced slowly from 2% in 2012 to 4% in 2013 for the most important segment of consumers: womenswear. The recovery began to occur in 2011, when exports increased, and the pace was maintained in the subsequent years (2012, 2013) around the same value (Appendix B). The clothing and fashion market recorded a clear recovery in 2013, and the partial statements to date (2015) show a further growth in 2014 (according to Euromonitor International, http://www.euromonitor.com/apparel-and-footwear-in-romania/report).

Unfortunately, only a small part of this increase is due to the capitalization on Romanian brand clothing products (such as apparel TinaR, La Femme, Nissa, and Secuiana); the largest share of the increase went to international brands, such as H&M, C&A, Zara, sportswear retailer Decathlon, and Intersport.

Secuiana Company is a Romanian company that produces garments for men and owns the Adam's by Secuiana retail chain of stores. The company achieved a turnover of LEI 27,515 in 2013, 11% higher than in 2012.

Other Romanian traditional manufacturers are these (FRD Center, 2015):

- Benrom is a producer owned by the Italian Group Benetton, which recorded in 2013 the turnover of EUR 70 million and had 100 employees.
- Modexim had in 2013 a turnover of EUR 4.5 million and 700 employees; it produces clothes for Steilmann, H&M, Max Mara, etc.

If we refer to the development of the clothing industry at a regional level, we observe an unequal distribution among eight macro regions. In Fig. 4.4, we observe that Bucharest Ilfov and the northern regions have the higher number of local units

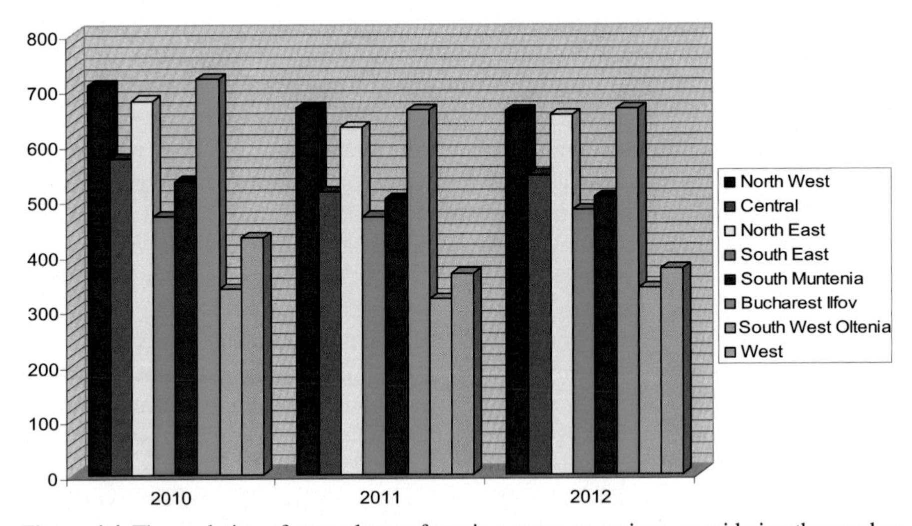

Figure 4.4 The evolution of apparel manufacturing sector on regions, considering the number of local units, 2010–2012.
Author's interpretation of statistical data.

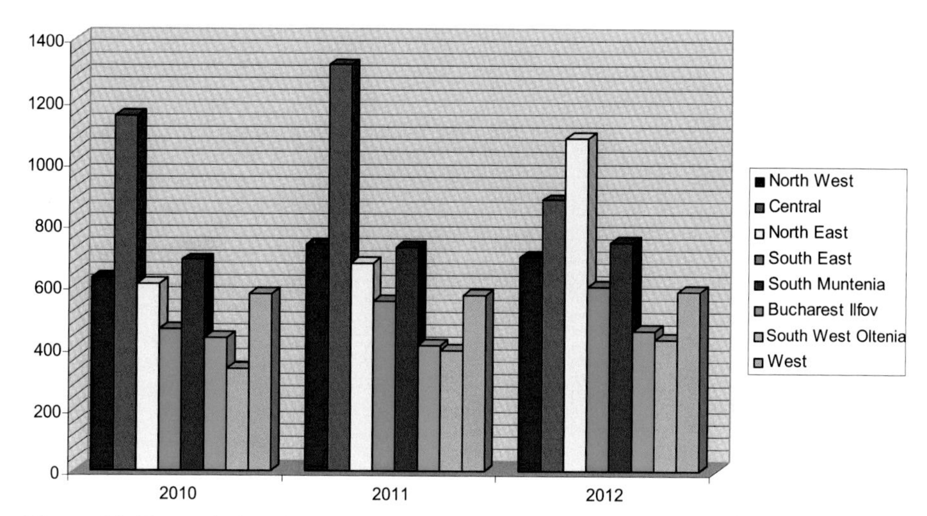

Figure 4.5 The evolution of apparel manufacturing sector on regions by turnover, 2010–2012 (million lei).
Author's interpretation of statistical data.

in the apparel sector, while the lowest level of units is registered in South West Oltenia. We can also observe that during 2010–2012, the rank of each region remained constant; none registered important changes (Romanian National Institute of Statistics).

If we make the same analysis, but take into account the turnover realized by the companies acting in the apparel sector, we can observe that the most profitable companies are in the Central and North East regions, and the least profitable is in South West Oltenia (Fig. 4.5). We also observe a decrease in turnover realized by the companies in the Central Region, while the same indicator registered an important increase for the companies from North East.

Regarding the companies with the most employed personnel, we can observe that companies from South Muntenia and North West Region have the highest number of employees (Fig. 4.6), while the Bucharest Ilfov Region is in last place.

Looking at the overall picture, we can say that though the Bucharest Ilfov Region has the highest number of units in the apparel sector; they have a low turnover and the lowest number of employees in the area.

According to European Commission Report, in *Forecasts for Romania Report* (http://ec.europa.eu/economy_finance/eu/countries/romania_en.htm, 2015), Romania is one of the European countries that recorded an economic growth of 2.8% and a lower unemployment rate of 6.8% in 2014. Also, the European Commission forecasts that Romania's economic growth will remain robust in 2015 and 2016, the exports will increase in 2015 and 2016 at a slow pace, and the unemployment will drop by 6.4% in 2016.

The 2014–2020 National Export Strategy, published in September 2013 by the Ministry of Economy-Department for Foreign Trade and International Relations,

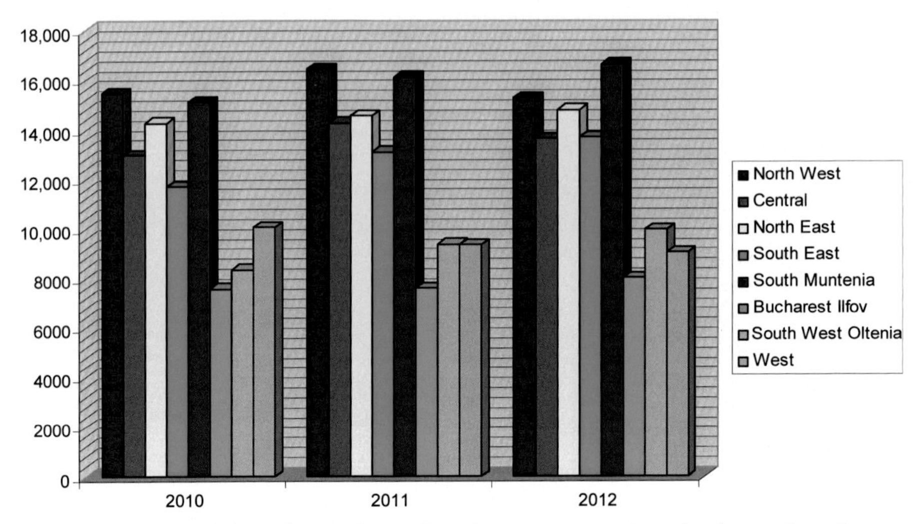

Figure 4.6 The evolution of apparel manufacturing sector on regions, by the number of employees.
Author's interpretation of statistical data.

provides a few strands with which to solve a number of problems facing manufacturing companies in the garment industry:

- diversification of production through increased flexibility and quick capacity to adapt to market requirements
- realizing products at lower costs without lowering their quality, which implies increased productivity, staff training, and production efficiency through production technologies modernization
- encouraging the creative process in the field to contribute to the development of brands, originality, style, and quality
- promoting companies holding nationally and internationally recognized brands through modern marketing policies, promotional offers, and creating customer advisory departments

In this context, we have reasons to believe in economic recovery of the clothing sector and its fashion component, driven by private consumption, public investment based on EU funds absorption, foreign capital inflows, and on a cheap, skilled labor force.

4.3.2 Facing the global market: Romanian products and their competitive advantage

Due to the obsolete technology, lack of a management IT based on real-time information, and a slow transition from traditional to digital marketing, most Romanian companies cannot compete with famous companies in domestic and foreign markets.

Romania is still predominantly involved in "lohn" production, even if a small number of companies have started to produce their own brands for the domestic market, particularly but not exclusively.

The lohn system or OPT (outward processing trade) means the subcontracting of some orders for producers and distributors from world-renowned companies and using a cheaper, skilled labor force from less developed countries. This business model has been used a lot and is still practiced in Romania in the clothing manufacturing industry and in fashion. In the medium- and long-term, the main disadvantage of the performer is that it loses its identity in the market, the goods it produces being sold under the name of the company that owns the brand. Also, through their contracts, big companies do not invest in buildings and equipment; they try to use existing physical resources, and this does not prevent them from transferring their production at any moment to another country that gives more favorable conditions (Plank and Staritz, 2014).

The demand is mostly conditioned by fashion trends and costs of manufacture in Romania, which directly influence prices and, therefore, sales. The profitability of each clothing enterprise depends on their strategy of production but also on their policy of sale: owning a chain of stores or requesting support from retailers.

Large companies have advantages in selecting clothes in accordance with their requirements, distribution and redistribution on store requests, and the financial strength to use the various marketing strategies and, in particular, digital marketing.

Instead, small businesses are looking for lower cost or no-cost methods (also called guerrilla marketing). In 2015, a small enterprise can start with online sales after registering its label to protect it from trademark infringement.

A small company can compete with a large company only by approaching a particular range of products that excel in quality and inventiveness and by addressing a specific market segment.

But, when the business grows up, the manager must expand production, invest in marketing, and acquire a store, two stores, and so on. At this level, when both innovation and sales are the keys of success, we can say that the products manufactured by the company led to the emergence of a real brand.

The global economic crisis has exacerbated the state of competitiveness in the field, and apparel firms have reacted in different ways: upgrading their products at a faster pace, approaching niche products, or looking for new export markets.

In Romania, given the lack of government financial support where apparel is considered to be a "sunset" industry (Plank and Staritz, 2014), small and medium firms have tried to access credit from banks to survive or to continue production under the lohn system.

A study conducted on 102 clothing firms in 2011 shows that most of them considered this business model to be capable of assuring stability during crisis, so they did not agree to give up or decrease lohn production (Popescu and Radu, 2011).

Larger companies were also affected by the crisis, and some of them disappeared; others shifted to higher value products or looked for other markets. But, product upgrade implies design development and equipment upgrade, meaning important investments.

The share of Romanian apparel firms that can successfully compete with international firms in the domestic market is very small. Some of them became groups of international firms, as is the case of FF Group and Azadea Group Romania.

FF Group Romania is part of Folli Follie and is one of the most important players in the national fashion retail industry (Euromonitor International, 2015). Under the name of *Famous Brands*, FF Group has introduced in its network of stores a multi-brand concept, which means it gives consumers "a collection of famous brands," from sporting goods to fashion and accessories, from important firms such as Nike, Folli Follie, Calvin Klein, Converse, and Calvin Klein Jeans. Subsequently, these stores have changed their name to "Collective," after the name used in the company's home country of Greece.

Romania clothing Azadea Group is part of Azadea Group, a company owned by the Daher brothers, originating in Lebanon, which operates in more than 15 countries. It also works with a multi-brand regime, and in the 409 shops it owns, it offers buyers almost 40 brands (Harb and Chaaya, 2015). In the area of fashion, it is an international retailer for brands such as Bershka, Calzedonia, Grain of Malice, Marina Rinaldi, MaxMara, Piazza Italia, Pull & Bear, Stradivarius, and Zara.

We could say that, from the two success models—one of manufacturing and commercialization through its owns chain stores, the other of trading several other brands through a chain of shops belonging to the retailer—the latter managed better on the market in recent years.

Where is the Romanian apparel industry placed among corporate strategies?

As usual, it is somewhere in the production area. With a low-cost and skilled labor force, Romanian clothing remained hidden under the umbrella of lohn system, which gives companies the opportunity to not close and to not dismiss personnel.

Since 2013, the strategy has begun to change; Romanian companies are trying to assert the market through their own efforts, without fully giving up the material support provided by OPT trade, as a result of contracts that are ongoing or due to managerial prudence.

Their attention is focused on traditional markets or less tackled markets, internal and external, avoiding a highly competitive environment such as Western Europe (Farnell, 2014).

Clustering strategy in the clothing industry was rethought due to globalization and technological progress. It is true that local or regional cooperation is very important for some activities, but this way of thinking has proved to be inadequate against the desire to remain competitive on the European market (Pellegrin et al., 2015).

Including all the activities of the fashion supply chain (production, transportation, sale stores, and marketing) in the business process leads to smaller production costs and therefore to an increasing competitiveness. Romania has developed a manufacturing sector comparable to other EU member states, but it has a smaller market services sector.

To remedy the situation of the Romanian clothing industry, the Ministry of Economy proposed a national strategy for 2014–2020 that contains several ways intended to sustain Romanian firms (http://www.minind.ro/strategia_export/SNE_2014_2020.pdf, 2013):

- transition from lohn system to direct exports
- modernization of production with emphasis on creation (design)

- guideline sales for e-commerce
- improving value chain efficiency, notably with regard to storage, transport, and logistics documentation, by training and better cooperation between manufacturers and service providers in the sector
- increased production levels at an annual rate of 3.4%

Also, the national strategy for 2014–2020 contains ways to promote the export of Romanian products and the actions to be taken in the short, medium, and long term in fashion design (see Appendix D).

Competitiveness is an attribute derived from efficiency factors essential for creating more revenue, thus, a concept related to the productivity. If the statistical databases available for Romania partially cover the needs for a competitiveness analysis, there are several European and world organizations that publish such reports annually (Şerbanel, 2014).

A study realized by the World Economic Forum reflects that, in 2014, the most important factors that affected Romania's competitiveness negatively were tax rates, corruption, tax regulations, access to financing, inefficient government bureaucracy, inflation, restrictive labor regulations, and policy instability. But, there is an important competitive advantage: a low-cost labor force. Though a competitive advantage, this affects workers in textiles and the clothing industry; the value added per employee placed Romania on the penultimate position before Bulgaria's DG Enterprise and Industry in 2012. Even so, Romania does not exceed 12% of the results of the country with the higher added value per employee (Schwab et al., 2014).

With a share in turnover of 91.66% from the manufacturing industry in 2012 (8,619,903,951/9,403,920,778 lei), the clothing industry is among the main branches of the Romanian manufacturing industry and has an important export potential.

Figs. 4.7 and 4.8 show the values of apparel exports in 2012 for several countries. As we can see, Romania is far behind from the point of view of the export value

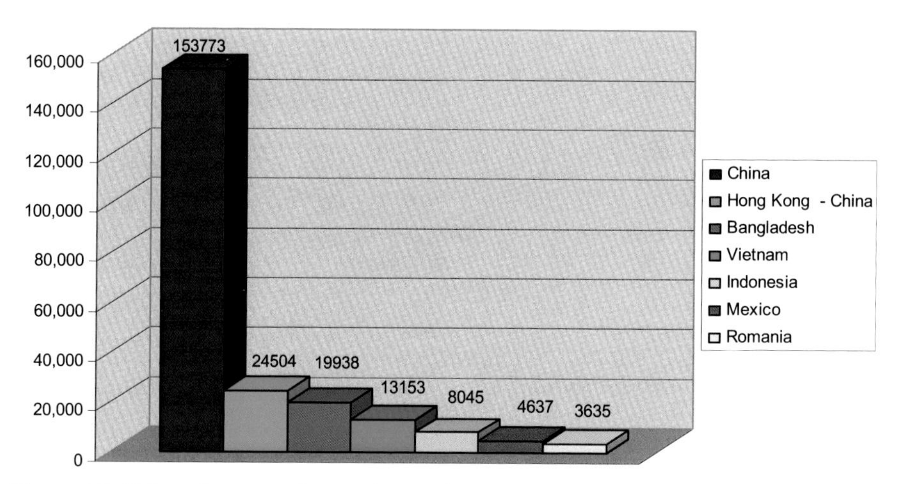

Figure 4.7 Value of exports of apparel sector in 2012 (million euro). Adapted from World Trade Organization (www.wto.org).

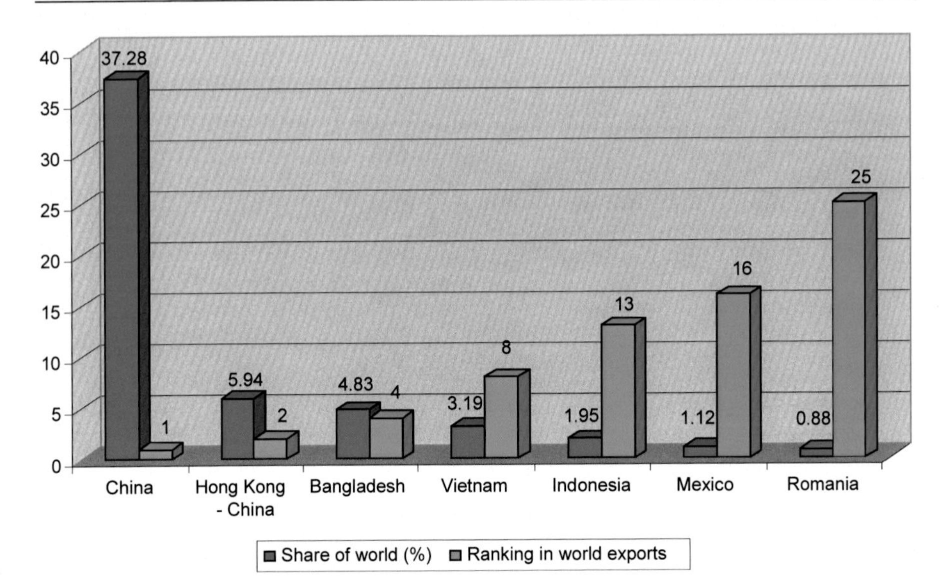

Figure 4.8 Share of world and ranking in world exports of apparel sector in 2012.
World Trade Organization (www.wto.org).

(EUR 3635 million) compared to the country with the higher export in the apparel sector (China with EUR 153,773 million), as well as the share of world (China: 37.28%, Romania: 0.88%).

There are Romanian companies with a tradition in the textile and clothing industry, and we extracted data on the evolution of two of them:

- SECUIANA SA (Fig. 4.9) registered a considerable decrease of turnover in 2010 and has seen a progressive increase up to 2013, with a total of 528 employees.
- FORMENS SRL (Fig. 4.10) had an upward trend throughout the period analyzed, and the number of employees increased from 530 in 2008 to 696 in 2013.

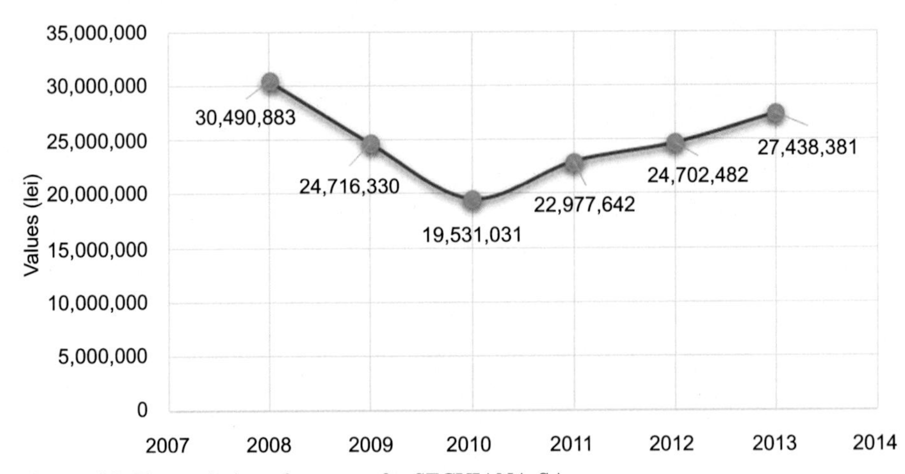

Figure 4.9 The evolution of turnover for SECUIANA SA.
Author's data processing from http://www.firme.info/secuiana-sa-cui3667158/.

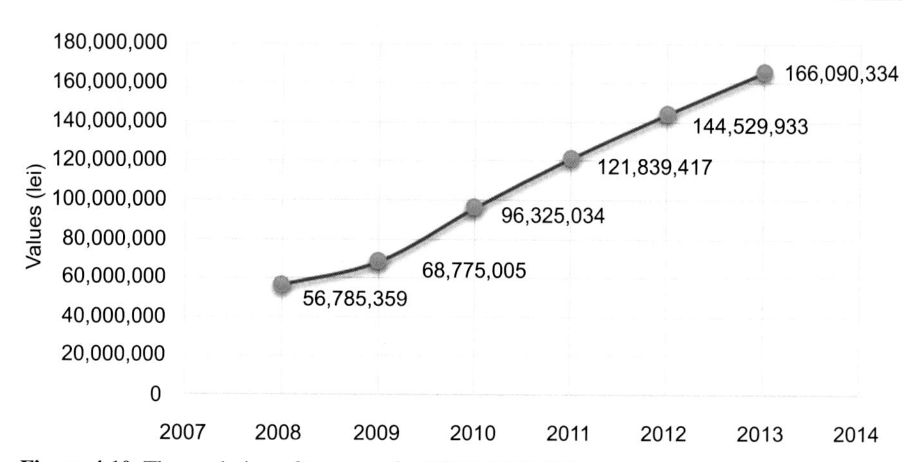

Figure 4.10 The evolution of turnover for FORMENS SRL.
Author's data processing from http://www.firme.info/formens-srl-cui12153083/.

As we said before, the difference between a successful company and another that fails in the market is made through creative activity, involving research and refurbishment.

Utility, quality, and comfort in use are the main features that should be taken into consideration when developing new products to respect the innovation perspective of a company.

Thus, Romanian designers were inspired by Romanian traditions in their collections and managed to impress the Western European fashion world with traditional blouses, dresses, and skirts that featured Romanian pure elements.

In Romania appeared new categories of textile and clothing that included products for military, agriculture, sport, and medical domains. Research studies were carried out mainly by the National Institute of Research and Development for Textiles and Leather.

During 2010–2014 the potential of IT&C was proved for textile and clothing companies. The companies oriented toward modernization of technologies and new strategies to design and commercialize new products. Another important challenge for companies was the problem of knowing the changes in customer behavior using social networks software. Finally, research centers have set up new IT&C services and improved marketing and communications.

The small-scale investments in the Romanian apparel industry, reaching only a weak growth of around 0.03% during 2010–2012, have been directed mainly toward two operations of the supply chain: production process improvements and research.

The companies' attention focuses on use of Internet services to promote their image and develop their online business. A set of measures to be applied between 2016 and 2020 are defined in the national strategy regarding Romania's Digital Agenda concerning advanced IT&C, and these include cloud computing and big data.

The surveys conducted on a sample of Romanian SMEs proved that management is highly convinced that success can be achieved by increasing Web visibility and quick adaptation to customer requirements, which involves a social media presence.

Business environment needs are supported by IT&C technologies through software packages that process unstructured data from social media: tools that may extract patterns, forecasts, and other useful information from this huge volume of data. This kind of data processing is made through BI and SNA software.

Knowing the changes in customer behavior using IT&C technology to collect and analyze the big data, provided especially from social networks, represented a huge challenge for companies.

4.4 Community-influenced decision-making: the answer is in the social cloud

Evolution of IT&C technologies in recent years has strongly influenced both production and distribution of apparel products worldwide.

Today, software and hardware companies' attention focuses on three main areas:

1. computer-aided design for manufacturing patterns and garments
2. using the Internet to create and sell custom designs
3. collecting and processing information from social networks to analyze customer preferences and make the best decisions for their satisfaction

This division does not mean a separation or exclusion of some ways in favor of others, but conversely, by cooperation and mutual support, every step forward in one direction is felt immediately in the other two.

We will briefly present the trends and achievements of the first two directions, and we will insist on the third, which we consider essential in the supply chain decision-making.

1. It's a fact, confirmed daily in all areas of consumption, that current consumers are quickly "bored" by products used in the accelerated and ever changing pace of life, and dress style is the first questioned; it is the first that loses the "time test."
   The life cycle of products is getting shorter, and new product launch or increasing the volume of products no longer represents a means of success in the fashion world (D'Amico et al., 2013).
   Therefore, companies have introduced, in trial design and, more importantly, in manufacturing, machinery driven by software (computer-aided manufacturing systems). The current trend is to make clothes tailored to each client, so-called personalized products at prices that will not be significantly higher than those of regular series.
   Concerning the relation with customers, the personalized production assumes operations that could be made in stores, which involves specialized personnel, equipment, and more space, such as choosing and adapting a fashion assortment, but also operations that can be achieved only in the production area.
2. The obvious solution to satisfy new consumer demands is through the Internet, its involvement not only in online sales but also in production.

Fashion sites are much promoted, first, and second, they are intensively accessed to learn trends, which leads to subsequently purchasing the patterns proposed.
But, the problem of sizes and colors still remains, as well as the serial productions that are harder to accept. Designing personalized clothing means a dialogue with the customer to find measures and preferences, the use of software for creating and presenting virtual models (e-prototyping), the purchase of materials and manufacture on demand (by computer-based design), and operations that increase the sales price of the clothing product.

An example of a company that has great achievements in the field of "dynamic-clothing" is Optitex, which provides a complete solution in the fashion and apparel industry by the collaboration between design and production (http://www.optitex.com/). This is a solution used by designers such as Hugo Boss and Tommy Hilfiger.

Among the worldwide distributors of Optitex technology is also SC Texco Industrietechnik SRL, with headquarters in Sibiu, Romania, founded in 2013. The turnover increased from 741,276 lei in 2013 to 7,205,177 lei in 2014, according to the data from Ministry of Public Finances (www.mfinante.ro).

Also, the IT company GeminiCAD Systems launched their software environment called Gemini Pattern Editor for designing products or collections and Gemini Tailoring Assistant, a 2-D product covering the two applications in garment design: made-to-measure clothes with automatic reshaping of the patterns to fit the measurement of the customer, and mass production that provides full automatic grading of a single product for different size tables (http://www.geminicad.com/apparel_textile/made_to_measure_at.php).

During 2009–2013, according to Ministry of Public Finances, the enterprise recorded an increase on turnover of 4.19% in 2012 (4,192,179 lei) in comparison to 2009 (2,340,598 lei) and had a slight decrease of 3.51% in 2013 (3,510,049 lei) (http://www.mfinante.ro).

The "ready-to-wear" system (prêt-à-porter), providing suitable clothing products to a range of customers, is globally losing ground in favor of the "made-to-measure," accomplished to fit each client individually. The made-to-measure system is placed in financial terms between mass produced clothes and ones that are made to order (Suciu and Olaru, 2012).

A variety of advanced technologies have given rise to an explosive phenomenon of "mass customization," introducing clients in the stages of creation and design of the model, resulting in articles that are made-to-measure with a high degree of matching in competitive market prices.

Experts in the field have noted the difference between the two terms frequently used in the customized apparel industry (Suciu and Olaru, 2012):

- Mass personalization is defined as tailoring customized according to the tastes and preferences of users.
- Mass customization offers customers the ability to create and choose the product within certain specifications (custom tailors).

By promoting the mass customization method of production and sales, a number of intermediaries, such as wholesaler and retailer, disappear from the supply chain.

Examples of companies that have launched in the area and successfully practice this model of production and sale are these:

- Tinker Tailor lets the customer create a one-of-a-kind dress, top, or skirt, starting with the silhouette, then move on to fabric and embellishments (http://www.tinkertailor.com/).
- Bow and Drape encourages a woman's self-expression (http://www.bowanddrape.com/home/).
- Indochino is one of the most popular online sellers of made-to-order menswear (http://www.forbes.com/sites/deborahljacobs/2013/07/01/made-to-order-fashion-goes-mainstream/).
- Black Lapel is a site where shoppers can provide their measurements and get customized style, fabric, and fit (http://www.forbes.com/sites/deborahljacobs/2013/07/01/made-to-order-fashion-goes-mainstream/).
- Tailor Store is a company offering made-to-measure clothing for both men and women.
- Piol allows the customer to design a dress with a threefold approach to fashion: color, style, and fit (http://www.oliviapalermo.com/piol-designing-for-the-future/).

3. In an era in which businesses are becoming "data businesses," the clothing industry is moving to big data–driven technology, which is playing a part in trend forecasting.

Social media information, sales reports, and other kinds of information are aggregated to help designers in their future projects and retailers to acquire clothes in trend (Marr, 2015).

4.4.1 Big data tools and techniques

In this section, we will briefly present a new type of database (NoSQL, used for storing big data), the software that allows for massive parallel computing (Apache Hadoop), and several applications for collecting unstructured data, such as Apache Flume.

Finally, the data filtered by MapReduce must be analyzed using analytic tools, and we consider that Gephi and NodeXL are two representative applications for SNA software; so we made a short presentation of their facilities.

NoSQL databases manage a wide variety of unstructured data described by several models: key value stores, graph, and document data (Mohamed et al., 2014). NoSQL data are implemented in a distributed architecture based on multiple nodes, such as Hadoop software.

Hadoop is an open-source software platform that enables the processing of large data sets in a distributed computing environment. So, the distribution of data is realized to individual machines, also called nodes, which locally process the information and store the required outputs.

Hadoop is compatible with any of the operating system families (Linux, Unix, Windows, or Mac OS) and can be run on a single- or multi-node cluster. Hadoop has the advantage that it can be used in a datacenter as well as in a cloud environment.

The most widespread solution is the open-source Apache Hadoop distribution, which includes several components (Frank, 2014):

- *Hadoop Distributed File System (HDFS)*: the storage component
- *MapReduce engine*: the processing component

- *Hadoop Common*: a module for libraries and utilities
- *Hadoop YARN*: the component responsible for managing and scheduling cluster resources

Fig. 4.11 shows a Hadoop cluster, which is a set of computers running HDFS and MapReduce, one of them being the master node, and the others are slave nodes.

- HDFS is designed to split input data files into blocks and distribute them across the cluster to be processed in parallel. This component also has the function of storing the information across the machines involved. The master node (also called name node) coordinates distributed blocks, known as data nodes.
- MapReduce is a programming framework used by Hadoop to distribute work around a cluster so that tasks can be run in parallel on different nodes. MapReduce is controlled by the JobTracker software that resides on the master node, and it monitors the TaskTracker applications, running tasks at each node (Frank, 2014).

The MapReduce system is based on the two components performed (Brust, 2012):

- Map reads data in the form of key/value pairs and processes tasks on nodes, avoiding network traffic.
- Reduce waits until the Map phase is completed and then combines all the intermediate values into a list, providing final key/value pairs as outputs that are written into HDFS.

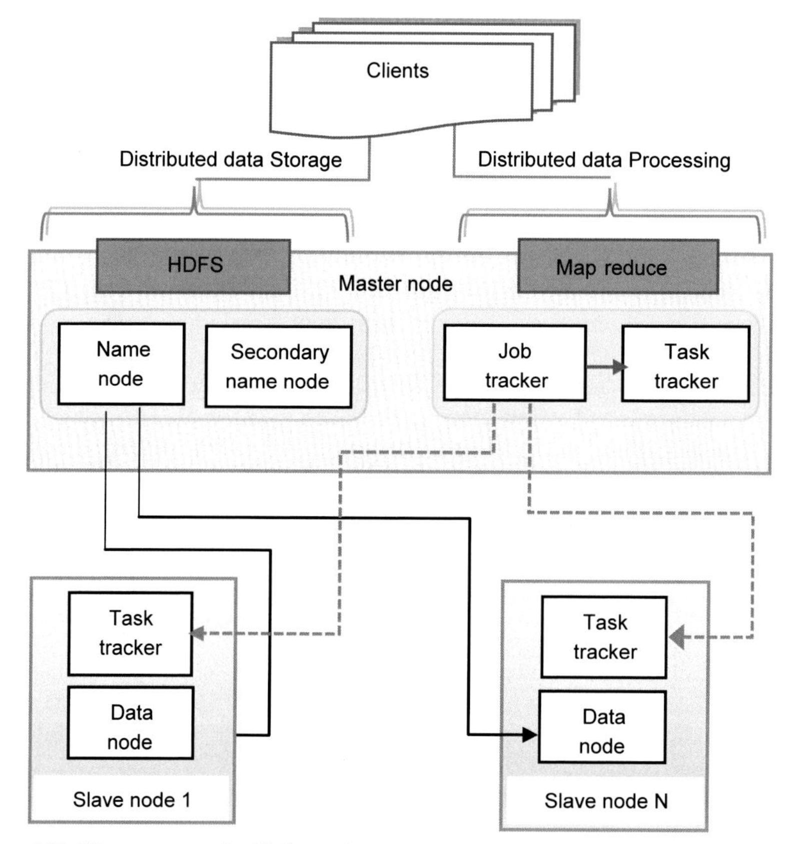

Figure 4.11 The structure of a Hadoop cluster.

The purpose of a secondary name node in this architecture is not to offer a backup of the primary node name, but to keep checkpoints in HDFS, helping the name node to function better. In a real configuration, the secondary name node is placed on a different machine from the primary name node in case of malfunction, so information could be recovered from the secondary name node (https://hadoop.apache.org/docs/).

Big data and clouds are two emerging technologies for processing tremendous datasets and applying them into business; through these, company leaders can know the true opinions of customers and make the best decisions leveraged on customer preferences.

SNA software is a combination of graphical network representation and statistical analysis, and the most representative applications in the domain are Gephi and NodeXL.

Gephi is a free, open-source interactive visualization and exploration platform for all kinds of networks and complex systems, capable of accommodating networks up to 50,000 nodes and 1,000,000 edges (Bastian et al., 2009).

NodeXL is a free, open-source template for Microsoft Excel that makes it easy to explore network graphs. It has a direct connection to social networks as Twitter, YouTube, and Flickr, and it collects the publicly available data and follows the relationships of users that have public accounts (Lieberman, 2014).

Concerning metric calculations, both tools allow the calculation of degree, centrality, clustering coefficient, and graph density of networks (Messarra, 2014).

Centrality identifies the most influential individuals in a social network, as well as the isolated persons. This measure shows the roles in a network: the leaders versus the isolates, the core of the network versus the periphery (Krebs, 2013).

Betweenness centrality represents the entity's position within the network and identifies the individuals who hold a favored position in collectivity (Krebs, 2013).

Closeness centrality is one of the most significant SNA metrics because it represents the speed of information distribution between nodes of the network. It measures the distance (through direct and indirect links) from a node to all others within the network (Borgatti, 2005).

These applications are successfully used to analyze pages and groups of Facebook and Twitter networks and emails. After using both SNA software and making a comparison of their features, we believe that NodeXL has the advantage of working with Microsoft Excel spreadsheets for data handling and allowing direct connection to social networks, while Gephi offers a better graphical module (Banica and Hagiu, 2015).

4.4.2 Big data on clouds for fashion

An increasing number of firms are choosing to use cloud computing technology; many brands and retailers consider that the solutions offered are helpful for managing the supply chain. It is necessary to connect the entire supply network to the clouds, from the manufacturer to customer, through the retailer and the financial institution of the settlement (Lutz, 2013).

But the big step has not been made in a sector as subjective and unpredictable as fashion. The next level in IT for business (meaning big data exploration to find useful information in all discussions, blogs, emails, and Facebook or Twitter pages) is yet a shy approach for the fashion domain. The first reason could be the costs, the second, the impact of new technologies, and the third could be mistrust—most designers and producers are more confident in their own intuition than a software report, even if it represents client feedback.

But if we look from the correct perspective, we realize that big data will change all aspects of fashion business in the years to come, from the ways to establish the preferred color for the next season to virtual personalized clothing and optimization of supply chains.

In our researches, we presented several scenarios of cloud-based big data, and we recommended hybrid clouds as the preferred option for the companies that may use private clouds to manage internal SQL data, while public clouds are intended to achieve and process the external data or archives, also called big data.

But, for our purpose, taking into account that SMEs are more numerous in the clothing and fashion industry, we consider that the optimal strategy is to preserve their internal structures (relational database management system, RDBMS; enterprise data warehouse, EDW) and use public clouds only for gathering and processing unstructured data.

The final objective is to analyze and extract meaningful insight from this labyrinth of data, to discover behavioral patterns of consumers, and to apply this information in management decisions (Banica and Hagiu, 2015).

In this section, we propose a model for integrating big data with an existing ERP system (or other information system), resulting a four-layer architecture based on a cloud Hadoop implementation. Thus, our model refers to a unified architecture, using Hadoop as a data integration platform, as is shown in Fig. 4.12.

The model includes four levels with these functions:

1. The first layer is designated for gathering structured and unstructured data for a company using nontraditional sources:
   a. Unstructured data originated from social networks, clickstreams, Websites, Weblogs, etc., and are collected with a software application such as Apache Flume.
   b. Structured data originated from external databases and are collected with a tool like Apache Sqoop or ETL.
2. The second layer refers to the parallel processing of big data using Hadoop: at the second level, Apache Hadoop is implemented to aggregate and process data coming from nontraditional sources.
3. The third level contains the existing enterprise applications and the data generated, which are structured data and constitute a consolidated data warehouse.
4. The fourth level is planned to analyze all data processed using analytics tools: business and modeling tools, statistical analysis, and exploring and visualization applications such as Gephi and NodeXL.

The proposed scenario requires that the first two levels are implemented using public clouds (Platform as a Service model), while the third level is considered to be the core of confidential and protected enterprise data, running in the on-premise datacenter or private cloud, using paradigms such as Software as a Service, Business Process as a Service, Storage as a Service, and Security as a Service.

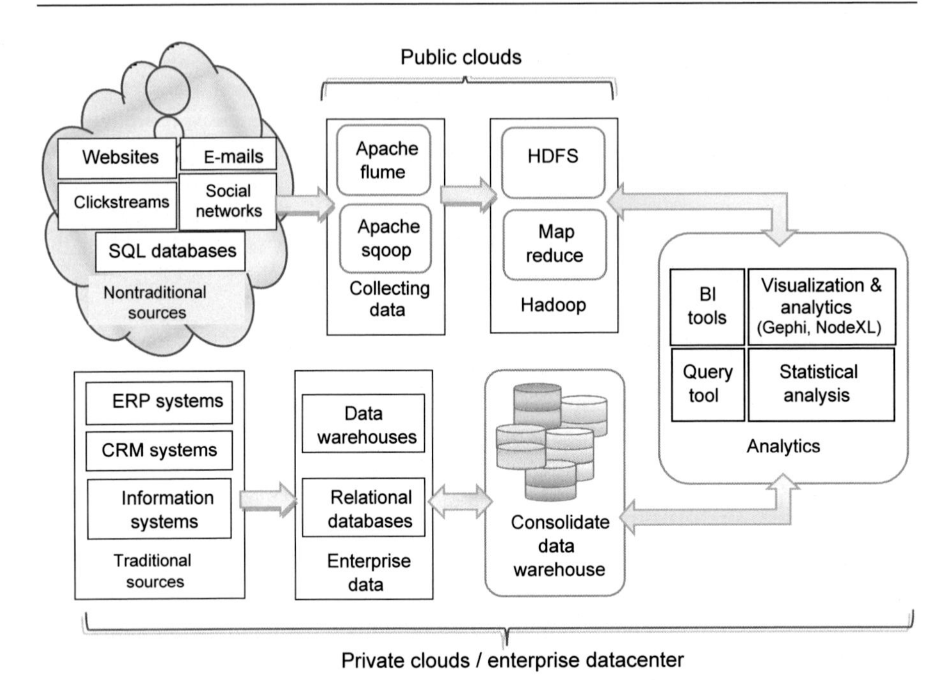

Figure 4.12 A model for integration of a big data–based cloud implementation to an existing ERP.

Some of the Hadoop distributions, such as Cloudera's CDH, IBM BigInsights, and Hortonworks Data Platform can be run in public clouds, such as AWS, Rackspace, Microsoft Azure, or IBM SmartCloud, as a service.

Where should the analytical tools be placed in this architecture?

We may consider the option of treating unstructured data (big data) separately from the company's internal data. But the main goal is to correlate all this information to have a complete perspective of the business, and in such conditions, we consider that the optimal solution is to install and launch the analytic software from the company's information system, having also the ability to access data processed by MapReduce functions and stored in an HDFS.

These are the features of the proposed conceptual model for clothing and fashion:

1. provides a powerful infrastructure at low subscription cost, by using cloud computing resources: the storage capacity, the security of hosted data, and the requested services
2. keeps the previous information system implemented in a private cloud or local datacenter as an extension, not a replacement
3. uses an open-source architecture (Apache big data platform) for collecting and processing the unstructured data from social media
4. provides a set of collaboration tools for all actors involved in the supply chain by using public cloud services
5. increases efficiency due to the ability to explore data using analytics and extracts interesting insights for each of the supply chain activities

4.4.3 Points of view and discussions about the influence of big data in a fashion supply chain

By leveraging the advantage of the flexibility, real-time information, and parallel processing, cloud-based big data is the first choice for SCM in the apparel and fashion industry. To be more specific, the related technologies improve the decision-making of each supply chain member (manufacturing, logistics, marketing, and retail):

1. From the manufacturer's point of view: by knowing the preferences of the target client, it is possible to perform short-term forecasting and adapt to ever-changing market trends, avoiding major risks. The manufacturer may own a chain of stores or work with retailers, and in such cases, the investment in IT tools for analysis of customer opinions is redirected to design and e-prototyping.

   The professionals in many industries are welcoming data analytics, but in the fashion industry the adoption is slower, and thus we observe some skepticism from fashion designers, who view their creations as a work of art and more difficult to be quantified, according to Professor Yilu Zhou from Fordham University, United States (Swayne, 2014).
2. For marketing activity: based on findings from social networks, blogs, emails, preferred products for specific customers, and the effects of a promotion campaign, management can identify new product opportunities, optimize pricing, and improve customer service.

   Exploring the network graphical representations, the marketers also identify the "brand ambassadors," known as loyal customers who are active on social media (Sharma, 2014) and try to attract them in future promotional campaigns.
3. For logistics activity: when integrated into a clothing company, this refers to the connection of locations with the flow of goods. By receiving information from social media related to traffic congestions and safety of routes, delays, or customer complaints, the management can make the best decisions regarding shipment planning, product delivery, and the security of the transport activity. Not all information is reliable; therefore management must design an intelligent and safe transport plan via visualizations and analysis of real-time data about the transport networks (International Transport Forum, 2015).
4. From the retailer's perspective: a retailer has direct connections with customers, so they are in the best position to know customer opinions and to want to know more. Usually, retailers have two closely related purposes: improving customer satisfaction and increasing sales opportunities. When the retailer is a department of the apparel company, it can use the data from social media about the latest fashion trends to provide a basis for decisions regarding future production.

   When the retailer is not integrated in the manufacturing company, it has the advantage of buying and selling products from different brands, so it needs to invest in software analysis that could offer a better perspective of the future marketplace.

   Using Hadoop, retailers may capture, analyze, and have a deep understanding of information coming from multiple channels, including fan sites, social network pages, emails, and blogs. By aggregating all relevant data about transactions with online behavior and social media activity, a unique repository can be built from multiple streams of data to obtain a general view of the transaction history and a "to do" planner.

The consumer has the opportunity to evaluate a new product or a new fashion collection, and she or he can use social media to exert a powerful influence over a retail brand's or product's success; also, the future client is able to choose a store or a commerce site, taking into account the slight differences in pricing that could influence their purchase decision. Cloud big data allows more price flexibility in comparison with traditional methods. Normally, the consumer is the main beneficiary of the new technologies, and he or she may express their opinion by a simple clickstream.

Even if they are not directly included in the fashion supply chain, the IT&C companies are indirectly involved because they need a market for the applications, as well as their services. Therefore, their offer could be the storage and processing of structured data in their clouds and/or gathering and analysis of large sets of unstructured data using the Hadoop implementation and analytics tools. This way, manufacturers and retailers can focus on management problems, instead of concerning themselves about hardware and software infrastructures, or about the implementation of software applications.

The cloud computing provider makes analytical reports available for all the supply chain members, while keeping the security and confidentiality of information at a high level. The analytical models offer an advantage to the decision-making process for manufacturing, marketing, merchandising, and pricing strategies.

In fashion's digital transformation, vendors' technology has an important role. There are various tools that can transform massive amounts of data into real business value in a very small time, and this can help fashion companies and retailers to understand, at any point in the product lifecycle, which trends are gaining and which are losing ground. These insights give them the possibility to reduce the risk of not selling the products they made by making changes to the design, production, or marketing before putting the goods on the market (Trites, 2013).

4.5 Conclusions

Fashion itself is a social phenomenon, so it is natural to find out what thousands of people think by exploring social networks and analyzing their likes, comments, and messages on Twitter or Facebook and to appreciate the sincerity in expressing what they like and dislike.

Forecasting consumer demand for fashion is a complex subject and has been the domain of elite groups and in-trend designers supporting successful brands. Big data will help to predict customer behavior to steer fashion houses in the right direction, but without ignoring the surprise factor and frequent retro trends and without replacing the creativity in design. Also, big data on clouds will contribute to the optimization of the supply chain and to the development of personalized clothing.

Instead fashion critics, manufacturers, and retailers will be more attentive to the "feelings" expressed on social media and their connection with sale reports. If they will ignore the views on social networks, their position in the marketplace will be taken by small- and medium-sized businesses working in the e-commerce sector, which is promoted by bloggers, fan groups, and celebrities with large social media followings.

In our future research, we intend to implement a small Hadoop cluster in the laboratory of the university and evaluate its performance working with unstructured data from social networks. Also, we will continue to evaluate analytic tools, comparing them against workloads with query, graphic interpretation, and dynamic approaches.

In conclusion, we underline that big data on clouds is a powerful tool that can improve the competitive edge of a clothing and fashion company—without exaggeration, this is a magic solution for success.

Appendix A: The evolution of the Romanian investments during 2008–2012

Table 4A.1 and Fig. 4A.1 records another indicator, which follows a similar trend as the turnover, namely the level of investments. We observe that the gross investments

Table 4A.1 Evolution of the investments in clothing sector during 2008–2012 (lei)

Year	2008	2009	2010	2011	2012
Gross investments	940,031,510	469,167,136	492,131,060	505,967,020	507,104,602
Net investments	522,182,524	255,150,620	286,997,322	296,966,074	331,586,941

Romanian National Institute of Statistics.

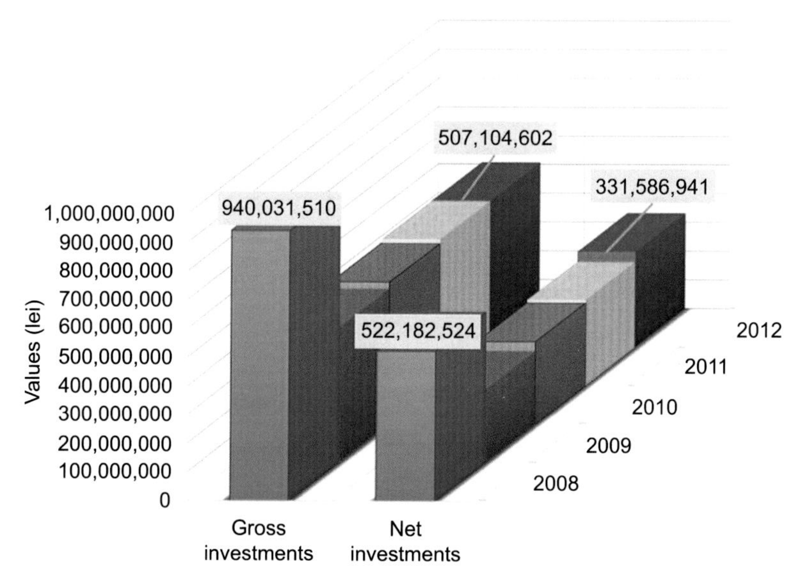

Figure 4A.1 Graphical evolution of the Investments in clothing sector during 2008–2012. Adapted from Romanian National Institute of Statistics.

registered a strong decline, from 47.6% in 2010, compared with 2008, and weak growth of around 0.03% in 2010–2012.

Appendix B: The evolution of Romanian exports and imports

From the assessment of Romania's international trade developments, in the 2010–2014 period it results that the exports increased continuously, and in 2014 they rose by over 40% compared to 2010. At the same time, imports rose by almost 25% in 2014 compared to 2010 (Table 4B.1 and Fig. 4B.1).

Table 4B.1 The evolution of Romanian export and imports (million euro)

Year	Exports (FOB)	Imports (CIF)	Commercial balance
2010	3113.36	3905.77	−792.41
2011	3774.29	4579.29	−805.00
2012	3755.77	4558.61	−802.84
2013	4130.18	4605.77	−475.58
2014	4371.64	4875.70	−504.06

Romanian National Institute of Statistics.

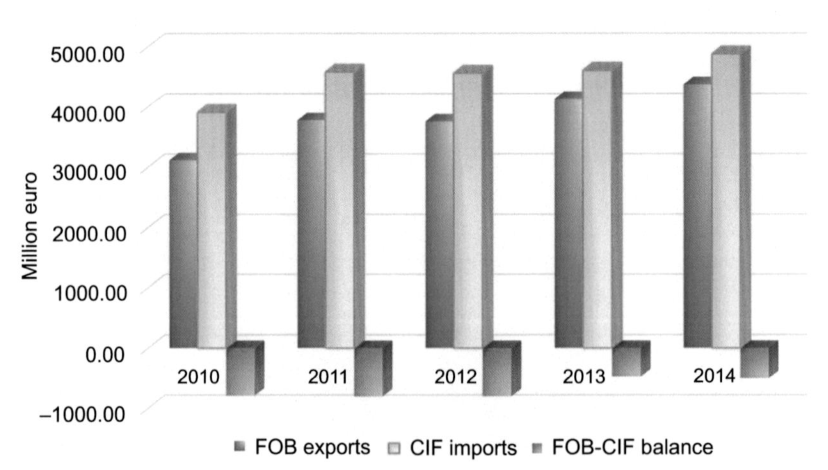

Figure 4B.1 The evolution of Romanian exports and imports during 2010–2014. Adapted from Romanian National Institute of Statistics.

Even if the exports increased constantly and in the first quarter of 2015, reaching 13,380.8 million euro, the imports also amounted to 14,753.1 million euro, so the commerce balance recorded a deficit of 1372.3 million euro, according to National Institute of Statistics.

Appendix C: The evolution of clothing sector exports during 2008–2012

The data from National Institute of Statistics, summarized in Table 4C.1 and plotted in the Fig. 4C.1, indicates a trend which records a peak in 2012.

We observe that the annual value of exports increased by 29.7%, which can be seen as an expansion of markets and an increase in the quality of exported products.

Table 4C.1 **Direct exports in Romanian apparel manufacturing sector (lei)**

Year	2008	2009	2010	2011	2012
Exports	4,577,909,374	4,432,142,031	4,816,871,283	5,478,323,263	5,939,615,616

Romanian National Institute of Statistics.

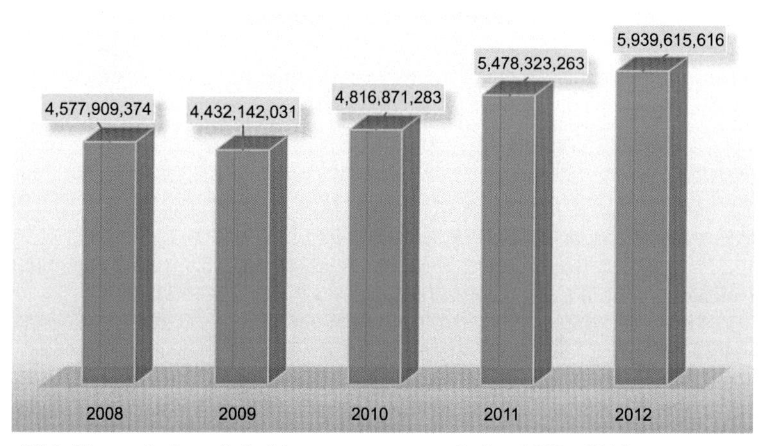

Figure 4C.1 The evolution of clothing sector exports during 2008–2012. Adapted from Romanian National Institute of Statistics.

Appendix D: The strategy to promote the Romanian exports

Internationally, in order to promote the export of a 100% Romanian product, it is proposed:

- Attaching particular importance to producers in Romania presence at international fairs, mainly: Paris, New York, Tokyo
- An interesting textile market could be United Arab Emirates, and it should be taken into account the presence at fairs held here
- Organizing economic missions abroad: contacts, appointments with clients in France, USA, China, Russia, stating that such missions should be managed by the Association in order to achieve tangible results

In terms of fashion design, business circles from this domain intend (http://www.minind.ro/strategia_export/SNE_2014_2020.pdf 2013):

On short term
- Developing a brand and communication strategy
- Transform the national values in internationally recognized brands
- Development of education, harmonization of the educational offer with labor market demand

On medium- and long-term
- Develop research and innovation in all fields
- Implementation of new business models based on innovation
- Create some design centers in the four regions of the country, with role of design workshops
- Interconnection, clustering, in order to ensure effective management of information, knowledge and business skills
- Develop public-private partnerships at national and international level
- Development of a stimulating framework for investment
- Development of own sales network in the extra community space, with the support of the Export Council
- Orientation toward new export markets

References

https://hadoop.apache.org/docs/. Date of access: 08.03.15.

Banica, L., Paun, V., Stefan, C., 2014. Big data leverages cloud computing opportunities. International Journal of Computers & Technology 13 (12), 5253–5263.

Banica, L., Pirvu, D., Hagiu, A., 2014. Intelligent Fashion Forecasting Systems/Springer, Chapter 9: Neural Networks Based Forecasting for Romanian Clothing Sector, pp. 161–194. http://dx.doi.org/10.1007/978-3-642-39869-8.

Bastian, M., Heymann, S., Jacomy, M., 2009. Gephi: an open source software for exploring and manipulating networks. In: International AAAI Conference on Weblogs and Social Media, San Jose, USA. Available at: https://www.aaai.org/ocs/index.php/ICWSM/09/paper/view/154/1009. Date of access: 16.04.15.

Bazavan, S., 2014. Zara's Owner Reports EUR 25 Mln Profit in Romania in 2013. Available at: http://business-review.eu/featured/zaras-owner-reports-eur-25-mln-profit-in-romania-in-2013-65164. Date of access: 24.03.15.

Bhosale, H., Gadekar, D., 2014. Big data processing using hadoop: survey on scheduling. International Journal of Science and Research (IJSR) 3 (10), 272–277. Available at: www.ijsr.net/archive/v3i10. Date of access: 14.04.15.

Borgatti, S., 2005. Centrality and network flow, Social Networks 27, pp. 55–71, Available at: http://www.analytictech.com/borgatti/papers/centflow.pdf. Date of access: 15.04.15.

Brust, A., 2012. CEP and MapReduce: Connected in complex ways. Available at: http://www.complexevents.com/2012/03/10/cep-and-mapreduce-connected-in-complex-ways/. Date of access: 12.03.15.

http://www.bowanddrape.com/home/. Date of access: 21.04.15.

http://www.c-and-a.com/ro/ro/corporate/company/. Date of access: 15.04.15.

D'Amico, S., Giustiniano, L., Nenni, M.E., Pirolo, L., 2013. Product lifecycle management as a tool to create value in the fashion system. International Journal of Engineering Business Management Special Issue on Innovations in Fashion Industry 1–6. http://dx.doi.org/10.5772/56856.

Dijcks, J.P., 2013. Big Data for Enterprise, Oracle White Paper. Available at: http://www.oracle.com/us/products/database/big-data-for-enterprise-519135.pdf. Date of access: 09.03.15.

Davenport, T.H., Dyché, J., 2013. Big Data in Big Companies. Available at: http://www.sas.com/content/dam/SAS/en_us/doc/whitepaper2/bigdata-bigcompanies-106461.pdf. Date of acesss: 19.03.15.

Euromonitor International, 2015. Market Research for Romania, Apparel and Footwear in Romania. Available at: http://www.euromonitor.com/apparel-and-footwear-in-romania/report. Date of access: 05.05.15.

Eurostat statistics, 2014. Available at: http://ec.europa.eu/eurostat/statistics-explained/index.php/File:V4_Use_of_cloud_computing_services_in_enterprises,_2014.png. Date of access: 07.04.15.

http://ec.europa.eu/economy_finance/eu/countries/romania_en.htm. Date of access: 08.09.15.

Farnell, J., 2014. An Integrated Industrial Policy for the Globalisation Era. Available at: http://ec.europa.eu/DocsRoom/documents/6313/attachments/1/translations/en/renditions/pdf. Date of access: 19.05.15.

Frank, L., 2014. Big Data Technology. Available at: https://datajobs.com/what-is-hadoop-and-nosql. Date of access: 27.03.15.

FRD Center, 2015. Garments Market in Romania 2015 – Demo Sector Brief. Available at: http://www.google.ro/url?sa=t&rct=j&q=&esrc=s&source=web&cd=7&cad=rja&uact=8&ved=0CEgQFjAG&url=http%3A%2F%2Fwww.frdcenter.ro%2Fassets%2FGarments-Market-Romania-2015-DEMO-Sector-Brief.pdf&ei=utmDVeTmD8XT7Qbjn4PACA&usg=AFQjCNGONRU9WyPA40iNovNTeyQxLI5NOg. Date of access: 29.03.15.

http://www.firme.info/. Date of access: 20.03.15.

http://www.forbes.com/sites/deborahljacobs/2013/07/01/made-to-order-fashion-goes-mainstream/. Date of access: 21.04.15.

Gupta, A., 2015. Forecasting the Fashion Future: Big Data Comes to Rescue Fashion Designers!. Available at: http://bigdata-madesimple.com/forecasting-the-fashion-future-big-data-comes-to-rescue-fashion-designers/. Date of access: 21.04.15.

http://www.geminicad.com/apparel_textile/made_to_measure_at.php Date of access: 19.03.15.

Harb, A.H., Chaaya, L.B., 2015. The Supply Chain's Strategic Importance for AZADEA Group available at: http://article.sapub.org/10.5923.j.mm.20150502.01.html. Date of access: 07.04.15.

International Transport Forum, 2015. Big Data and Transport: Understanding and Assessing Options. Available at: www.internationaltransportforum.org/pub/pdf/15CPB_BigData.pdf. Date of access: 07.09.15.

Iuga, V., Simion, I., 2015. Main Results of the Annual Survey for Romania of Global CEO Survey. Available at: http://www.google.ro/url?sa=t&rct=j&q=&esrc=s&source=web&cd=2&ved=0CCgQFjABahUKEwillNbE9JHHAhUBnCwKHZLNBwg&url=http%3A%2F%2Fwww.retail-fmcg.ro%2Fwp-content%2Fuploads%2F2015%2F07%2Fceosurvey15.pdf&ei=H_3BVeWIDYG4sgGSm59A&usg=AFQjCNEos5vDMklPiWNvTfNQdol30r_u6g. Date of access: 07.05.15.

Kotlik, L., Greiser, C., Brocca, M., 2015. Making Big Data Work: Supply Chain Management. https://www.bcgperspectives.com/content/articles/technology_making_big_data_work_supply_chain_management/.

Krebs, V., 2013. Social Network Analysis, a Brief Introduction. Available at: http://www.orgnet.com/sna.html. Date of access: 07.09.15.

Lieberman, M., 2014. Visualizing Big Data: Social Network Analysis, Digital Research Conference. Available at: https://www.google.ro/?gws_rd=ssl#q=Lieberman%2C+M.%2C+2014%2C+Visualizing+Big+Data:+Social+Network+Analysis%2C+Digital+Research+Conference%2C+. Date of access: 09.09.15.

Lutz, E., 2013. The Cloud Addresses Challenges in the Clothing Industry. Available at: http://www.supplychaindigital.com/logistics/3115/The-Cloud-addresses-challenges-in-the-clothing-industry. Date of access: 07.04.15.

Manyika, J., Chui, M., Brown, B., Bughin, J., Dobbs, R., Roxburgh, C., Byers, H.A., 2011. Big Data: The Next Frontier for Innovation, Competition, and Productivity. McKinsey Global Institute. Available at: www.mckinsey.com/mgi. Date of access: 09.03.15.

Marr, B., 2015. Ralph Lauren's Smart Shirt and the Future of Fashion. Available at: http://data-informed.com/ralph-laurens-smart-shirt-and-the-future-of-fashion/. Date of access: 27.03.15.

Mentzer, J., DeWitt, W., Keebler, J., Min, S., Nix, N., Smith, C., Zacharia, Z., 2001. Defining Supply chain management. Journal of Business Logistics 22 (2), 1–25. Available at: http://onlinelibrary.wiley.com/doi/10.1002/j.2158-1592.2001.tb00001.x/pdf. Date of access: 09.09.15.

Messarra, N., 2014. Introduction to Social Graph and NodeXL. Available at: http://nasri.messarra.com/introduction-to-social-graph-and-nodexl/. Date of access: 19.04.15.

Ministry of Public Finances of Romania, Available at: http://www.mfinante.ro. Date of access: 24.04.15.

Mohamed, A.M., Altrafi, O.G., Ismail, M.O., 2014. Relational vs. NoSQL databases: a survey. International Journal of Computer and Information Technology 3 (3), 598–601.

Mooi, E., Sarstedt, E., 2011. A Concise Guide to Market Research, Chapter 9 Cluster Analysis. Springer, pp. 237–284. http://dx.doi.org/10.1007/978-3-642-12541-6_9.

Naslund, D., Williamson, S., 2010. What is management in Supply chain management? - a critical review of definitions, frameworks and terminology. Journal of Management Policy and Practice 11 (4), 11–28.

Nagurney, A., Yu, M., 2010. Fashion Supply chain management through cost and time minimization from a network perspective. In: Choi, T.-M. (Ed.), Fashion Supply Chain Management: Industry and Business Analysis, 2011. IGI Global, Hershey, PA, pp. 1–20.

http://www.oliviapalermo.com/piol-designing-for-the-future/. Date of access: 21.04.15.

http://www.optitex.com/. Date of access: 15.03.15.

Pellegrin, J., Giorgetti, M.L., Jensen, C., Bolognini, A., 2015. EU Industrial Policy: Assessment of Recent Developments and Recommendations for Future Policies. Available at: http://www.europarl.europa.eu/thinktank/en/document.html?reference=IPOL_STU(2015)536320. Date of access: 27.03.15.

Plank, L., Staritz, C., 2014. Economic up- and Downgrading in Romania: From Full-package to OPT– and Back?, Global Competition, Institutional Context, and Regional Production Networks: up- and Downgrading Experiences in Romania's Apparel Industry. Available at: http://www.oefse.at/fileadmin/content/Downloads/Publikationen/Workingpaper/WP50Global_Competition.pdf. Date of access: 25.05.15.

Popescu, D., Radu, C., 2011. Strategies to increase the competitiveness of industrial products in the context of the global economic crisis. Annals of DAAAM & Proceedings of the 22nd International DAAAM Symposium 22 (1), 1123–1124. Available at: http://www.daaam.info/Downloads/Pdfs/proceedings/proceedings_2011/1123_Popescu.pdf. Date of access: 24.05.15.

Report from Economist Intelligent Unit Report, 2014. The Future of Business: Supply Chains. Available at: http://www.economistinsights.com/business-strategy/analysis/future-business-supply-chains/fullreport.

Romanian National Institute of Statistics, Available at: http://www.insse.ro. Date of acesss: 04.03.15; 24.04.15.

Rosenbush, S., Totty, M., 2013. How Big Data Is Changing the Whole Equation for Business. Available at: http://www.wsj.com/articles/SB10001424127887324178904578340071261396666. Date of access: 24.05.15.

Rouse, M., 2014. Big Data Analytics. Available at: http://searchbusinessanalytics.techtarget.com/definition/big-data-analytics. Date of access: 08.04.15.

Schwab, K., Sala-i-Martín, X., 2014. The Global Competitiveness Report 2014–2015, World Economic Forum. Available at: www3.weforum.org/WEF_GlobalCompetitivenessReport_2014-15.pdf. Date of access: 28.03.15.

Şerbanel, C.I., 2014. Romanian textile Industry and its competitive advantage. SEA - Practical Application of Science II (2 (4)), 2014, Available at: http://sea.bxb.ro/Article/SEA_4_45.pdf. Date of access: 28.04.15.

Sharma, N., 2014. Sphere of Influence - the Importance of Social Network Analysis. Available at: http://search.pb.com/baynote/socialsearch?q=Sharma+Sphere+of+Influence&cn=pitneybowes&cc=www&mode=gsa. Date of access: 10.09.15.

Shen, B., 2014. Sustainable fashion supply chain: lessons from H&M. Sustainability Journal 2014 (6), 6236–6249. http://dx.doi.org/10.3390/su6096236. Available at: www.mdpi.com/journal/sustainability. Date of access: 29.03.15.

Socher, R., Perelygin, A., Wu, J., Chuang, J., Manning, C., Ng, A., Potts, C., 2013. Recursive deep models for semantic compositionality over a sentiment tree bank. In: Conference on Empirical Methods in Natural Language Processing (EMNLP 2013). Available at: http://nlp.stanford.edu/sentiment/. Date of access: 27.05.15.

Suciu, M.C., Olaru, S., 2012. Research Final Report. Available at: http://scoalapostdoctorala.ase.ro/doc/olaru.pdf. Date of access: 25.05.15.

Swayne, M., 2014. Big Data May Be Fashion Industry's Next Must-have Accessory. Available at: http://news.psu.edu/story/338772/2014/12/17/research/big-data-may-be-fashion-industrys-next-must-have-accessory. Date of access: 18.04.15.

The 2014—2020 National Export Strategy, 2013. Ministry of Economy, Department of Foreign Trade and International Relations. Available at: http://www.minind.ro/strategia_export/SNE_2014_2020.pdf. Date of acesss: 28.04.15.

Trites, D., 2013. Big Data: The Next Big Trend in Fashion, SAP Business Trends. Available at: http://scn.sap.com/community/business-trends/blog/2013/09/12/big-data-the-next-big-trend-in-fashion. Date of access: 21.04.15.

http://www.tinkertailor.com/ Date of access: 21.04.15.

World Trade Organisation, http://www.wto.org. Date of access: 02.03.2015.

Using artificial neural networks to improve decision making in apparel supply chain systems

P.C.L. Hui, T.-M. Choi
The Hong Kong Polytechnic University, Kowloon, Hong Kong

5.1 Introduction

Apparel retailers offer a great variety of apparel products to consumers. They change the product styles and colors very frequently in order to meet the dynamic consumer preferences. However, offering a large product variety creates business challenges as costs are higher and the respective operations are more difficult. In addition, time becomes an important key factor. It is commonly agreed that apparel products have the following characteristics which pose challenges to proper supply chain management.

1. The products have short life cycles. The selling period can be as short as a week or two.
2. The demand is dynamic and changes very frequently with respect to style and color. It may be influenced by factors such as weather or celebrities.
3. The product demand is difficult, if not impossible, to predict.
4. The consumer normally makes a buying decision when he or she is confronted with the product at the point of purchase (Christopher et al., 2004).

In the apparel industry, roughly speaking, there are basically two extreme operations modes: one is the "ready-to-order" mode in which the apparel product is produced in good shape and ready for selling (Fig. 5.1), and the other is "make-to-order" in which the apparel product is made based on the user's request (Fig. 5.2). Different operations modes have different apparel supply chain structures and dynamics.

In the apparel supply chains under the ready-to-order operations mode, the product design is initiated by the company after predicting the major customer preference and changing style and color trends. After the design of apparel products is finalized, production starts based on the design specification and the demand forecast. Once the product quality meets the design requirements and international standards, such as

Design → Production → Distribution → Retailing

Figure 5.1 Apparel supply chain for the ready-to-order operating mode.

Individual user's request → Design → Production → Deliver to customer directly

Figure 5.2 Apparel supply chain for the make-to-order operating mode.

Information Systems for the Fashion and Apparel Industry. http://dx.doi.org/10.1016/B978-0-08-100571-2.00005-1

color fastness and shrinkage, all finished products would be shipped to the centralized distribution center and the distribution center will allocate the needed quantities to individual retail outlets based on their demand forecasts. Size assortment (eg, XS, S, M, L, and XL) of each apparel product for each retail outlet are varied based on their demographic characters. Typical industrial examples include many mass market labels and volume retailers.

In the apparel supply chain under the make-to-order operations mode, the product design is initiated by an individual user's requests, such as body fitting and material selection, the production is tailormade and the finished garment could be delivered to the individual customer directly. Typical examples of this kind of apparel supply chain include the traditional tailors for men's suit, ladies' wedding dresses, and health-care apparel products. Of course, the mass customization program, initiated by many apparel brands, also belongs to this category.

This chapter mainly focuses on the ready-to-order operations mode in apparel supply chain management because it is the most commonly adopted one in the fashion industry and it involves complicated decision making, such as prediction of inventory level, determination of customer's buying preference, sales forecasting, and optimization of resources allocation. In the other section of this chapter, we discuss what specific decision process is being made in the apparel supply chain and the possible application of artificial neural networks (ANN) to improve the respective decision making in the apparel supply chain. Finally, the limitations of applying ANN for apparel supply chain management and possible research areas for further development are also discussed.

5.2 Decision process involved in the apparel supply chain

The apparel supply chain aims to provide the right fashion products to satisfy the market needs, with the lowest possible cost, the fastest speed, and the maximized profit, simultaneously. From a certain perspective, one can view the fashion product as the final outcome of the apparel supply chain system and the customers' preference and demographical information are the inputs into the system. Based on the works of Burns and Bryant (2002), the major steps involved in the apparel supply chain systems include: (1) product design and development, which is used to determine whether the supply chain can provide the right, and feasible, product to the market; (2) demand forecasting, which is critically important for driving the sourcing and inventory planning of the apparel supply chain (Choi, 2013); (3) ordering and replenishment, which relates to the inventory quantity decision and ordering timing; it also affects the inventory cost and the flexibility of the supply chain system; (4) price negotiation, which usually relates to the transaction cost between supply chain partners and may also relate to lead time; (5) quality control, which affects price and also product quality; and (6) information sharing, which is useful to enhance supply chain performance and lower the risk for all supply chain members from inherent problems such as the bullwhip effect. To a certain extent, one may argue that information sharing is most crucial because all operations decisions made in the apparel supply chain system are related to information availability, ie, how much information is shared among supply chain members and the degree of information transparency. For example, apparel manufacturers could better forecast their demand and develop a better production plan if they are given the point-of-sale data from their retail customers. In addition, suppliers could use manufacturers' production schedules to better prepare their own production and inventory control, which would ultimately help ensure reliable supplies to the manufacturers.

It is crystal clear that every supply chain member has to clearly define the specific nature of the decision they must make. Each supply chain member needs to collect pertinent information from different relevant sources. For example, fashion designers collect the customers' preferences from trade shows, fashion books, fashion blogs, journals, magazines, fashion gurus, and a variety of other important sources in fashion. Through the process of collecting the right information, each supply chain member will identify several possible paths of action, or alternatives for their selection. For example, apparel manufacturers would identify different fabric sources which are available and suitable for producing their fashion products to satisfy the retail buyers' material requirements and design specifications. In going through the evaluation process, each supply chain member should be able to put the available alternatives in a priority order with respect to their own ultimate goal. For instance, apparel manufacturers would evaluate the fabric sources based on their track record, their reputation, their fabric quality and their post-sales services to determine the priority list of fabric suppliers. After the priority list is set, each supply chain member might select the best one or a combination of alternatives for "real implementation."

As another example, in fashion retailing, a fashion retailer might rely on the past sales history and inventory on hand to determine the inventory replenishment policy and the needed stock level required for each fashion item in the coming week. However, the inventory policy can be a dynamic one as it will be affected by the market situation. For example, when the sales performance of some fashion items is below expectations, an attractive discount offer should be made to attract more customers to buy them and the inventory replenishment scheme will be adjusted accordingly. Of course, the fashion retailer should let the supplier know this change as soon as possible so that the whole supply chain decision is synchronized.

As a remark, in such an entire decision-making process, information gathering, decision making, and alternatives evaluations are all very tedious and complicated. It is not easy to work out in a short period of time. In the fashion industry, consumer demands are very volatile and the fashion designers may fail to deliver fashion items which meet the customers' preference in terms of style, color, and size. Facing these constraints, apparel supply chain members attempt to use advanced information technologies, such as artificial neural networks (ANNs), for improving the decision quality, and achieving an accurate response to the market.

Essentially, ANNs are "artificial intelligence (AI) models" which can be viewed as the biologically inspired software applications designed to simulate the way human brains process information. It is commonly believed that the first conceptual ANN model was developed in the 1940s. In fact, ANNs can help gather knowledge by detecting the patterns and relationships in data and learn (and are trained) through experience. An ANN consists of a large number of artificial neurons, which are connected with the carefully fine-tuned coefficients (weights), which constitute the *input layer, hidden layer, and output layer*. The structure of ANNs is commonly known as a *multilayered perceptron*, ie, a network of many neurons. In each layer, every artificial neuron has its own weighted inputs, transfer function, and one output. The performance of a neural network is determined by: (1) the transfer functions of its neurons, (2) the learning rule, and (3) the layer structure itself. The weights are the adjustable variables and the weighed sum of the inputs would constitute the "activation of neurons." The activation signal is passed through the transfer function to produce a single output of the neuron. Transfer functions introduce nonlinearity to the ANN. During training, the interunit connections are optimized with respect to the data, until the error in predictions is minimized and the network reaches the prespecified level of desirable accuracy. Notice that one popular measure on optimizing the weights of an ANN is known as *back-propagation*. Training the network (ie, adjusting the weights) will generate the error (ie, the difference between the desired result and the guess). The error, however, must be fed backward through the network. The final error ultimately adjusts the weights of all the connections, ie, optimizes them, for the real implementation in the ANN. Once the ANN is trained and tested with the right weights decided, it can be given to predict the output.

With its long development history and wide range of applications, the ANN model represents a promising modeling technique, especially for data sets having highly nonlinear relationships, which are frequently encountered in fashion business processes. Notice that in terms of model specification, ANNs require no knowledge of the data source but, since they often contain many weights that must be estimated, they require a large training data set. In addition, ANNs can combine and incorporate both literature-based and experiment-based data to solve problems. The popular application domains of ANNs include classification and clustering, pattern recognition, forecasting, and analytical modeling.

In the coming section, some applications of ANN in apparel supply chain systems are discussed and how it may outperform the traditional approach is illustrated.

5.3 Applications of ANN in apparel supply chain to improve their decision

5.3.1 Data mining of customer's buying preference

In fashion retailing, fashion retailers need to collect a large amount of sales-related transactions data and the customer shopping history. These data sets can help the fashion retailer identify customer behaviors and discover customer shopping patterns and trends to improve the quality of customer service and to achieve better customer retention and satisfaction. The problem faced by many fashion retailers is that much information on specific customers is missing (ie, information is incomplete) and the amount of such relevant data and information is huge. It is difficult for fashion retailers to understand the purchase behaviors of individual customers in good detail and to track changes in purchases as affected by the environmental variables, such as economy changes and fashion trend changes. Thus, data mining is one of the approaches that can help identify the buying preference of potential customers and allow the fashion marketing team to target them for better customer service and retention.

Data mining mainly contains data analysis and discovery algorithms to produce significantly important patterns and/or rules over a large amount of data (Fayyad et al., 1996, Lungu and Bara, 2012). ANN is one of the popularly applied data-mining tools that can be used to find the hidden patterns from the given data. ANN comprises three components: the architecture or the model, the learning algorithm, and the activation functions. ANNs are being trained to recognize and retrieve patterns so that both proper pattern recognition and accurate functions estimation can be obtained. With this dual nature, ANNs effectively serve as the tool to implement data mining.

In most cases of data mining, the ANN adopts the back-propagation algorithm and conducts learning on a feed-forward neural network. In the work of Velu and Kashwan (2015), the ANN is used to classify the customers into different groups with specific buying behaviors observed over a long period of time. The different groups are formed based on their buying of a specific product, buying frequency of the product and preference of the product in the presence of "similar product availability." The study also classifies the customers with respect to the level of royalty (regular) and the ad hoc purchase pattern (random). The input variables for the grouping of customers include the frequency of product buying, preference of the product, and regular and random purchase. The output variables for classification of customers are high-valued, medium-valued, and low-valued customer classes, respectively. A three-layer back-propagation ANN is constructed to model the relationship between such input variables and output variables in the work of Velu and Kashwan (2015). After training the ANN with the experimental data, it was found that the customer classification is highly accurate. This study shows that the ANN is a promising tool to be deployed in predicting the customer's buying preference.

5.3.2 Sales forecasting

Compared to other retailing businesses, it is well known that sales forecasting is a very difficult task in fashion retailing because fashion product demand is basically "crazy" and rather unpredictable. It also fluctuates with the ever-changing preference of consumers and the fashion product's life cycle is very short (Choi and Sethi, 2010; Choi et al., 2013). In addition, the sales of fashion products are strongly affected by seasonal factors, fashion trend factors, and many other variables, such as weather, pricing, marketing strategy, item features, and macroeconomic trends (Thomassey, 2010). Thus, fashion retailers may need to keep a large amount of stock to meet their customers' demands. Meanwhile, it is known that an apparel supply chain system includes upstream members, such as fiber manufacturers and fabric producers, and downstream members, like designers, distributors, wholesalers, and retailers. The bullwhip effect (Lee et al., 1997), which depicts the information distortion problem, has a strong impact on the apparel supply chain system. Improving sales forecasting could help reduce the bullwhip effect, which will also directly enhance the efficiency of the apparel supply chain system (Liu et al., 2013). It makes sales forecasting even more important in apparel supply chains.

In traditional fashion sales forecasting, it relies on the commonly seen statistical methods, such as linear regression, moving average, weighted average, exponential smoothing, double exponential smoothing, Bayesian analysis, ARIMA, and SARIMA (Box et al., 2008). However, such statistical methods may fail to achieve a desirable forecasting outcome (eg, in terms of accuracy) as fashion sales are affected by multiple factors, such as fashion trends and seasonality, and exhibit a highly irregular pattern (Choi et al., 2011).

In an influential study (Frank et al., 2003), the ANN model is used for conducting fashion retail sales forecasting. The three-layers' back-propagation feed-forward ANN architecture is implemented to learn the patterns of sales of garments in the past to forecast sales in the future. The ANN has 10 neurons at the input layer, 30 neurons at the hidden layer, and one neuron at the output layer. The author employs 198 sales weeks' data for the training of the ANN. Ten sales weeks' data are used for testing the ANN's performance. Nine sales weeks' data are used to verify the forecasting performance of the ANN by comparing the forecasted data with the actual sales data. In the computational result, it is found that the ANN performs best compared with various other statistical forecasting models. It is thus proven that the ANN is an effective tool in fashion sales forecasting, especially when there are a large number of sales data for the ANN's learning.

5.3.3 Prediction of stock level

As we mentioned earlier, fashion products have a short life cycle. It is difficult to spot the fashion trend quickly and to decide the right inventory level. In addition, making a timely decision is critical yet not easy. It is known that poor inventory decisions lead to serious overstocking or understocking problems. A commonly known method to deal with uncertainty in demand and supply is to keep a safety stock. However, it is expensive as safety stock is used as a buffer. To deal with this issue, the effective estimation of the amount of required safety stock for an individual fashion product is required. For the company to properly plan its inventory investment, it is also important to predict the amount of required safety stock and advanced information technology, such as ANN, may help.

In the work of Li et al. (2008), the back-propagation algorithm of a multilayer feed-forward network ANN is built for forecasting of an inventory level (Fig. 5.3). The ANN consists of neurons and the connections between neurons. Their proposed ANN is also composed of positive propagation and back-propagation. In positive propagation, the state of any neuron in a layer will only affect the state of neurons in the next layer. If the expected outcome cannot be found in the output layer, the network enters into back-propagation. According to the error signal obtained from positive propagation, the network adjusts the connection weight values between neurons for all layers to find out the best network output (Zhang, 2004, 2011; Wang and Sun,

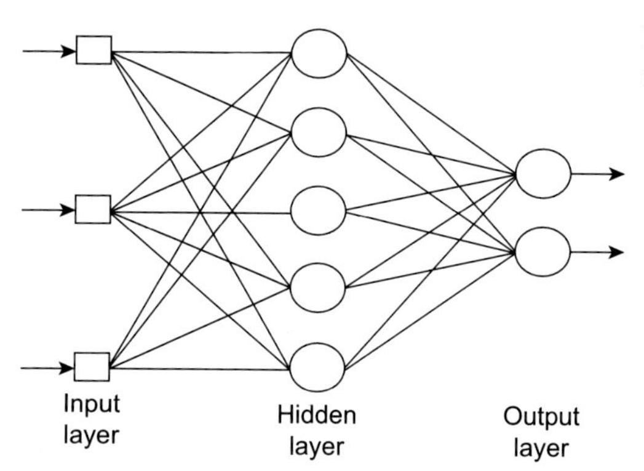

Figure 5.3 Multilayer feed forward network.

2005; Bao and Kai, 2004; Karhunen and Oja, 1997; and Yuan, 2006). Twenty important impact factors are considered to predict the stock level. These factors could become the input values of each neuron at the input layer. Such factors include population size for the product sale region, average wages for purchasing in that region, product sales quantity and price, the previous month's inventory of the product, potential competitive influence, potential customer demand, market share, market entry barriers, marketing capacity, etc. Many factors are qualitative and the method of expert scoring can be used to forecast the possible safety stock levels.

As a remark, the forecasting of safety stock level by ANNs can be treated as a kind of nonlinear mapping from inputting the scores of each factor to exporting the final forecasting values of the inventory level of the product. For instance, in a specific ANN, there are a total of 20 neurons at the input layer and only one neuron at the output layer in the three-layer back-propagation neural network structure. The value of the output neuron ranges from 0 to 1. The higher the score, the lower the inventory level. The number of neurons at the hidden layer can be determined by the calculation speed and the prediction accuracy requirements. In the work of Li et al. (2008), seven neurons are selected for the hidden layer. After training the back-propagation neural networks with the actual score of each factor assessed by experts, the error of predicting the inventory level generated by the ANN, as reported in the work of Li et al. (2008), lies between 1.2% and 5%, which is highly accurate.

5.3.4 Optimization of marker planning in apparel production

In the apparel supply chain, apparel production is important for transforming the fashion design concept into the physical fashion product. Apparel production here covers fabric cutting, sewing, and finishing processes. Before the cutting process, the entire fashion design would be decomposed into various pieces of garment parts with block patterns which will be placed on the marker. After receiving the needed production information, such as sizes, styles, colors, and order quantities, the marker maker will need to calculate the required quantities of "fabric lays" to prepare the optimized marker (the best arrangement of block patterns onto "fabric lays" so as to minimize the fabric wastage) for cutting. In the calculation of the required quantities of fabric lays, some factors, such as the size distribution, the fabric type and characteristic, the cutting system's knife length, the number of fabric layers, and the spreading table's length would be considered, and they all affect the productivity of apparel production. Such data patterns are usually nonlinear and complex in nature and thus the ANN is one of the important tools to ensure optimization of marker planning.

In the work of Kayar and Ozel (2008), a three-layer back-propagation feed-forward ANN architecture is employed for achieving optimization of the marker plan. There are seven neurons at the input layer, 10 neurons at the hidden layer, and one neuron at the output layer. The input neurons include the ones for gender, sleeve length, size distribution, and fabric width. The output neuron only represents the length of marker. The back-propagation training algorithm is used to train the ANN. After training and testing, the ANN is deployed for yielding the optimal marker length for a particular apparel production order. This ANN system has been tested with 77 markers and it is found that the target and the tested output are approximately the same. It thus proves that the ANN method can be used to help optimize the marker planning process very effectively.

5.4 Conclusion and limitations of using ANNs in apparel supply chain systems

The apparel supply chain system includes many different functions with a final objective to satisfy customer requirements with the most efficient and effective manner. The process in the apparel supply chain usually starts from fashion design in which the appropriate design works which fit the target customers' preferences are created. The next stage is to transform the design works into physical fashion products by selecting the appropriate fabric sources and efficient production methods. Within a short period of time, the finished products would need to be delivered to the right retailing points at the right time for selling to consumers. During the entire apparel product life cycle, there are various decisions to be made by each supply chain member of the apparel supply chain system. For example, based on the customers' preference and fashion trends revealed via data-mining tools, fashion designers would provide the best possible fashion design for the target customers. Apparel manufacturers could use data to identify the best marker plan for improving productivity in apparel production. Apparel retailers need to conduct sales forecasting and use it to determine the appropriate stock level for the apparel products. To achieve scientifically sound and promising decisions, former studies have examined how the ANNs could be used to help. In fact, the supervised learning algorithms in the ANN models can help reveal complex nonlinear relationships from data and they can also be easily implemented as decision support systems for real-world applications. Thus, in this paper, we have discussed how the ANNs can be used to help.

However, there are still some potential problems in using the ANN models for decision making in apparel supply chain systems. Firstly, the training time required by the ANN is nontrivial and it increases according to the complexity or size of data sets. Such long training time becomes a major hurdle to the real-world applications of ANNs. To be specific, since the fashion industry is driven by the "fast fashion" model, quick decision making is always desirable. As a result, for many fashion supply chain systems, using ANNs requires the members to equip themselves with state-of-the-art high-power computing devices. The next problem is that the performance of the ANN is often unsatisfactory when the ANN has to work with: (1) noisy data, (2) a large number of inputs, and (3) small training data sets. Unfortunately, in many cases, historical sales data are highly noisy, which means that the risk of overfitting in the ANN is high.

To deal with the above-mentioned problems, some unsupervised learning algorithms might be used in the ANN. They could learn by following some representative particular data patterns that reflect the statistical structure of the overall data patterns. Compared with the supervised learning scenario, there are no explicit target outputs or "environmental evaluations" associated with each input (Dayan, 1999). This approach helps reduce the computational time required. As a remark, a method such as the self-organizing map and the adaptive resonance theory-based models are commonly used in implementing the unsupervised learning algorithm. Next, to address the concerns on the limitations of ANN (eg, its ability to handle noisy datasets, etc.), one may consider some newer methodologies, such as the evolutionary neural network (ENN) model and the extreme learning machine (ELM) (Huang et al., 2006; Zhu et al., 2005; Sun et al., 2008; Yu et al., 2012).

For future research, it will be interesting to see how apparel companies implement the ANN-based tools in real-world software applications and integrate them with their existing applications. It will also be interesting to explore other application domains. For example, color forecasting is an important part of fashion design and sales forecasting, yet the current industrial practices are highly subjective and are based on guesses by individual designers or "experts." The use of ANN can hence potentially improve the accuracy of forecasting. Last but not least, one may also consider how ANN models may work with other methods to establish hybrid models. Expectedly, the hybrid models can take advantage of their component methods and achieve a better performance than just using ANN alone (Choi et al., 2014).

References

Bao, Y.Y., Kai, X.M., 2004. Evaluation of the investment risk of high tech project based on neural network model. Journal of Hefei University of Technology 27 (7), 851–854.

Box, G.E.P., Jenkins, G.M., Reinsel, G.C., 2008. Times Series Analysis: Forecasting and Control, Wiley Series in Probability and Statistics, fourth ed. John Wiley & Sons, Hoboken, NJ, USA.

Burns, L.D., Bryant, N.O., 2002. "The Business of Fashion Designing, Manufacturing, and Marketing. Fairchild Publications, Inc., New York.

Christopher, M., Lowson, R., Peck, H., 2004. Creating agile supply chains in the fashion industry. International Journal of Retail & Distribution Management 32 (8), 367–376.

Choi, T.M., Sethi, S., 2010. Innovative quick response programs: a review. International Journal of Production Economics 127 (1), 1–12.

Choi, T.M., Hui, C.L., Yu, Y., 2011. Intelligent time series fast forecasting for fashion sales: a research agenda. In: Proceedings of the International Conference on Machine Learning and Cybernetics, pp. 1010–1014. Guilin, China, July 2011.

Choi, T.M., 2013. Optimal apparel supplier selection with forecast updates under carbon emission taxation scheme. Computers and Operations Research 40, 2646–2655.

Choi, T.M., Hui, C.L., Yu, Y. (Eds.), 2013. Intelligent Fashion Forecasting Systems: Models and Applications. Springer, New York, NY, USA.

Choi, T.M., Hui, C.L., Liu, N., Ng, S.F., Yu, Y., 2014. Fast fashion sales forecasting with limited data and time. Decision Support Systems 59, 84–92.

Dayan, P., 1999. Unsupervised learning, appeared in Wilson, R.A., Keil, F. (Eds.), The MIT Encyclopedia of the Cognitive Sciences. Published by MIT, pp. 1–7.

Fayyad, U., Piatetsky-Shapiro, G., Smyth, P., 1996. From data mining to knowledge discovery in databases. AI Magazine. ISSN: 0738-4602 17 (3), 37–54.

Frank, C., Garg, A., Raheja, A., Sztandera, L., 2003. Forecasting women's apparel sales using mathematical modeling. International Journal of Clothing Science and Technology 15 (2), 107–125.

Huang, G.B., Zhu, Q.Y., Siew, C.K., 2006. Extreme learning machine: theory and applications. Neurocomputing 70 (1–3), 489–501.

Karhunen, J., Oja, E., 1997. A class of neural network for independent component analysis. IEEE Transaction on Neural Networks 18 (3), 486–504.

Kayar, M., Ozel, Y., 2008. Using neural network method to solve marker making "calculation of fabric lays quantities" efficiency for optimum result in the apparel industry. In: Proceedings of 8th WSEAS International Conference on Simulation, Modelling and Optimization, Santander, Cantabria, Spain, September 23–25, 2008, pp. 219–223.

Lee, H.L., Padmanabhan, V., Whang, S., 1997. Information distortion in a supply chain: the bullwhip effect. Management Science 43 (4), 546–558.

Li, T., Zhang, S., Xiong, F., 2008. A novel inventory level forecast model based on artificial neural network. In: Proceedings of 2008 International Conference on Multimedia and Information Technology, pp. 82–85.

Liu, N., Ren, S., Choi, T.M., Hui, C.L., Ng, S.F., 2013. Sales forecasting for fashion retailing service industry: a review. Mathematical Problems in Engineering 2013, 738675, 9 pages.

Lungu, I., Bara, A., 2012. Improving decision support systems with data mining techniques. In: Advances in Data Mining Knowledge Discovery and Applications – Chapter 18. In Tech Publisher, Croatia, ISBN 978-953-51-0748-4, pp. 397–418.

Sun, Z.L., Choi, T.M., Au, K.F., Yu, Y., 2008. Sales forecasting using extreme learning machine with applications in fashion retailing. Decision Support Systems 46 (1), 411–419.

Thomassey, S., 2010. Sales forecasts in clothing industry: the key success factor of the supply chain management. International Journal of Production Economics 128 (2), 470–483.

Velu, C.M., Kashwan, K.R., 2015. Artificial neural network based data mining technique for customer classification for market forecasting. International Journal of Advancements in Computing Technology 7 (1), 24–31.

Wang, Y.W., Sun, C.H., 2005. Study of risk indices and fuzzy integrative evaluation of construction project investment. Mathematics in Practice and Theory 35 (8), 16–21.

Yu, Y., Choi, T.M., Hui, C.L., 2012. An intelligent quick prediction algorithm with applications in industrial control and loading problems. IEEE Transactions on Automation Science and Engineering 9 (2), 276–287.

Yuan, Z., 2006. Research on improved BP neural network. Information Technology (2), 89–91.

Zhang, C.M., 2004. Application of neural network to investment risk evaluation of high-technical projects. Journal of Henan University of Science and Technology 25 (4), 36–38.

Zhang, X.H., 2011. A neural network synthetic evaluation method for the investing risk of the high technical project. Information Transaction 20 (5), 608–611.

Zhu, Q.Y., Qin, A.K., Suganthan, P.N., Huang, G.B., 2005. Evolutionary extreme learning machine. Pattern Recognition 38 (10), 1759–1763.

Smart systems for improved customer choice in fashion retail outlets

B. Pan
Seamsystemic Design Research, London, UK

6.1 Context overview

6.1.1 The rising consumer awareness, knowledge, and access to fashion information

Social media and the online community of Facebook, Instagram, Twitter, Pinterest, YouTube, commentary blogs, and sourcing apps form a powerful and open fashion forum. These global and freely accessible platforms cross social demographics, market sectors, and geographic borders for discussion, review, validation, endorsement, recommendation, and referral. This 21st-century phenomenon has become mainstream [3] and is not only bypassing but often brokering traditional channels and closed circuits of fashion industry buyers and press for trend prescription, buying selection, and editorial endorsement. Where fashion information used to be privileged property, reserved for industry and trade for a minimum of 6–12 months ahead of consumer access, it is now not only widely available to consumers simultaneously, but it is also competing through the same channels with layman commentators who validate fashion information, commanding at times many more followers online for the attention and endorsement of consumers.

6.1.2 The unequal fashion system relationships between brand/retailers, manufacturers, and consumers

Diversification of market sectors, regions, and lifestyles prompts unprecedented product proliferation [4] on one hand, and free and easy access to infinitely multiplied sources and channels of fashion information on the other. Together, the result is a dramatically increased opportunity for consumers to encounter fashion information, where in the past there were mainly three channels: consumer printed press, brand advertisement/catalogs, and physical department stores. Theoretically, consumers should therefore be more knowledgeable about fashion products and brands and have much improved access and options to far wider choices in the retail market place, enabling them to make more informed decisions about their purchases.

In reality, however, as the fashion retail market sees increasing polarization toward luxury and value chains [5], larger brands/retailers are leveraging their financial scale

Information Systems for the Fashion and Apparel Industry. http://dx.doi.org/10.1016/B978-0-08-100571-2.00006-3

advantage with an ever more aggressive and dictatorial approach to market push-out [1–2, 6]. Global market share and supply chain sources are monopolized by large brands/retailers to gain preferential treatment for commercial properties, distribution channels, and media coverage; and price wars create undercut and subsequently wipe out smaller independent brands/retailers as well as specialized manufacturers that previously provided diversity and quality in mid-price product offerings that bridge the gaps between the extremes of luxury and value chains.

Consequently, instead of having meaningfully increased and substantially improved selections in products and brands, consumer choices are actually severely reduced. This is due to the common scenario where most prime shopping locations and retail premises in/around top-tier cities (which serve as global fashion shopping destinations) are dominated by the same or similar offering of standard products, whether in the luxury or value market sector, with little room for stylistic variety, creative artistry, or real personal choice beyond the recognizable seasonal palette from the same household name brands or value retailers. Consumers are therefore experiencing choice fatigue, disenchantment, and increasingly, distrust for large brands [7] in the wake of numerous serious social and environmental scandals of unethical practice involving high-profile brands/retailers in recent years, such as the 2013 Rana Plaza garment factory building collapse in Bangladesh.

6.1.3 *Social, economic, and environmental issues facing the contemporary fashion retail consumption model*

Moreover, the favorable profits made by large-brand houses/retailers are not distributed across the generally low-margin/low-paid supply chain manufacturers. While a GBP £39 jacket purchased from a name-brand mass market fashion retailer in 1991 would cost £75 with an average inflation of 2.8% per year in the UK (according the Bank of England inflation calculation), a similar looking jacket in the same retail store is now priced at £29 some 25 years later (ie, over 60% less than what it would be in comparable material, labor, logistics, and associated supply chain costs). So how and why are the retailers able to make their sums work to reverse 2.5 decades of economic appreciation?

With a basic calculation estimate (Fig. 6.1), clear numeric evidence would indicate that a continuous drive to move supply chain works to ever poorer regions of the world to use cheaper subcontracted manufacturers and achieve more competitive production prices plays a core part in the unequal distribution of commercial gain.

UK recommended retail price (RRP)	£29 (GBP)	= Consumer purchase price
−20% value added tax (VAT)	£24	= Pre-tax retail price
Minimum retail markup 2X	£12	= Wholesale price
Minimum wholesale markup 2X	£6	= Ex-factory price
Manufacturer's margin 2X	£3	= Garment gross cost
Typical material cost at 50% of garment	£1.50	= Garment cut & sew cost
Minimum ½ Day (4–5 h) to cut/sew a typical jacket	£0.30–£0.375	= Sewers maximum hourly pay

Figure 6.1 Retail to production wage reverse calculation table.

An RRP £29 piece of merchandise, excluding the UK value added tax, is GBP £24. By a simple calculation of fashion retail standard pricing structure of two times the minimum wholesale to retail markup, and another two times margin to wholesale price, leaves a manufacturing ex-factory price of £6/piece, where the manufacturer would make another two times margin on gross garment cost. Since material cost typically composes 50% of garment gross cost (at £3 in this case), this leaves the cut/sew cost at £1.50. As a jacket would typically take no less than half a day to cut, sew, and finish (factoring in a minimum of 8–10 h work day), this would equate to an average worker's wage of £0.3–0.375 per hour. And there are many more items in the large retail chain merchandise offering where prices start from less than half of the example jacket.

While customers (end consumers) appear to benefit from the perceived "value for money" in the short-term gain from this enormous price drop, quality and satisfaction also drop significantly through extremely short product life-span afforded by a cheap purchase intended for short-term use. More importantly, the economic, environmental, and social consequences of casual overconsumption propelled by the large-brand house/mass retailers' marketing push presents an enormous hidden cost to all on a much greater scale. In the United Kingdom, for instance, this contemporary phenomenon of the fast fashion trend, now also known as "throwaway fashion," where inconceivably low prices encourage fashion garments to be treated as disposables, is causing an unprecedented crisis of global waste increase. Statistics show that British consumers are now buying four times more than they did in the 1980s (of which 40% are from value retailers), and yet, they are discarding an average of 35 kg of clothing per person per year (mostly nonbiodegradable)—almost as much as is purchased per person per year in total. Clothing/textile waste thus rose from 7% to 30% of total waste in the United Kingdom within the last 5 years [8]. As a result, Britain will run out of landfill space on its soil in about 5–7 years if the problem persists at the current rate. This is while already exporting a large proportion of its waste.

It is therefore sufficient to draw from this overview, by evidence and deduction, that the following issues facing the fashion market place of today's beckon structural change:

1. While technological advances through open-source digital medium have broadened the forum of information and commentary for consumers, they have not necessarily resulted in widened access for customers in the fashion market place due to the control of product distribution channels by large-brand/retail corporations.
2. Increased speed and quantity in product mixes from the similar types of "fast fashion" large retailers/brand houses that dominate market shares have neither provided real, improved choices nor satisfaction for consumers. This is illustrated by the ratio of garments that are purchased versus discarded annually.
3. Ultimately, only at the very end of the long fashion system supply chain workings do customers have any real impact on the fashion system in the decisions they make. These consists of two choices only: to buy or not to buy (including returns).

There is one more decision point that they encounter after the sale: to keep or to throw away, although at this point the decision impact is already removed from the brand houses/retailer workings.

Beyond a social commentary level, there is currently little to no interaction in the relationship between the brand house/retailer and the customers other than in the point of sales transaction. The key question for this chapter is therefore, "How do we improve real consumer choice in fashion retail outlets?"

6.2 Research parameter

6.2.1 *Fashion industry and market supply chain as a system*

As per research previously carried out by our team, this chapter continues to examine the complex fashion industries as a "fashion system." This includes all processes output to the market place and their relationship with the end retail consumers (collectively referred to as customers). The research separates the activities (research, design, sourcing, development, production, distribution, marketing, etc.) of brand houses/retailers/manufacturers (collectively producers) as push-based supply, back-end processes of the system. Retail sales, promotion, and customer service activities are the pull-based demand, front-end processes of the system. The fashion system refers to all fashion supply chain (FSC) workings of the global apparel industries and market place [9]. And the garment is defined as the object of the fashion system that is constantly piped through the entire FSC and is the ultimate output to the customer.

6.2.2 *Customer choice in conventional fashion retail offerings*

There is much research on consumers' purchase decision-making, widely agreed to be a five-stage process [10] that consists of problem identification, information search, evaluation of alternatives, purchase decision, and post-purchase decision. From a customer's point of view, the sequential thought process that correlates to the five-stage decision process are shown in Fig. 6.2.

This process, illustrated in Fig. 6.2, is in fact a "search and sieve" funneling process that ultimately affords a choice of "yes" or "no" in each garment style assessed by a customer, with fixed attributes of each garment in the search as developed and pushed out by producers. To find the desired garment style, customers are expected to carry out and reiterate this same process multiple times, and each time to arrive at the

	Consumer decision-making	Customer shopping thought process	Garment in quest
1	Problem identification	I have need/want for garment type A with a, b, c description ➢ with varying degrees of time pressure for obtaining it	Describe a, b, c details/features
2	Information search	Where can I find/shop garment type A ➢ search internet and/or survey stores	Shortlist options for comparison
3	Evaluation of alternatives	Like or don't like/ fit or not fit ➢ short list potential candidates for trying	Alternatives versus features matched
4	Purchase decision	How much I like it versus how much I am willing to spend on it ➢ to buy or not to buy	All or some features matched at what price
5	Post-purchase decision	Love it or not so sure it serves style/function requirement ➢ to wear/keep or to return/discard.	To buy or not to buy

Figure 6.2 Customer decision-making thought process table.

closed-end answer of "yes" "no." And if the answer arrived at is "no," customers are then expected to start the search and sieve process again until one suitable item is eventually encountered.

For the customer, this is a laborious and time-consuming process where patience or time can easily run out and result in giving up (this equals a loss of sales opportunity for producers). Even when an item came close to suiting a customer's desire, there is typically no communication channel between customers and producers for inquiry or feedback, and at no point in each search and sieve process is there any opportunity for simple modification or adaptation to be incorporated. Therefore, even when a short-listed garment came close to matching the search criteria, one minor garment feature discrepancy or dislike could ultimately direct the decision to not buy.

For the producer, this is a highly inefficient method of pushing out product ranges and mixes that had undergone long and costly design and development processes—ultimately commanding a 50/50 chance of each item being selected and purchased by the customer. Ultimately, of the hundreds or thousands of stock keeping units (SKUs) designed, developed, sourced, prescreened out (dropped from collection), or preselected (wholesale purchased) to be produced and eventually funneled through the producer's own long and costly supply chain processes, there is only one point of encounter with its customer at the very end of the full supply chain, with a maximum 50% chance of each SKU favored and selected by the customer. Multiply this percentage chance of sales on the scale of mass production and fast fashion unit quantity and SKU styles, then the probability decreases infinitely. Thus there is little wonder that a 50–60% sell-through ratio renders an industry norm, and even in the most successful case of brand/retailers where seasonal sell-through reaches an enviable 80–85% average [11], it is measured again over 10,000 styles created. Multiply that by units of garment produced, then the 15–20% unsold items translate to millions of units of unsold garments, seasonally. This enormous volume of pre- or post-sale unwanted textiles and garments is threatening the environment at an unprecedented pace and scale. And it is but the output pumped through by one successful global retailer. The infinitely more waste generated by a whole industry, including most retailers with much higher percentages of unsold items that result in garment and textile waste, is thus colossal.

6.2.3 Mass customization retail strategies increase actual customer choice

Traditional organizational models for the fashion system and their supply chains, customarily aligned to be based on product and/or supplier market characteristics, now insufficiently support the required target market customer focus [12]. Mass customization (MC) strategies have therefore been identified not only to provide alternatives but also competitive advantages to brand houses and retailers in an age of differentiation, individualization, and diversification of consumer lifestyles, tastes cultures, and personal requirements.

Contemporary examples of MC successfully implemented as a marketing pull strategy for consumer interests range from top-end luxury brands such as Louise Vuitton offering personalized initials and color choice monograms on handbags, to mass market brands such as Converse's "design your own sneakers" where customers are given choices on 10 attributes for each pair of customized sneakers they wish to purchase. These are cases that offer actual increased and improved options for customer's choices in each purchase, instead of the finite two options of yes (to buy) or no (not to buy). However, although market acceptance for MC offerings from the example mainstream brands is high, it is by no means a common offering to customers in the fashion retail market place.

A key reason for the underuse of MC strategies is the inherent inertia of a highly fragmented fashion system, where key drivers of the fast and perpetually moving fashion wheel are the large retailers/brand houses whose brand emphasis and financial investment are generally placed in their downstream marketing and retail promotional expenditures, as opposed to the upstream development and innovation activities.

The prerequisite for better serving consumer demands and preferences in a dynamic and fast-moving market place requires brand houses/retailers to systematically collect, analyze, and make use of customer-centric data. The data must be incorporated into a fully integrated structure reengineered from the traditional FSC processes, starting right from supply chain upstream of design and sourcing, all the way down to sales, marketing, and after-sales customer relationship management (CRM). To meaningfully make use of continuously monitored and collected customer information for the purpose of streamlining and improving producer's relevant decision-making with regard to customers preferences, Pan's Collective Activity Map for a mass customized FSC model [13] identified and integrated producer's preselection points and customer's choice points in the cyclical FSC design-production processes.

6.3 Model proposition

6.3.1 Fashion system framework reconfiguration

6.3.1.1 Principles of modularization

Modularity is a strategy for organizing complex components, products, and processes efficiently [14] within a system that has embedded hierarchical coordinates [15]. A modular system comprises the modules or units that can be designed independently yet can function together as an integrated whole. It allows the separation, recombination, mixing, and matching of components within a set of rules defined by the system architecture. Modularity can be achieved by classifying and partitioning information into visible design rules and hidden design parameters early on in the design process, and by communicating these parameters and rules to all parties involved.

A number of industries have trialed and errored modularity as a design and manufacturing strategy for MC, for instance, in automobiles, consumer electronics, and personal computer [16] industries. Theory and practice concurs that thorough implementation of modularity leads to four key competitive advantages for companies: (1) faster product evolution [17], (2) managing complexity, (3) lower costs, and (4) increased product variety and strategic flexibility.

The three key parameters for the visible design rules of modularization are architecture, module, and interface. Architecture is the overall framework through which modules fit together to become a whole. It denotes the choice of modules to be combined and the individual function of each. Modules may have hidden design parameters and decisions made within the individual module that do not affect other modules in the architecture assembly. So long as a set of common standards sets the rules and criteria for testing module conformity and measuring module performance against one another, the modules remain compatible within the overall architecture. The interface provides detailed definition on how the modules connect and correlate to fit with one another as a whole. It has the attribute of an aggregate made from a number of different components.

6.3.1.2 Architecture

Architecture specifies which modules it contains, what their functions are, and how the functional elements are arranged. The architecture for the fashion object is the garment style. Each garment is composed of a set of hierarchical elements that can be separated and/or joined up with elements of other garments. Each garment style belongs to a product family of garment type by its common object attributes, sharing a common platform in the architecture of modular design. Platform is defined by a collection of assets shared by a set of products [18] (eg, components, processes, knowledge, and relationships [19]). More specifically, it is a set of subsystems purposefully planned and developed so as to form a common structure, from which a stream of derivative products can be designed and produced efficiently. Platform architecture enables increased external variety for customers and reduced complexity for producers [20].

6.3.1.3 Modules

A module is an individual component that forms the smallest SKU in the new MC model presented by this research. The garment architecture is composed of multiple garment modules that possess common components and option-related components [21]. Each component style belongs to a component type. Following this logic of hierarchy, each component type forms a subproduct family and is a subclass of a garment style. The new design logic of this order lies in systematically and hierarchically classifying all garment components.

There are three principle parameters to which modules must conform. First, all modules that belong to the same subproduct family must possess a key common attribute. Second, all modules from the same subproduct family must share and follow the same rules for the modular architecture. Finally, all modules must inherit a set of built-in specifications. These design and technical specifications can be altered, disassembled, and reassembled provided that the specific interface between components remains coherent.

6.3.1.4 Interface

Interfaces describe the standard and detail of how modules fit together, connect, and communicate as an operation. Conforming components can be replaced, reused, and/ or exchanged by way of interface. Each component in each object class has an interface where it joins up with other elements to make a complete garment style. Lines, cuts, shapes, and measurements are built into the module design, and the interface enables module interchangeability during and after both production or retail purchase.

6.4 Deploying smart systems compilation of customers' choice through modular customization model

The ongoing research, analysis, development, and application carried out by this research team present key conditions, parameter, processes, components, classification, and specification of the fashion system. Three main types of information are identified to be crucial to the efficient, effective, economic, and sustainable operation of its perpetual and cyclical output: (1) customer information, (2) product information, and (3) the flow of information.

The integrated data system of this new modular customization model is a central database that stores, manages, and enables access of various use cases in the FSC front-end and back-end to input, extract, and interpret collated information. Its three main apparatus include a data bank where information collected and generated is stored, an application server where the business logic of the operation is determined, and a front-end user interface through which different use cases (eg, designers, manufacturers, retailers, and customers) input, access, and utilize relevant information.

6.5 Conclusion

6.5.1 Redefining customer choice in customized fashion design and retailing

A redefinition of choice is necessary to answer the key question asked at the beginning of this chapter, "How do we improve real consumer choice in fashion retail outlets?" If a customer is looking for a "little black dress" in spring when it is not festive cocktail party season, to be confronted with dozens of colorful dresses is not to be offered an actual choice at all. To be offered a variety of materials, shapes, cut, and detailing of non-full-length black dresses would be to have actual choices worth considering for the customer. And, to be presented with a few dress shape options, as well as the choice of various lengths, sleeves, neckline styles, and/or embellishments, would be to substantially improve real customer choice.

The new modular customization model presented by this research builds in multiple design options for each garment into the upstream supply chain processes, enabling product development management to be streamlined in both human and material resources, while product options and variety can be multiplied in production, distribution, and retail.

6.5.2 Reorienting producers' push supply chain to customers' pull demand chain

By deploying the smart system that applies modular customization throughout the FSC, three structural changes and subsequent benefits to the existing fashion system are introduced (see the comparison of a current conventional retail model and a new modular customization model in Fig. 6.3):

1. Customers are offered increased and improved real choices for the garment in their search, and they have a more effective and efficient way to purchased items that more accurately match their search criteria (at new model choice point A); thus they are more likely to be committed to and satisfied with their purchase. This reduces both the chance of garment return (thus benefiting the producers) and the rate of garment disposal (thus reducing environmental impact).
2. Producers are able to increase percentage chances of garments being sold due to their ability to modularly customize to match varied customers' needs and wants, without having to start from the beginning of the design development processes. This not only postpones, but in fact reduces, the heavy dependence on fast-moving, trend-based, blind decisions made in the upstream FSC processes well ahead of product time-to-market. Also, it could potentially improve the overall sell-through ratio as compared to the current, conventional push-through approach of brand houses/retailers.
3. Through the implementation of this modular customization model, reduction of waste in both human and material resources can begin right at the upstream of the FSC, from creative conceptualization to design development, to sourcing and production, to retail product and service provision for creative customization and low-environmental impact up-recycling.

6.5.3 Toward a new target for resource efficiency, consumption sustainability, and customer satisfaction

Despite the increasingly challenging global retail environment for small and medium enterprises (SME) independent brand houses/designers/retailers, given the right message and method, SMEs often have more operational agility, business flexibility [22], and are more in touch with direct consumer information and feedback. Therefore, they have opportunities to be more directly responsive and consistently influential to their target market clientele, particularly as consumers grow more skeptical toward large corporations.

The British ethical and sustainable fashion brand DEPLOY demi-couture is one such SME independent designer brand/retailer specialized in high-end design tailoring

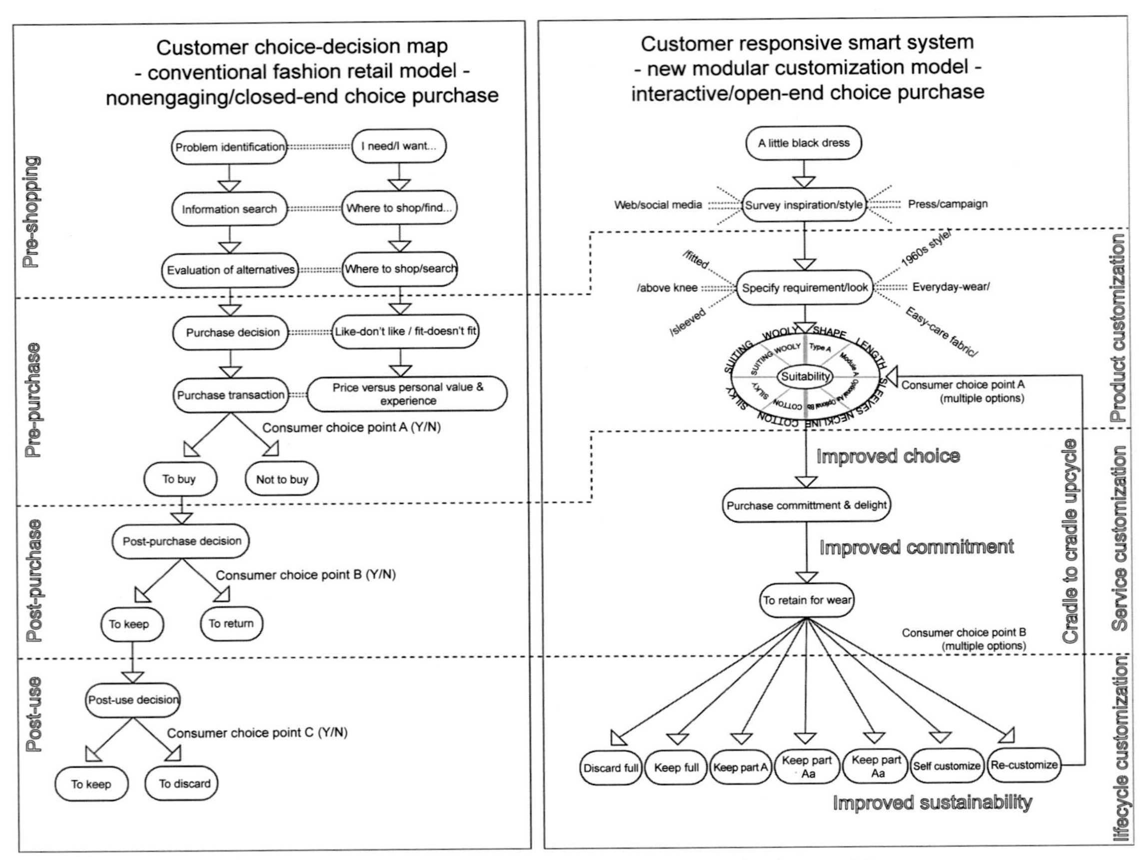

Figure 6.3 Customer choice comparison in conventional and new modular fashion customization models.

that contributed toward the innovation and pilot of this research. DEPLOY utilized the theory and experimented with the practice of the new Modular Fashion Customization model both in its FSC processes and in the market place. The brand objective is to drive changes in fashion and lifestyle consumption through providing higher levels of quality product and service customization right from the upstream processes of design, sourcing, and development. And DEPLOY has indeed seen direct results in substantially reducing waste in its own FSC workings of development and production, as well as increasing customer choice and product life cycle, thus affording long-term customer satisfaction.

The possibilities for this new model and its manifestation in DEPLOY demi-couture's prosumer relationship with its global clientele presents two significant benefits: (1) gaining improved margins through streamlined process efficiency as opposed to through constant squeeze on material and labor costs resulting in reduced quality; and (2) engaging retail customers' brand interest and loyalty through real and improved choice and thus their voluntary participation in ethical and sustainable fashion consumption. While the personal impact made to individual customers and its supply chain partners has been measurably high, the scale of DEPLOY's brand reach is evidently still limited. It is therefore the objective of this research team to encourage more brand/retailers, large or small, to deploy this smart system approach to reap more ethical and sustained collective benefit for all parties of the fashion system.

Further studies and collaborative development incorporating mobile applications (apps), CRM, and bundling technology would enhance prospects for the Modular Fashion Customization model developed. However, although a smarter and more accurate option matrix for both producer's design and customer's choice could be automated by further technological advances, the research team would argue that meaningful long-term customer relationship in fashion clothing retail can ultimately be fostered through a combination of physical (human, personal, and experiential) and virtual (technology, online, and media) interaction.

References

[1] Birnbaum. Birnbaum's global guide to winning the great garment war. 5th ed. Hong Kong: Third Horizon Press; 2005.

[2] Jimenez, Koslum. Fashion law: a guide for designers, fashion executives, and attorneys. New York: Bloomsbury; 2014.

[3] Stephen, Galak. The complementary roles of traditional and social media publicity in driving marketing performance. 2010 [A Working Paper, 2010/97/MKT, INSEAD Working Paper Collection].

[4] Bakewell, Mitchell. Generation Y female consumer decision-making styles. Int J Retail Distrib Manag January 2003;31:95–106.

[5] Retail Gazette. Cost polarisation and the divergent retail sector. November 21, 2014.

[6] Aguelov. The dirty side of garment industry. Boca Raton (FL, USA): CRC Press; 2016.

[7] Farrer, Fraser. Conscience clothing: polarisation of the fashion textile market. 2008. http://eprints.port.ac.uk/id/eprint/18551.

[8] Allwood, Laursen, Malvido de Rodriguez, Bocken. Well dressed? Cambridge, UK: University of Cambridge Institute of Manufacturing; 2006.
[9] Pan, Holland. A mass customised supply chain for the fashion system at the design production interface. J Fash Mark Manag 2006;10(3):345–59.
[10] Jeddi, Atefi, Jalali, Poureisa, Haghi. Consumer behavior and consumer buying decision process. Int J Bus Behav Sci 2013;3(5):20–35.
[11] Miller. Fashion's new fast lane. Forbes.com; September 12, 2006. http://www.forbes.com.
[12] Rinnebach. Redefining fashion business models: today's challenges, tomorrow's competitive edge. Düsseldorf, Germany: Kurt Salmon; 2014.
[13] Pan. A mass customisation implementation model for the total design process of the fashion system. In: Choi, editor. Fashion supply chain management: Industry & business analysis. Hershey (PA, USA): IGI Global; 2012.
[14] Sanchez, Mahony. Modularity, flexibility, and knowledge management in product and organization design. Strateg Manag J 1996;17:63–76 [Special Issue: Knowledge and the Firm].
[15] Baldwin, Clark. Managing in an age of modularity. Harv Bus Rev 1997;75(5).
[16] Asan, Polat, Sedar. An integrated method for designing modular products. J Manuf Technol Manag 2004;15(1):29–49.
[17] O'Grady. The age of modularity. USA: Adams & Steele Publishers; 1999.
[18] Robertson, Ulrich. Planning for product platforms. Sloan Manag Rev Summer 1988;39.
[19] Muffatto, Roveda. Developing product platform: analysis of the development process. Technovation 2000;20:617–30.
[20] Sanderson, Uzumeri. Managing product families: the case of the Sony Walkman. Res Policy 1995;24(5):761–82.
[21] Danese, Romano. Improving inter-functional coordination to face high product variety and frequent modifications. Int J Oper Prod Manag 2004;24(9):863–85.
[22] Battisti, Perry. Walking the talk? Environmental responsibility from the perspective of small-business owners. Corp Soc Responsib Environ Manag 2013;18(3):172–85 [Special Issue: Critical Research in Sustainability Debate].

Intelligent procurement systems to support fast fashion supply chains in the apparel industry

7

D.A. Serel
Ipek University, Ankara, Turkey

7.1 Introduction

Retailers selling fashion apparel need to carry inventories of their offerings as the market demand for fashion goods is usually highly uncertain. Inventories help retailers cope with demand estimation errors and unforeseen surges in demand. Shortages of products not only result in lost sales in the current sales period but also influence the retailer's future profits negatively. The customers dissatisfied because of a stockout experience may be less willing to visit the retail store or website again in the future, which leads to a decrease in the retailer's future sales. Nonetheless some successful fashion retailers, such as Zara and H&M, manage to operate with low levels of inventory while keeping shortage occurrences under control. As carrying high levels of inventory increases the risk of leftovers and consequently reduces the retailers' profits caused by markdowns on selling price in the clearance season, retailers operating with lower stocking levels without sacrificing customer service levels can achieve higher profits and have competitive advantage. The key to reducing inventory while preserving availability of product to the customers at an acceptable level is to have a better match between supply and demand. A possible approach to reduce mismatch between supply and demand is to use more accurate demand forecasts to determine stocking quantities. There exist various ways to improve demand forecasts, including application of advanced statistical techniques and qualitative market research methods. The length of time between the preparation of demand forecast and the occurrence of actual demand is considered to be related to the magnitude of error associated with the demand forecast. In this chapter we review analytical models for enhancing the effectiveness of inventory decisions by taking into account the relationship between the time of the procurement order and the demand forecast error. Our discussion focuses on a subset of the operations management literature that explores how to incorporate multiple supply modes into the procurement system of a retailer selling seasonal fashion goods with uncertain demands. Intelligent use of demand forecasts in choosing appropriate stocking levels of fashion apparel is an important part of information systems used by firms in the clothing industry.

Given that error in forecasting uncertain customer demand for fashion apparel decreases as the time interval between ordering and selling season shortens, naturally the retailer will try to order its merchandise as late as possible in order to escape the costly

Information Systems for the Fashion and Apparel Industry. http://dx.doi.org/10.1016/B978-0-08-100571-2.00007-5

consequences of grossly misestimated demand. However, waiting for demand forecasts to be improved over time may lead to an increase in procurement expenditure. The supply price of the merchandise will generally be higher when the suppliers are asked to complete delivery of the products within a shorter time period. Consider the offshore sourcing practice of American manufacturers and retailers. It is generally the case that the product can be sourced from suppliers in the emerging economies, such as China and Cambodia, at a relatively low cost. However, the geographical distance between these countries and the retailers in the US, combined with the preference for the low-cost shipping route (that is, ocean containers), necessitates that the procurement order be issued to the offshore supplier several months before the start of the selling season. On the other hand, local suppliers in the US for the same product can accommodate purchase orders with shorter delivery time requests as the products can be shipped by railway or trucks to their destinations. However, these local suppliers will likely charge a higher supply price than the offshore suppliers since their labor and other production costs will generally be higher than those of offshore suppliers.

Thus, the decision problem for the fashion retailer is whether to place its order for goods with the suppliers (1) at an earlier time based on a low purchase cost but coarse demand forecast, or (2) at a later time based on a high purchase cost but more accurate demand forecast. Academic researchers in the operations management area have developed various mathematical models to tackle this problem. This chapter will provide a summary of these models, which can serve as a useful reference for professionals involved in the design and analysis of fast fashion supply chains.

In the manufacturing of products with volatile demands, the use of improved and timely demand forecasts facilitated by reduced lead times in production and sourcing has been referred to as quick response (QR) manufacturing. For application of quick response manufacturing ideas, the textile apparel industry has proved to be an especially suitable setting in the last three decades. A well-known industry example showing the potential value of the QR approach is the American skiwear manufacturer Sport Obermeyer, discussed in detail in Fisher and Raman (1996). Confronted with highly volatile demand for skiwear products, such as parkas, pants, suits, sweaters, jackets, and various accessories, the company had to decide production quantities of these products by taking into account its production capacity. To prepare for a particular selling season, the production at Sport Obermeyer was completed in two subsequent periods. In the first period, production was completed without any retailer orders. Thus, based on an initial demand forecast, the company decided the size of the first production batch for each product. Before the start of the second production period, some retailers began to place orders with the company. In the second period, production quantities were decided by taking into account these initial orders received from retailers. These initial orders were interpreted by Sport Obermeyer as early signals for the final demand for the products. The greater the retailer orders placed at the beginning of the season for a product, the higher the demand would be for that product in the whole season that Sport Obermeyer predicted. For example, Fisher and Raman (1996) stated that after observing just 20% of seasonal demand for one product line, it was possible to forecast the demand in the remaining part of the season with greater accuracy. Utilizing a more accurate demand forecast, the company could have reduced the risks of overstocking and understocking, which would translate into a reduction in inventory-related costs. Fisher and Raman (1996) developed an algorithm that can be used to determine appropriate production quantities in the two periods. Using real sales figures at the company, they found that the company could have increased its profits by 60% had it implemented their algorithm for the production decisions rather than relying on its own ad hoc decision process.

Another example for integrating local and offshore production within a quick response framework in the apparel industry is the VF Corporation, which supplies jeans to Wal-Mart retail stores in the US. The routine orders for the products are met by the plants outside the US that entail a long lead time; for urgent replenishment orders placed by Wal-Mart, the company's two plants located in the US are used to achieve faster delivery performance (Jin, 2004).

In this chapter we provide an overview of seasonal inventory models that facilitate incorporation of more accurate demand forecasts into the procurement systems of fast fashion retailers. Different aspects of the quick response manufacturing and ordering systems have been investigated in the literature. Our discussion will concentrate on a group of closely related research articles that share a common framework. We can divide the papers reviewed in this chapter into two main categories.

The main idea in the papers belonging to the first category is to divide the selling season for a fashion retailer into two periods and place two separate stocking orders instead of a traditional single order. Note that the decision problem we consider can be seen from the perspective of a manufacturer or a retailer. There is an uncertain customer demand for a product and if the decision maker is a manufacturer, it needs to decide the size of production batches in the two periods. If the decision maker is a retailer, it needs to decide the size of purchase orders in the two periods. Essentially, this problem is a variant of the well-known single period (newsvendor) inventory problem in the operations management literature. Instead of the single procurement order issued prior to the selling season in the classical newsvendor problem, in the new setting there is an additional ordering opportunity that can be utilized by the retailer after observing actual sales in the early season. Using this new demand information, the retailer has a chance to adjust its stocking quantity to achieve a better match against market demand in the remaining part of the season. As described above in the context of the Sport Obermeyer case, the model of Fisher and Raman (1996) is an example for the papers in the first category.

The papers in the second category also utilize new demand information to improve the stocking decisions, but the demand forecast update is made before the start of the selling season. Thus, instead of actual sales observations for the product in consideration, the new demand information comes from other sources such as customer surveys or observed sales of related products. The fashion retailer again has two purchasing opportunities. The first order is placed based on a rough demand forecast. The second order is decided later, based on a revised demand forecast. In this group of papers, the retailer does not have a chance to reorder the product during the selling season.

The papers reviewed in the next two sections have built upon one of the two settings described above, and explored a variety of issues surrounding this practical decision problem. In Section 7.2, we look into the models that allow reordering during the selling season. In Section 7.3, we go over the models that assume that all procurement orders are placed prior to the selling season. The chapter concludes with a recap of the main ideas in Section 7.4.

7.2 Two-period models with reordering during the selling season

To illustrate the use of demand forecast updating in the two-period setting, we first briefly review the mathematical optimization model presented in Fisher and Raman (1996). It is assumed that there are n products and two separate purchase orders can be placed for each product. There are two selling periods. Let x_{i1} be the initial order for product i, $i = 1, \ldots, n$. Thus, inventory on hand at the beginning of period 1 is x_{i1}. The vector x_I contains the inventory values x_{i1}. After observing the demand for product i in the first period, D_{i1}, the purchase order for the second period is placed, which is delivered immediately, that is, lead time for the second order is zero. Let x_i be the total number of units produced (or ordered) for product i in the two orders combined, $x_i \geq x_{i1}$, $i = 1, \ldots, n$. Also let D_i be the total number of units demanded for product i in the two periods. The vectors D_I and D contain the demand values for all products in period 1 only and the total demand in the two periods, respectively. Similarly, let x be the vector containing the total production quantities x_i. Fisher and Raman (1996) assume that demands for a fashion product in two periods are dependent and the probability density function (pdf) of D_i given the observed value of D_{i1} is represented by $h_i(D_i|D_{i1})$. The unit cost of overproducing product i is shown by O_i, and the unit cost of underproducing product i is shown by U_i. Thus, for product i, each unit of leftover stock costs O_i, and each unit of demand shortage (insufficient stock) costs U_i. It is also assumed that total number of units produced (or ordered) in period 2 for all products combined is limited to the shared production capacity, K. There is no capacity constraint associated with initial production.

The fashion retailer in this setting tries to minimize the total cost of overproducing and underproducing by first optimally choosing the initial production x_I, and then, based on the demand observation D_I, by determining the optimal total production x. This two-stage optimization model can be expressed as:

$$\underset{x_I \geq 0}{\text{Minimize}} \quad \text{EC}(x_I) = E_{D_I} \min_{x \geq x_I} \quad E_{D|D_I} \quad \sum_{i=1}^{n} \left[O_i(x_i - D_i)^+ + U_i(D_i - x_i)^+ \right]$$

$$s.t. \quad \sum_{i=1}^{n} (x_i - x_{i1}) \leq K,$$

where $(y)^+ = \max(y,0)$. Fisher and Raman (1996) show that the expected cost function $\text{EC}(x_I)$ is convex. In this model the constrained capacity in the second period causes the initial order to be positive. If the capacity in the second period was unlimited, no initial order would be placed in the optimal solution. Fisher and Raman (1996) also investigate the case where the first and second period production quantities are required to be greater than certain minimum production levels.

Other two-period models in the literature consider that the difference in unit purchase costs for the initial and second order may cause the initial order to be positive. In this case lower purchase cost associated with initial order to some extent counteracts the higher demand forecast error. Milner and Rosenblatt (2002) develop such a single-product model in which the buyer places the orders for both the first period and second period prior to the selling season but has the flexibility to adjust the second order upward or downward at the end of the first period by incurring an order adjustment penalty. As the demands for first and second periods are correlated, the adjustment to the second-order quantity is decided at the beginning of period 2 according to the revised demand distribution for period 2. Milner and Rosenblatt (2002) take into account holding and shortage (loss of goodwill) costs in both periods in their formulation. Let D_i be the demand in period i, and Q_i be the order quantity for period i decided prior to the selling season, $i = 1, 2$. The conditional distribution of second period demand is shown as $F_{D_2|D_1}(\cdot)$. The holding cost and shortage cost for period i are denoted by h_i, and π_i, respectively, $i = 1, 2$. It is assumed that unit purchase cost associated with both Q_1 and Q_2 is c, but at the beginning of period 2 the buyer can obtain additional units at unit cost c_H and also the buyer can cancel any portion of Q_2 at unit cost $c - c_L$; that is, a rebate c_L is offered by the supplier for each unit not demanded by the buyer at the beginning of period 2, $c_L \le c \le c_H$. Let u be the order quantity for period 2 after adjustment. Denoting the unit revenue by p and defining $S = Q_1 + Q_2$, Milner and Rosenblatt (2002) derive the total expected profit function as:

$$
\begin{aligned}
\mathrm{EP}(Q_1, S) = E_{D_1}\{&p\min(Q_1, D_1) - cQ_1 - h_1(Q_1 - D_1)^+ - \pi_1(D_1 - Q_1)^+ \\
&+ \max_{u \ge 0} E_{D_2|D_1}\big[p\min(Q_1 - D_1 + u, D_2) - (c - c_L)(S - Q_1) - c_L u \\
&- (c_H - c_L)(u - S + Q_1)^+ - h_2(Q_1 - D_1 + u - D_2)^+ \\
&- \pi_2(D_2 - Q_1 + D_1 - u)^+\big]\}.
\end{aligned}
$$

Milner and Rosenblatt (2002) prove that the expected profit function $\mathrm{EP}(Q_1,S)$ is concave. They also show that, depending on demand realization in period 1, D_1, the optimal adjusted order in the second period, u is given by:

$$
u = \begin{cases}
v_2(D_1) - (Q_1 - D_1) & \text{if} \quad C(S) < D_1, \\
Q_2 & \text{if} \quad B(S) < D_1 \le C(S), \\
v_1(D_1) - (Q_1 - D_1) & \text{if} \quad A(Q_1) < D_1 \le B(S), \\
0 & \text{if} \quad D_1 \le A(Q_1),
\end{cases}
$$

where $v_1(D_1) = F^{-1}_{D_2|D_1}\left(\frac{p+\pi_2-c_L}{p+\pi_2+h_2}\middle|D_1\right)$, $v_2(D_1) = F^{-1}_{D_2|D_1}\left(\frac{p+\pi_2-c_H}{p+\pi_2+h_2}\middle|D_1\right)$, and $C(S)$, $B(S)$, and $A(Q_1)$ are thresholds that depend on S and Q_1. To determine the optimal initial order quantities, define $H(x,y) = -F_{D_2|D_1}(y-x|D_1=x)$, and let q_1 and s satisfy the following equations:

$$(\pi_1+h_1)F_{D_1}(q_1) + (\pi_2+h_2+p)\int_{-\infty}^{A(q_1)} F_{D_1}(x)\frac{\partial H(x,q_1)}{\partial x}dx = \pi_1,$$

$$\int_{B(s)}^{C(s)} F_{D_1}(x)\frac{\partial H(x,s)}{\partial x}dx = \frac{c_H-c}{p+\pi_2+h_2}.$$

Milner and Rosenblatt (2002) show that if $q_1 < s$, then optimal $Q_1 = q_1$ and optimal $S = s$. Otherwise, $Q_1 = q_1 = s$ satisfies the following equation at the optimal solution:

$$(\pi_1+h_1)F_{D_1}(s) + (\pi_2+h_2+p)\int_{-\infty}^{C(s)} F_{D_1}(x)\frac{\partial H(x,s)}{\partial x}dx = \pi_1 + c_H - c.$$

The backup agreement between a catalog company and a manufacturer discussed by Eppen and Iyer (1997) is another example for a two-period ordering system with demand forecast update. The reordering during the season may be filled from a backup which is set aside earlier. After observing early sales in the season, the catalog company may decide not to order an arbitrary portion of the backup quantity by paying a cancellation penalty to the manufacturer. The products ordered before the season and products ordered from the backup cost the same. Eppen and Iyer (1997) also take into account possible customer returns during the season in inventory decisions, and assume lost sales if a stockout occurs in the first period. They show that an order-up-to policy is optimal in both periods.

Some variants of two-period models have been investigated in the literature. Next, we describe the main characteristics of these models. The papers reported in this section are classified according to a group of modeling assumptions in Table 7.1.

7.2.1 Multiple products with heterogeneous Poisson demands

Bradford and Sugrue (1990) explore the case where demand for each product during the arbitrary time interval $[0,t]$ is distributed as Poisson with unknown but stationary mean rate λt. They assume that the parameter λ follows a gamma distribution across the individual products, and it is re-estimated at the end of the first period based on observed demands for products in the first period. These assumptions cause demand distribution for a product conditional on its period 1 demand to be distributed as negative binomial. The problem is to determine optimal orders to be placed at the beginning of each period so that total expected profit in two periods is maximized.

Table 7.1 **Features of papers reviewed in Section 7.2**

Paper	Number of products	Number of reorders	Channel coordination	Price-dependent demand	Time of late (postponed) orders	Lead time for late orders
Bitran et al. (1986)	Multiple	More than 1			Fixed	
Bradford and Sugrue (1990)	Multiple	1			Fixed	
Eppen and Iyer (1997)	Single	1			Fixed	
Fisher et al. (2001)	Single	1			Flexible	Positive
Fisher and Raman (1996)	Multiple	1			Fixed	
Lau and Lau (1997)	Single	1			Fixed	Positive
Lau and Lau (1998)	Single	1			Flexible	
Li et al. (2009)	Single	1			Flexible	Positive
Matsuo (1990)	Multiple	More than 1			Fixed	
Milner and Kouvelis (2002, 2005)	Single	1			Flexible	Positive
Milner and Rosenblatt (2002)	Single	1			Fixed	
Petruzzi and Dada (2001)	Single	1		Yes	Fixed	
Raman and Kim (2002)	Multiple	More than 1			Fixed	
Subrahmanyan and Shoemaker (1996)	Single	1		Yes	Fixed	Positive
Zhou and Wang (2012)	Single	1	Yes		Fixed	

7.2.2 Price-sensitive demand

Assuming negative binomial distribution for demand with a price-dependent distribution parameter, Subrahmanyan and Shoemaker (1996) use dynamic programming to find optimal prices and order quantities in a two-period problem. They also present numerical results for a three-period problem in which lead time for the second order is one period.

Petruzzi and Dada (2001) study a two-period problem in which expected demand decreases linearly in demand. Using an additive demand model, they consider a setting in which orders for both periods are placed before the first period. After observing demand in period 1, the buyer can cancel a part of the order previously given for period 2, or can add new units to this standing order to increase the stocking level for period 2. In addition to these stocking decisions, the buyer also needs to choose the optimal selling price in each period. Petruzzi and Dada (2001) show that finding the optimal solution to this problem can be reduced to a search for one decision variable.

7.2.3 Flexible timing of the order placed after demand forecast update

All models described so far assume that the lengths of the two periods are fixed, and the time of the second purchase order is not a decision variable. Some researchers have relaxed this assumption and allowed the time of second order to be a decision variable. Milner and Kouvelis (2002) develop a continuous time model in which demand follows a Brownian motion with an unknown drift and known variance. The entire selling season is divided into two time segments. By observing demand in the first time segment, the buyer updates its demand forecast for the second segment using a Bayesian approach. Milner and Kouvelis (2002) assume that there is a fixed lead time for second order, and in case of stockout, excess demand is lost. They also assume that the buyer follows a $(s(\tau), Q_2(\tau))$ ordering policy, which is defined as: if inventory level at time τ equals $s(\tau)$, an order of $Q_2(\tau)$ is placed. To minimize the cost in the second segment (approximately) when an order is placed at time τ, they present two equations that $s(\tau)$ and $Q_2(\tau)$ should satisfy for a given τ.

In a similar research, Milner and Kouvelis (2005) look into a Martingale demand process in which demand evolves over time as a result of shocks which can be, for example, the effects of advertising, competitive product offerings, changes in the business environment, etc. Milner and Kouvelis (2005) argue that the Bayesian demand model in their 2002 paper is applicable to fashion products for which mean demand need only be estimated. They argue that Martingale model of demand evolution is suitable for innovative products for which demand may be determined by exogenous factors. In computing inventory holding cost, they make the simplifying assumption that ending inventory over any time segment is positive, that is they ignore the possibility of demand shortage when calculating the amount of inventory carried.

Fisher et al. (2001) propose a model in which if demand before the time of reordering exceeds the initial inventory, excess demand is backordered subject to the condition that it is possible to fill backorders from replenishment. They show that expected cost in the time interval between reordering time and the end of the season is neither convex nor concave in the inventory position. Hence, finding the optimal solution is complicated. Fisher et al. (2001) propose a solution heuristic for a given reorder time, and suggest finding the optimal reorder time by solving the model by varying reorder time.

Li et al. (2009) generalize the model in Fisher et al. (2001) to the case where inventory holding costs and backorder costs may be time-dependent. They show that only under certain conditions, the optimal policy associated with the timing of second order is the $(s(\tau), Q_2(\tau))$ policy explored in Milner and Kouvelis (2002).

7.2.4 Supply chain coordination

While many researchers have approached the stocking problem in a quick response environment from the perspective of the fast fashion retailer, it is also important to consider the supplier's production costs and the overall supply chain efficiency. Because the supplier's production costs may depend on the lead time allowed for delivery by the retailer, and in a decentralized system the supplier and the retailer try to maximize their own profits, it is natural to search for production policies that maximize the total supply chain profit, that is, the combined profit of the retailer and the supplier.

Zhou and Wang (2012) develop a two-period model in which a buyer may order from a manufacturer at the beginning of each period. The retailer's order at the beginning of the second period is made based on actual demand information collected from the first period. There is a fixed ordering cost for the buyer, and a fixed production setup cost for the manufacturer. The manufacturer does not carry inventory and determines the wholesale price for the product. The authors propose an improved revenue-sharing contract that can coordinate this supply chain and prevent the profit loss caused by double marginalization which is associated with the simple wholesale price contract.

7.2.5 More than one demand forecast revision

It is possible to allow demand forecast update to be made multiple times during the selling season while continuing to produce the products between these updating times. Raman and Kim (2002) extend the two-period multiproduct model in Fisher and Raman (1996) to a general n-period setting in which production capacity is limited and possibly different in each of the n production periods. In addition to overstocking and understocking costs incurred at the end of the season, there are also inventory holding costs incurred for each period and caused by carrying items produced in earlier periods in stock.

Bitran et al. (1986) study a similar problem in which products belong to a group of families. Products in the same family share the same production setup and consume the same amount of resources per unit. Although there is no setup cost associated with a switchover within a family, switching production from one family to another leads to a setup cost. Bitran et al. (1986) make the following key assumptions: (1) demands for products in a family are distributed as multivariate normal, (2) Bayesian approach is used to update demand distribution parameters at the beginning of each period, (3) holding costs are incurred in each of the production periods, (4) each product family is set up exactly once, (5) production capacity in each period is constrained, and (6) demands occur in the last period of the model. The problem is to determine when (in which period) to produce a family as well as how much of each product in a family to produce. Bitran et al. (1986) propose a hierarchical procedure to find an approximately optimal solution. Matsuo (1990) adapts the discrete time model of Bitran et al. (1986) to a continuous time setting and proposes a heuristic sequencing procedure to specify the starting and finishing times of production of each family.

7.2.6 Mid-season replenishment without forecast updating

Lau and Lau (1998) study the two-period problem in which demand forecast for period 2 is not updated after observing demand in period 1. The order placed for the second period arrives in zero lead time. They provide a solution procedure for determining order quantities. They also consider the extensions of fixed setup cost for second order, and time of second order as a decision variable. In a related research, Lau and Lau (1997), working with normal and beta distributions, present numerical results for the setting where lead time for the second order is positive, and demands in the two periods are dependent.

7.3 Multiple-order models with all orders placed before the selling season

Due to the length of the lead time required for delivery, it might not always be possible to place and receive an order within a selling season that has a short duration. Nonetheless, valuable demand information can still be obtained prior to the selling season without observing actual sales of a product. For example, there may be some similar products of which actual sales are observed, and these sales may be correlated with the sales of the product that have not yet been observed. Color of fashion garments is an attribute that has an important effect on sales in a particular season. Demands of fashion apparel with the same color, say red, will usually be correlated (Choi et al., 2003). A similar positive dependence can be expected among demands for clothes made of organic and green cotton. Sales of other fashion products in a closely related category generate useful demand signals for retailers. In the music industry, the entry position of single song releases in the Top 100 charts provides an early signal regarding sales of the full album that will be released later (Choi et al., 2003; Seifert et al., 2015).

In this section we focus on information systems supporting fast fashion businesses in which demand signals are incorporated into the procurement decisions before the start of the selling season. A more detailed explanation of models in this group can be found in Serel (2014). We adopt the framework of presentation in Serel (2014) but expand the coverage of literature by discussing a larger set of references that extend the basic setting in a variety of directions.

To illustrate the basic ideas, we start with the model of Choi et al. (2003) that considers two purchasing opportunities before the start of the season. Suppose the first order Q_1 is placed at time 1 and, based on a revised demand forecast, the second order Q_2 is placed later at time 2. According to initial demand forecast at time 1, demand is assumed to be distributed as normal with an unknown mean m and a known variance σ_1^2. It is also assumed that the unknown mean m is normally distributed with mean μ_1 and variance d_1.

Because actual demand for the product cannot be observed before the season starts, the retailer tries to estimate potential demand via other ways such as early order commitments, market research techniques, expert judgments, fashion shows, and sales observations of other related products. The demand information collected is translated into an observation from the demand distribution, say x, which is used to derive posterior distribution of the parameter m by using Bayes' rule. It follows that the predictive distribution for demand, which is used in determining the additional order size at time 2, is normal with mean $\mu_2 = \left(\mu_1\sigma_1^2 + xd_1\right)/\left(\sigma_1^2 + d_1\right)$ and variance $\sigma_2^2 = \sigma_1^2 + d_2$, where $d_2 = \left(\sigma_1^2 d_1\right)/\left(\sigma_1^2 + d_1\right)$.

We define τ as the salvage value per unit of leftover inventory, π as the unit shortage cost (loss of goodwill) incurred when demand is unfulfilled, p as the selling price, c_1 as the unit purchase cost at time 1, and c_2 as the unit purchase cost at time 2, $c_1 < c_2$. Assuming Q_1 is given and μ_2 has been observed, we first find the optimal additional order at time 2. The expected profit at time 2 for given values of Q_1 and μ_2 is

$$\begin{aligned}\mathrm{EP}_2(Q_1, Q_2) = {} & pE[\min(Q_1 + Q_2, D)] + \tau E[Q_1 + Q_2 - D]^+ \\ & - \pi E[D - Q_1 - Q_2]^+ - c_2 Q_2\end{aligned}$$

where D is the demand during the selling season. Recalling that mean of the demand distribution is μ_2 and variance of the demand distribution is σ_2^2, the optimal Q_2 maximizing $\mathrm{EP}_2(Q_1,Q_2)$ is given by:

$$Q_2^* = \max\left\{0, \mu_2 + \left(\sigma_1^2 + d_2\right)^{0.5}\Phi^{-1}(s) - Q_1\right\},$$

where Φ^{-1} (.) is the inverse cumulative distribution function (cdf) of a standard normal variable, and s is the well-known critical fractile solution of the standard newsvendor problem, that is, $s = \frac{p + \pi - c_2}{p + \pi - \tau}$. Thus disappointing new market information leading to a low value for μ_2 will imply that optimal policy at time 2 will be not to place a new order.

We now substitute the optimal Q_2 value into the expected profit function $\mathrm{EP}_2(Q_1,Q_2)$. Q_2^* is positive if $\mu_2 > Q_1 - \left(\sigma_1^2 + d_2\right)^{0.5}\Phi^{-1}(s)$. The expected profit at time 2 when $Q_2^* > 0$ is

$$\begin{aligned}J_1(Q_1,\mu_2) = {} & (p - c_2)\mu_2 + (\tau - c_2)\left(\sigma_1^2 + d_2\right)^{0.5}\Phi^{-1}(s) \\ & - (p + \pi - \tau)\left(\sigma_1^2 + d_2\right)^{0.5}\Psi\left(\Phi^{-1}(s)\right) + c_2 Q_1,\end{aligned}$$

where $\Psi(u) = \int_u^\infty (z-u)\phi(z)dz$ is the unit loss function, and $\phi(z)$ is the pdf of a standard normal variable, respectively. On the other hand, if $\mu_2 \leq Q_1 - (\sigma_1^2 + d_2)^{0.5}\Phi^{-1}(s)$, $Q_2^* = 0$, and expected profit in this case is

$$J_2(Q_1, \mu_2) = p\mu_2 + \tau(Q_1 - \mu_2) - (p + \pi - \tau)(\sigma_1^2 + d_2)^{0.5}\Psi\left(\frac{Q_1 - \mu_2}{\sqrt{\sigma_1^2 + d_2}}\right).$$

By using these expressions, we obtain the expected profit function at time 1, $\mathrm{EP}_1(Q_1)$, as:

$$\mathrm{EP}_1(Q_1) = \int_{-\infty}^{Q_1-(\sigma_1^2+d_2)^{0.5}\Phi^{-1}(s)} J_2(Q_1, \mu_2)g(\mu_2)d\mu_2 + \int_{Q_1-(\sigma_1^2+d_2)^{0.5}\Phi^{-1}(s)}^{\infty} J_1(Q_1, \mu_2)g(\mu_2)d\mu_2 - c_1Q_1.$$

Thus expected profit at time 1 is computed using the probability density function of μ_2, $g(\mu_2)$, which is a normal pdf with mean μ_1 and variance $\sigma^2 = d_1^2/(\sigma_1^2 + d_1)$. By adapting the proof of Lemma 1 in Choi et al. (2003), it can be shown that the expected profit function $\mathrm{EP}_1(Q_1)$ is concave in Q_1. This result implies that the optimal order quantity at time 1 Q_1, if positive, satisfies the following first-order condition:

$$(p + \pi - c_2)\Phi(\varepsilon) + (c_2 - c_1) - (p + \pi - \tau)\int_{-\infty}^{\kappa} \Phi\left(\frac{Q_1 - \mu_1 - \gamma\sigma}{\sqrt{\sigma_1^2 + d_2}}\right)\phi(\gamma)d\gamma = 0,$$

where $\Phi(.)$ is the cdf of a standard normal variable, $\kappa = Q_1 - (\sigma_1^2 + d_2)^{0.5}\Phi^{-1}(s)$, $\gamma = \frac{\mu_2 - \mu_1}{\sigma}$, and $\varepsilon = \frac{Q_1 - (\sigma_1^2 + d_2)^{0.5}\Phi^{-1}(s) - \mu_1}{\sigma}$. A method to evaluate the expected profit for any given Q_1 value is described in Serel (2009).

We remark that Yan et al. (2003), working with a different type of forecast update process, study a similar problem and derive a similar result for the initial order quantity. The setting in which all purchase orders are placed by the retailer before the season starts has also been used in a number of papers in the literature. We now summarize the salient features of selected papers falling in this category. The main features of models discussed in this section are shown in Table 7.2.

7.3.1 Flexible timing of the purchase orders

As the demand forecast may improve and the purchase cost per unit may increase simultaneously over time, a natural extension is to treat the time of the second order as a controllable variable. Choi et al. (2004) assume that there are multiple supply delivery modes with faster deliveries implying higher costs. A faster delivery means the

Table 7.2 Features of papers reviewed in Section 7.3

Paper	Number of products	Number of reorders	Channel coordination	Price-dependent demand	Constraint on resources	Random purchase cost for late orders
Chen et al. (2006)	Single	1	Yes			
Chen et al. (2010)	Single	1	Yes	Yes		
Choi (2007)	Single	1		Yes		
Choi (2013a)	Single	1				
Choi (2013b)	Single	1				
Choi et al. (2003)	Single	1				Yes
Choi et al. (2004)	Single	1				
DeYong and Cattani (2012)	Single	1				
Donohue (2000)	Single	1	Yes			
Gurnani and Tang (1999)	Single	1				Yes
Huang et al. (2005)	Single	1				
Kim (2003)	Single	More than 1				
Ma et al. (2012)	Single	1				
Miltenburg and Pong (2007a)	Multiple	1				
Miltenburg and Pong (2007b)	Multiple	1			Yes	
Serel (2009)	Single	1		Yes		Yes
Serel (2012)	Multiple	1			Yes	
Song et al. (2014)	Multiple	1			Yes	Yes
Wang et al. (2012)	Single	More than 1				
Yan et al. (2003)	Single	1				
Zheng et al. (2015)	Single	1			Yes	

retailer can postpone the purchase order to a later time period and utilize a more improved demand forecast. Formulating a Bayesian forecast update model with normal demand distribution, Choi et al. (2004) determine the optimal time and quantity of a single order using a dynamic programming model. Choi (2013b), assuming there is a carbon emission tax proportional to the delivery time, investigates a more general variant of this supplier selection problem. Kim (2003) explores a problem similar to Choi et al. (2004) in which demand in the season is assumed to be correlated with a market information variable. This correlation is assumed to increase as the season approaches, and hence variance of the conditional distribution of demand given the market information gets smaller over time. The time interval between the time of first order and the selling season is divided into subperiods and the retailer has an opportunity to place an order in each of these subperiods. Kim (2003), formulating a dynamic programming model, shows that the profit-to-go function in each subperiod is concave in the current inventory level and demand indicator.

Wang et al. (2012) investigate the issue of determining the time and quantity of orders based on a Martingale model of forecast evolution (MMFE) approach, in which demand forecasts in successive periods are linked by random adjustment terms. The authors consider two different forecast updating methods. In the additive model, the demand forecast in a period is updated by adding an adjustment term to the forecast in the previous period. In the multiplicative model, the ratios of successive forecasts are an exponential function of the adjustment term. The adjustment terms are independent and normally distributed. The forecast in the last period is considered as the realization of market demand in the selling season. There is no salvage value or shortage penalty. The authors study both the single-order problem and the multiple-order problem. They show that a state-dependent base stock policy is optimal in the multiple-order problem in which the retailer is not constrained to placing a single order.

7.3.2 Price-sensitive demand

While selling price is generally considered to be an exogenous factor in the literature on intelligent fast fashion procurement systems, it is also possible to treat price as a decision variable when customer demand changes in response to a change in selling price. Choi (2007) develops a two-stage model in which initial order quantity is decided based on a tentatively selected selling price. Following collecting information on price sensitivity of customers, the retailer determines the final stocking level and selling price. Serel (2009), given a number of possible realizations for the random purchase cost at time 2, investigates the problem of selecting the optimal price and stocking quantity associated with each purchase cost value. Chen et al. (2010) look into the issue of coordinating a manufacturer–retailer supply chain when a linear additive price-dependent demand model is applicable.

7.3.3 Purchase cost uncertainty

In some situations, purchase cost at time 2 may not be known with certainty at time 1 when the initial order is placed. Gurnani and Tang (1999) assume that purchase cost at time 2 is a random variable with a discrete probability distribution. The same assumption is also made in Choi et al. (2003) and Serel (2009). Thus the optimal ordering decision at time 2 depends on both the demand signal and observed purchase cost. In general, the initial order may have a positive optimal value only when the unit purchase cost at time 1 is less than the *expected* unit purchase cost at time 2.

Song et al. (2014) explore a multiproduct problem with uncertain purchase prices for products at stage 2. The initial order at time 1 is subject to resource constraints such as budget, capacity, or space. However, any amount can be ordered at time 2. Song et al. (2014) propose a solution algorithm for this problem.

7.3.4 Limited supply at time 2

Zheng et al. (2015) solve the two-order problem by adding a constraint on the size of the second order. Assuming that the amount that can be ordered by the retailer at time 2 is limited to a predetermined upper bound, Zheng et al. (2015) show that order quantity at time 1 decreases with this bound, and derive expressions for the order quantity when conditional distribution of demand at time 2 given demand forecast at time 1 is normal or lognormal. Miltenburg and Pong (2007b) study a multiproduct problem with limited supply capacity at both time 1 and time 2.

7.3.5 Order cancellation flexibility

If the new market information indicates a limited sales potential for the product, the retailer may prefer to cancel a part of the first order at time 2. In some models, this cancellation flexibility is available to the retailer, and the impact of this option on the optimal ordering policy is explored. Huang et al. (2005) derive the optimal policy when the retailer incurs both fixed and variable costs to adjust the initial order size at time 2. Serel (2009) presents similar results for the case of stochastic purchase cost at time 2. DeYong and Cattani (2012) study a problem in which there are two possible demand distributions for the product, pointing to low and high demand levels in the season. The demand distribution that will be observed in the season is learned at time 2. Given the possibility of modifying the initial order at time 2, DeYong and Cattani (2012) determine the optimal ordering policy at time 1 and time 2.

Sethi et al. (2007) investigate the impact of order cancellation flexibility in a setting where there exists a customer service level constraint. They find that the effect of service level constraint on optimal total order quantity and optimal expected profit is similar to that in the problem without order cancellation flexibility (see Section 7.3.14).

Ma et al. (2012) focus on a retailer having a loss-averse objective function, and study the scenario with the possibility of order cancellation at time 2 in their paper. They find that the retailer's order quantity at time 1 increases if order cancellation is allowed.

7.3.6 Purchase budget constraint

Serel (2012) looks into the problem of optimally allocating a given monetary budget between orders at time 1 and at time 2. Comparing the optimal initial orders in the budget-constrained and budget-unconstrained problems, Serel (2012) shows that, provided the budget exceeds a threshold value, the optimal initial order quantity in the constrained problem will always be greater than or equal to that in the unconstrained problem.

7.3.7 Multiproduct problem

The models discussed in Section 7.2.5 explore how to schedule production of multiple products when there are inventory holding costs associated with producing products before the selling season. Assuming that carrying products ordered at time 1 in inventory during the preseason time interval does not cause any additional inventory holding costs, Serel (2012) extends the single-product model of Choi et al. (2003) to the case of multiple products with correlated demands, and also explores the ordering policy under a budget constraint. Miltenburg and Pong (2007a,b) present a two-stage multiproduct model with forecast updating in which order quantities for products are limited by production capacities in both the first and second stages. Song et al. (2014) study a multiproduct problem with random purchase costs and multiple resource constraints.

7.3.8 Coordinating the supply chain

Donohue (2000) looks into the issue of coordinating a supply chain where the manufacturer decides the quantities produced using the two production modes. The first production mode is slow but costs less, and the second production mode is fast but expensive. The manufacturer charges different wholesale prices for the retailer's initial order and the second order that relies on improved demand forecast. It is also assumed that the retailer can return unsold items back to the manufacturer at the end of season at a prespecified buyback price. Thus the retailer's decisions are the order quantities at time 1 and at time 2. Donohue (2000) finds efficient values of the wholesale prices and the buyback price that coordinate this supply chain.

Chen et al. (2006) consider the setting where the manufacturer decides the production quantity at time 1, and the retailer decides its order quantity at time 2 after demand forecast revision. Thus the retailer cannot order more than the production quantity determined by the manufacturer at time 1. If the retailer orders less than the production quantity, the manufacturer's loss due to overproduction is partially compensated by the retailer. As in Donohue (2000), the retailer can also return unsold items back to the manufacturer at the end of season at a prespecified buyback price. Chen et al. (2006) find values of the wholesale price, the proportion of overproduction loss shared by the retailer, and the buyback price that coordinate this supply chain.

Chen et al. (2010) extend the setting in Chen et al. (2006) to the case of price-dependent demand. A linear additive demand function is used to represent the relationship between demand and selling price. A parameter of this linear demand function, which is treated as random at time 1, is observed at time 2 so that the demand forecast is improved. Based on improved forecast, the retailer has to decide both the order quantity and selling price at time 2. Chen et al. (2010) show that a three-parameter risk and profit sharing contract coordinates this supply chain.

Yang et al. (2011) explore coordination of a two complementary suppliers, one retailer supply chain in which each supplier supplies a particular component required in the assembly of the finished product. The lead times of components are different, and only one order can be placed for each component. An improved demand forecast is available when the component with shorter lead time is ordered. Yang et al. (2011) propose a buyback contract with cancellation penalty to coordinate this assembly system.

7.3.9 Loss-averse retailer's problem

Instead of the risk-neutral decision maker case usually assumed in the literature, Ma et al. (2012) solve the problem of a loss-averse retailer. In this problem, the retailer has a reference target profit. If the realized profit at the end of season is above this target, the retailer's utility function is the same as the risk-neutral utility function. However, if the realized profit is under the target, the retailer's utility is assumed to decrease by an amount proportional to the difference between the realized and target profits. Ma et al. (2012) show that loss-averse utility function is concave in the initial order quantity, and derive the first-order condition for this order quantity. They observe in their numerical experiment that the order quantity at time 1 decreases as the penalty coefficient for not reaching the target profit increases.

7.3.10 Order policy for product components

When a product is assembled from different components which have different supply lead times, demand expectations for the product may be different when the orders for the components are issued at different times. Basically, the components that have longer lead times need to be ordered based on less accurate demand information. Yang et al. (2011) study such a two-component assembly system and find that the system can be coordinated by allowing the retailer to return leftover inventory to the supplier, and charging the retailer a linear cancellation penalty if a part of the order placed for the component with longer lead time is canceled.

Thomas et al. (2009) explore the interaction between an original equipment manufacturer (OEM) and a contract manufacturer (CM) supplying two complementary components with different lead times to the OEM. The components are assembled by the OEM to produce the finished product. The OEM provides the first coarse demand forecast to the CM when the CM starts producing the long lead time component. If the OEM provides the improved second demand forecast to the CM, the CM can use this forecast to determine the production quantity for the short lead time component. The OEM places orders for both components after the demand for the product is realized; the deliveries of components are limited by the production decisions made earlier by the CM. Thomas et al. (2009) study when it is beneficial to the OEM to share the updated demand forecast information with the CM, and to partially share the CM's overproduction cost.

Zhang et al. (2013) investigate a multiproduct system in which it is possible to customize a common component according to each product. There are two alternative strategies for producing the products. In the first alternative, product-specific components are used to produce all products at time 1. In the second alternative, referred to as the quick response system, the common component produced at time 1 is differentiated for each product at time 2. While the unit product cost in the quick response system is higher than that in the first alternative, stocking quantities for products are determined based on more accurate demand information in the quick response system. Zhang et al. (2013) find that the quick response system yields higher profits under higher demand uncertainty, holding cost, and shortage penalty cost.

7.3.11 Order policy with consideration of carbon footprint

The undesirability of carbon emissions due to their negative effect on the environment has caused the carbon footprint of transportation activities to be explicitly considered in some inventory optimization models in the literature. Choi (2013a) studies a dual-sourcing problem with offshore and domestic suppliers. Although the offshore supplier's unit production cost is lower, the carbon footprint tax charged on the transportation from the offshore supplier is higher because the fashion products sourced from the offshore supplier have to be transported over a longer distance. Given that an improved forecast is used for ordering from the domestic supplier, Choi (2013a) derives the optimal mix of offshore and domestic orders when a tax on carbon footprint is included in the model.

Choi (2013b) incorporates the carbon emission tax scheme into the problem of selecting the best supply delivery mode among a set of delivery modes with different lead times (shipping distances), which is investigated in Choi et al. (2004). The shorter lead time supplier is more expensive but also results in a smaller carbon tax. The optimal supplier is found by solving a dynamic programming model. Choi (2013b) explores the effect of linear and quadratic carbon taxation schemes on the retailer's profit margin and the optimal supplier choice, and also finds that the existence of a local supplier that accepts returns from the retailer with a partial refund at the end of season increases the chance that this local supplier is selected as the most preferred supplier.

7.3.12 Demand learning by precommitted orders

Extending a model proposed by Weng and Parlar (1999) which does not involve forecast updating, Tang et al. (2004) analyze a problem in which a retailer offers a discounted selling price to its customers if they commit to purchase the product before the selling season starts. Customers may decide to wait to purchase the product at regular price during the selling season. It is also assumed that there is another competing brand of which customers may switch to the brand with discounted price. This brand switching rate depends on the discount rate applied on the selling price. The demands for brands are distributed as bivariate normal. Given that total mean demand for the two brands is constant, by using the precommitted orders information, the retailer can improve the demand forecast for the selling season. The retailer's problem is to decide the optimal discount percentage on selling price and, after collecting the precommitted orders, the stocking quantity for the season. McCardle et al. (2004) extend the single-firm model of Tang et al. (2004) to the case of two competing retailers of which demands are distributed as bivariate normal. They find that if it is optimal for one firm to offer a preorder option, there is a unique equilibrium in which both firms offer preorder option to customers.

Zhao and Stecke (2010) study the effect of precommitted orders on the profit of a retailer when the customers have uncertain valuations for the product. Depending on the selling season profit margin for the product and the customers' valuations, they identify when each of the following three strategies is optimal for the retailer: no preorder option offered, moderate discount for preorders, and deep discount for preorders. In a companion paper, Prasad et al. (2011) consider risk-averse customers who may even buy the product at a higher price in the preorder period in order to avoid the risk of stockout in the regular season. Customers are classified according to whether they are informed or uninformed of the advance purchase offering. Both customer segments are assumed to have the same valuation probability density function.

Li and Zhang (2013) study a similar problem in which there exist heterogeneous customers who have different but certain valuations for the product. The customers behave strategically when deciding to place a preorder; they take into account the possibility of a lower price in the regular selling season and also the risk of unavailability of the product in the regular season. Li and Zhang (2013) find that demand information obtained via preorders may actually decrease the retailer's profit because it enables the retailer to improve the availability of product, which reduces the retailer's ability to charge a high preorder price to strategic customers. They also find that even offering a price guarantee, which compensates early purchasers in case of a later price decrease, may not prevent the decrease in the retailer's profit completely.

Li et al. (2014) develop a model in which customers have uncertain valuations for the product, and preorder customers can return the product to the retailer at a refund. They derive the retailer's optimal pricing and refund rate decisions in the three scenarios: no preorder option offered, full refund for returns, and partial refund for returns.

7.3.13 Two-period problem with exogenous market information update

Cheaitou et al. (2014) study a two-period problem in which the buyer places orders for each period before the start of the first period. At the end of the first period, the buyer can return part of the available inventory to the supplier, and the demand forecast for the second period is updated not based on observed sales in the first period but based on exogenous market information. It is also possible to order an additional amount at the beginning of the second period. Assuming unmet demand in the first period is backordered, Cheaitou et al. (2014) show that optimal ordering policy is characterized by order-up-to and salvage-up-to levels.

7.3.14 Optimal policy under a service level constraint

Sethi et al. (2007) extend the basic model to the case where total order quantity is required to meet a specified service level constraint, which is stated as, for each demand signal observed, the probability that total order quantity exceeds demand in the season must be greater than a specified value. Sethi et al. (2007) show that optimal total order quantity is nondecreasing and the optimal expected profit is nonincreasing with respect to the target service level for each observed demand signal. They also show that the buyback contract studied by Donohue (2000) also coordinates their service-constrained supply chain, thus extending the service-unconstrained setting of Donohue (2000) to the case with a constraint on service level.

7.3.15 Second order based on perfect demand information

Placing the second purchase order after observing the demand fully can be regarded as a special case of the two-order problem with demand forecast update. Given that the unit cost of the second order is higher than that of the first order, the retailer now needs to balance this difference in purchase cost with the risk of leftover inventory created by the first order placed before observing the random demand. In this newsvendor problem with emergency order, if the second-order size is not constrained by a capacity limit, the shortage risk is completely eliminated. Extensions of this problem, such as the case of price-dependent demand (Agrawal and Seshadri, 2000; Serel, 2015), strategic customers (Yang et al., 2015), capacitated reactive production (Cattani et al., 2008; Chung et al., 2008), and carbon footprint constraint (Arikan and Jammernegg, 2014) have been studied in the literature.

7.3.16 Use of demand signals in a multiperiod problem with service level constraint

Bensoussan et al. (2011) explore a multiperiod problem in which the retailer aims to achieve a specific service target. In their analysis, they consider two types of service targets separately: (1) the long-term in-stock probability must exceed a specified minimum value, or (2) the long-term fill-rate target. The retailer commits to a minimum order per period, and also a supplemental order is possible in each period based on observed demand signal in that period. The dependence of demand signals across periods is reflected by a positive recurrent Markov chain. No inventory is carried from period to period. As solving the problem with multiple ordering opportunities within the planning cycle is complicated, they formulate a two-stage single-period problem, and show that solution of this problem is asymptotically optimal for the original problem.

7.4 Conclusion

In this chapter, we have discussed how production/stocking decisions in the fast fashion apparel industry can be improved by utilizing up-to-date demand information in intelligent procurement systems. Responding quickly to customer information and using more accurate demand forecasts in determining seasonal stock levels can significantly reduce the overstocking and understocking costs. We have classified the inventory models with forecast update based on whether the new demand information for a fashion product is obtained by observing initial sales of the product in the selling season, or by collecting other forms of market information that provide useful signals regarding the demand for the product. Distinguishing features of a variety of analytical models have been described.

We remark that, due to our focus, we have not discussed some other analytical models related to the fast fashion retail industry. For example, some researchers have studied competition between retailers that have quick response replenishment capability and the retailers that do not have this capability (Caro and Martinez-de-Albeniz, 2010; Lin and Parlakturk, 2012; Wang et al., 2014; Wu and Zhang, 2014). Cachon and Swinney (2011) have studied the joint effect of "trendy" product design and quick response manufacturing on the profitability of a retailer serving strategic customers. Models that allow multiple price changes in the selling season have been described in a number of papers (eg, Sen and Zhang, 2009; Caro and Gallien, 2012; Talebian et al., 2014).

While many different aspects of fast fashion supply chains in the apparel industry have been investigated in the literature, future research may build upon the existing models to further explore fast fashion production and retailing systems in new directions. For example, some models can be extended to multiretailer distribution systems by considering logistics and transportation costs. Another important factor that may be incorporated into the previous analytical formulations is the effect of environmental taxes and regulations and carbon footprint. Sustainable product design and corporate social responsibility are the issues attracting growing attention in the business world recently. Analyzing inventory and product design decisions in fast fashion supply chains in light of these issues may yield interesting findings and insights for practitioners and policy makers. Development of intelligent procurement models for the fast fashion industry is expected to continue as the retailers adopting quick response manufacturing ideas and systems succeed in the marketplace.

References

Agrawal, V., Seshadri, S., 2000. Impact of uncertainty and risk aversion on price and order quantity in the newsvendor problem. Manufacturing & Service Operations Management 2, 410–423.

Arikan, E., Jammernegg, W., 2014. The single period inventory model under dual sourcing and product carbon footprint constraint. International Journal of Production Economics 157, 15–23.

Bensoussan, A., Feng, Q., Sethi, S.P., 2011. Achieving a long-term service target with periodic demand signals: a newsvendor framework. Manufacturing & Service Operations Management 13, 73–88.

Bitran, G.R., Haas, E.A., Matsuo, H., 1986. Production planning of style goods with high setup costs and forecast revisions. Operations Research 34, 226–236.

Bradford, J.W., Sugrue, P.K., 1990. A Bayesian approach to the two-period style-goods inventory problem with single replenishment and heterogeneous Poisson demands. Journal of the Operational Research Society 41, 211–218.

Cachon, G.P., Swinney, R., 2011. The value of fast fashion: quick response, enhanced design, and strategic consumer behavior. Management Science 57, 778–795.

Caro, F., Gallien, J., 2012. Clearance pricing optimization for a fast-fashion retailer. Operations Research 60, 1404–1422.

Caro, F., Martinez-de-Albeniz, V., 2010. The impact of quick response in inventory-based competition. Manufacturing & Service Operations Management 12, 409–429.

Cattani, K.D., Dahan, E., Schmidt, G.M., 2008. Tailored capacity: speculative and reactive fabrication of fashion goods. International Journal of Production Economics 114, 416–430.

Cheaitou, A., van Delft, C., Jemai, Z., Dallery, Y., 2014. Optimal policy structure characterization for a two-period dual-sourcing inventory control model with forecast updating. International Journal of Production Economics 157, 238–249.

Chen, H., Chen, J., Chen, Y., 2006. A coordination mechanism for a supply chain with demand information updating. International Journal of Production Economics 103, 347–361.

Chen, H., Chen, Y.F., Chiu, C.-H., Choi, T.-M., Sethi, S., 2010. Coordination mechanism for the supply chain with lead-time consideration and price-dependent demand. European Journal of Operational Research 203, 70–80.

Choi, T.M., 2007. Pre-season stocking and pricing decisions for fashion retailers with multiple information updating. International Journal of Production Economics 106, 146–170.

Choi, T.M., 2013a. Local sourcing and fashion quick response system: the impacts of carbon footprint tax. Transportation Research Part E: Logistics and Transportation Review 55, 43–54.

Choi, T.M., 2013b. Optimal apparel supplier selection with forecast updates under carbon emission taxation scheme. Computers & Operations Research 40, 2646–2655.

Choi, T.M., Li, D., Yan, H., 2003. Optimal two-stage ordering policy with Bayesian information updating. Journal of the Operational Research Society 54, 846–859.

Choi, T.M., Li, D., Yan, H., 2004. Optimal single ordering policy with multiple delivery modes and Bayesian information updates. Computers and Operations Research 31, 1965–1984.

Chung, C.S., Flynn, J., Kirca, O., 2008. A multi-item newsvendor problem with preseason production and capacitated reactive production. European Journal of Operational Research 188, 775–792.

DeYong, G.D., Cattani, K.D., 2012. Well adjusted: using expediting and cancelation to manage store replenishment inventory for a seasonal good. European Journal of Operational Research 220, 93–105.

Donohue, K.L., 2000. Efficient supply contracts for fashion goods with forecast updating and two production modes. Management Science 46, 1397–1411.

Eppen, G.D., Iyer, A.V., 1997. Backup agreements in fashion buying – the value of upstream flexibility. Management Science 43, 1469–1484.

Fisher, M., Rajaram, K., Raman, A., 2001. Optimizing inventory replenishment of retail fashion products. Manufacturing & Service Operations Management 3, 230–241.

Fisher, M., Raman, A., 1996. Reducing the cost of demand uncertainty through accurate response to early sales. Operations Research 44, 87–99.

Gurnani, H., Tang, C.S., 1999. Optimal ordering decisions with uncertain cost and demand forecast updating. Management Science 45, 1456–1462.

Huang, H., Sethi, S.P., Yan, H., 2005. Purchase contract management with demand forecast updates. IIE Transactions 37, 775–785.

Jin, B., 2004. Achieving an optimal global versus domestic sourcing balance under demand uncertainty. International Journal of Operations & Production Management 24, 1292–1305.

Kim, H.S., 2003. A Bayesian analysis on the effect of multiple supply options in a quick response environment. Naval Research Logistics 50, 937–952.
Lau, A.H.L., Lau, H.S., 1998. Decision models for single-period products with two ordering opportunities. International Journal of Production Economics 55, 57–70.
Lau, H.S., Lau, A.H.L., 1997. Reordering strategies for a newsboy-type product. European Journal of Operational Research 103, 557–572.
Li, C., Zhang, F., 2013. Advance demand information, price discrimination, and preorder strategies. Manufacturing & Service Operations Management 15, 57–71.
Li, J., Chand, S., Dada, M., Mehta, S., 2009. Managing inventory over a short season: models with two procurement opportunities. Manufacturing & Service Operations Management 11, 174–184.
Li, Y., Xu, L., Choi, T.M., Govindan, K., 2014. Optimal advance-selling strategy for fashionable products with opportunistic consumers returns. IEEE Transactions on Systems, Man, and Cybernetics: Systems 44, 938–952.
Lin, Y.-T., Parlakturk, A., 2012. Quick response under competition. Production and Operations Management 21, 518–533.
Ma, L., Zhao, Y., Xue, W., Cheng, T.C.E., Yan, H., 2012. Loss-averse newsvendor model with two ordering opportunities and market information updating. International Journal of Production Economics 140, 912–921.
Matsuo, H., 1990. A stochastic sequencing problem for style goods with forecast revisions and hierarchical structure. Management Science 36, 332–347.
McCardle, K., Rajaram, K., Tang, C.S., 2004. Advance booking discount programs under retail competition. Management Science 50, 701–708.
Milner, J.M., Kouvelis, P., 2002. On the complementary value of accurate demand information and production and supplier flexibility. Manufacturing & Service Operations Management 4, 99–113.
Milner, J.M., Kouvelis, P., 2005. Order quantity and timing flexibility in supply chains: the role of demand characteristics. Management Science 51, 970–985.
Milner, J.M., Rosenblatt, M.J., 2002. Flexible supply contracts for short life-cycle goods: the buyer's perspective. Naval Research Logistics 49, 25–45.
Miltenburg, J., Pong, H.C., 2007a. Order quantities for style goods with two order opportunities and Bayesian updating of demand. Part I: no capacity constraints. International Journal of Production Research 45, 1643–1663.
Miltenburg, J., Pong, H.C., 2007b. Order quantities for style goods with two order opportunities and Bayesian updating of demand. Part II: capacity constraints. International Journal of Production Research 45, 1707–1723.
Petruzzi, N.C., Dada, M., 2001. Information and inventory recourse for a two-market, price-setting retailer. Manufacturing & Service Operations Management 3, 242–263.
Prasad, A., Stecke, K.E., Zhao, X., 2011. Advance selling by a newsvendor retailer. Production and Operations Management 20, 129–142.
Raman, A., Kim, B., 2002. Quantifying the impact of inventory holding cost and reactive capacity on an apparel manufacturer's profitability. Production and Operations Management 11, 358–373.
Seifert, M., Siemsen, E., Hadida, A.L., Eisingerich, A., 2015. Effective judgmental forecasting in the context of fashion products. Journal of Operations Management 36, 33–45.
Sen, A., Zhang, A.X., 2009. Style goods pricing with demand learning. European Journal of Operational Research 196, 1058–1075.
Serel, D.A., 2009. Optimal ordering and pricing in a quick response system. International Journal of Production Economics 121, 700–714.

Serel, D.A., 2012. Multi-item quick response system with budget constraint. International Journal of Production Economics 137, 235–249.

Serel, D.A., 2014. Flexible procurement models for fast fashion retailers. In: Choi, T.M. (Ed.), Fast Fashion Systems – Theories and Applications. CRC Press: Taylor & Francis, Leiden, pp. 59–75.

Serel, D.A., 2015. Production and pricing policies in dual sourcing supply chains. Transportation Research Part E: Logistics and Transportation Review 76, 1–12.

Sethi, S.P., Yan, H., Zhang, H., Zhou, J., 2007. A supply chain with a service requirement for each market signal. Production and Operations Management 16, 322–342.

Song, H.-M., Yang, H., Bensoussan, A., Zhang, D., 2014. Optimal decision making in multi-product dual sourcing procurement with demand forecast updating. Computers and Operations Research 41, 299–308.

Subrahmanyan, S., Shoemaker, R., 1996. Developing optimal pricing and inventory policies for retailers who face uncertain demand. Journal of Retailing 72, 7–30.

Talebian, M., Boland, N., Savelsbergh, M., 2014. Pricing to accelerate demand learning in dynamic assortment planning for perishable products. European Journal of Operational Research 237, 555–565.

Tang, C.S., Rajaram, K., Alptekinoglu, A., Ou, J., 2004. The benefits of advance booking discount programs: model and analysis. Management Science 50, 465–478.

Thomas, D.J., Warsing, D.P., Zhang, X., 2009. Forecast updating and supplier coordination for complementary component purchases. Production and Operations Management 18, 167–184.

Wang, T., Atasu, A., Kurtulus, M., 2012. A multiordering newsvendor model with dynamic forecast evolution. Manufacturing & Service Operations Management 14, 472–484.

Wang, T., Thomas, D.J., Rudi, N., 2014. The effect of competition on the efficient-responsive choice. Production and Operations Management 23, 829–846.

Weng, Z.K., Parlar, M., 1999. Integrating early sales with production decisions: analysis and insights. IIE Transactions 31, 1051–1060.

Wu, X., Zhang, F., 2014. Home or overseas? An analysis of sourcing strategies under competition. Management Science 60, 1223–1240.

Yan, H., Liu, K., Hsu, A., 2003. Optimal ordering in a dual-supplier system with demand forecast updates. Production and Operations Management 12, 30–45.

Yang, D., Choi, T.M., Xiao, T., Cheng, T.C.E., 2011. Coordinating a two-supplier and one-retailer supply chain with forecast updating. Automatica 47, 1317–1329.

Yang, D., Qi, E., Li, Y., 2015. Quick response and supply chain structure with strategic consumers. Omega 52, 1–14.

Zhang, J., Shou, B., Chen, J., 2013. Postponed product differentiation with demand information update. International Journal of Production Economics 141, 529–540.

Zhao, X., Stecke, K.E., 2010. Pre-orders for new to-be-released products considering consumer loss aversion. Production and Operations Management 19, 198–215.

Zheng, M., Shu, Y., Wu, K., 2015. On optimal emergency orders with updated demand forecast and limited supply. International Journal of Production Research 53, 3692–3719.

Zhou, Y.W., Wang, S.D., 2012. Supply chain coordination for newsvendor-type products with two ordering opportunities and demand information update. Journal of the Operational Research Society 63, 1655–1678.

Intelligent demand forecasting systems for fast fashion

Brahmadeep, S. Thomassey
University Lille Nord of France, ENSAIT-GEMTEX, 2 allée Louise et Victor Champier, Roubaix, France

8.1 Introduction

For most industries, sales forecasting is a key factor for success in the supply chain and inventory management. It is especially true in the textile/apparel industry where the lead time is very long compared with the lifespan of the products. However, many other constraints make the sales forecasting very complex in this competitive market, such as seasonal sales that are very sensitive to the weather, the very volatile consumer demand, the new collections that provide new products without any historical data, the huge variety of products, and the many exogenous factors that disturb the sales. For these reasons, sales forecasting of apparel products has led to many works in the literature (Choi et al., 2014). Among the models suitable for the textile/apparel industry, some authors have investigated the fast fashion concept, which raises further issues. Indeed, fast fashion retailing requires more specific sales forecasting systems and, consequently, makes the forecasting challenge more difficult. For instance, the limited lifespan and the lack of historical data lead the researchers to implement advanced models with faster and more robust learning process. The forecasting models proposed in the literature specifically for fast fashion demonstrate promising results in fast fashion conditions compared to forecasting models used as benchmark. However, these models generally deal with a focused issue for a specific aim, such as short-term forecasts (Du et al., 2015) or the relationship between time cost and forecasting error for SKU forecasts (Yu et al., 2011). Thus, it seems interesting to investigate the real enhancement provided by one of these advanced techniques on the whole fast fashion supply process (ie, including the long-term forecast before the beginning of the sales, the short-term forecast during the season of sales, and the strategy of replenishment of the stores). Consequently, as per the literature review, we implement a two-stage forecasting system that carries out long- and short-term forecasts, combined with a replenishment simulation of the stores. The forecast accuracy and also the residual inventory and the total sales are the three performance indicators used to evaluate the quality of the models.

This chapter is organized as follows:

The next section deals with the main features of fashion and fast fashion retail and more particularly the requirements of sales forecasting. It describes the specificities of the fast fashion products that should absolutely be taken into account in the forecasting systems.

Information Systems for the Fashion and Apparel Industry. http://dx.doi.org/10.1016/B978-0-08-100571-2.00008-7

Section 8.3 describes the forecasting methods that exist in the literature to respond to the constraints of the fashion and the fast fashion sales.

Then, Section 8.4 proposes an intelligent system based on sales forecasting and replenishment modules. This system is composed of the long-term forecasting model, a short-term forecasting model, and a store replenishment model. The system is implemented on real data from a French retailer.

The last section concludes with the limitations of this study and suggests further perspectives that should be considered in the near future.

8.2 Fashion and fast fashion sales forecasting

In fashion industry, the consumer demand can be considered to be very volatile (Sen, 2008; Choi, 2007). Indeed, the offer is very large and the consumer need has to be satisfied immediately, or the consumer switches to another brand or store. Thus, companies have to keep a high service level and a very competitive price.

Generally, traditional brands ensure the low price of their products with supplies in low-cost countries that generate a long lead time. The service level is obtained with inventories in warehouses that are often difficult to manage and are expensive.

Thus, to deal with these constraints, companies have to rely on efficient and accurate sales forecasting systems. These systems should be perfectly suited to the requirements of the apparel market. First, it is important to know the product, the sales features, and how the retailer will use the forecasts (Armstrong, 2001). Then, the forecasting method can be wittingly selected.

The most significant features of a sales forecasting system for fashion products have been listed and detailed in Thomassey (2014). To be specific, the sales forecasting system has to take into account these features:

- Forecast horizon. Generally, the supply strategy of distributors or retailers is based on two steps: a supply in a long-term horizon and replenishment in a short-term horizon (Thomassey, 2010). Thus, two forecast horizons are generally required for fashion products: a long-term horizon (ie, six months to one year) to plan the sourcing and the production, and a short-term horizon (ie, a few weeks) to replenish the warehouse if necessary and to adjust the orders and deliveries of local stores.
- Aggregation level. The short lifespan and the reference changing for each collection require the company to aggregate the data. The main issue is then to select the right level and criteria for the aggregation. Usually, aggregated data (eg, by product family) are used for long-term forecasts, whereas SKU-level data is required for short-term forecasts.
- Seasonality. The sensitivity to the seasons is changing according to the considered apparel products. For instance, a pullover is logically more sensitive than underwear. This sensitivity should be integrated into the forecasting system, especially for long-term forecasts.
- Exogenous factors. Garment sales are impacted by many exogenous factors. These factors could be very difficult to manage since they are sometimes not controlled or even unknown. Indeed, some variables are not available (eg, competitor data) or predictable (eg, weather

data) and thus cannot be integrated into the forecasting system. Their impact also fluctuates in time and magnitude; hence it is impossible to exactly identify and quantify them (De Toni and Meneghetti, 2000). However, explanatory variables are essential to model the clothing sales, and if possible the most relevant ones should be integrated into the computation of the forecast (Thomassey, 2010).

In the competitive fashion market, the fast fashion concept emerges as a new success strategy for some companies. Indeed, fast fashion consists in offering in each season a larger number of articles produced in smaller series and continuously changing the assortment of products displayed in the stores (Caro and Gallien, 2010). Consequently, fast fashion companies have to offer a shortened lead time as well as an improved design to attract customers (Cachon and Swinney, 2011). However, this success is reliant on taking into account several novel operational challenges associated with the implementation of fast fashion (Caro and Gallien, 2010).

The most relevant example of a fast fashion company investigated in literature concerns the Zara brand of the Inditex group. This company is able to achieve in 2 weeks the whole cycle from design to sales in the store (Ghemawat and Nueno, 2003). In Caro and Gallien (2007), a comparative study between the supply chain of Zara and the traditional retailers is proposed. It appears, for instance, that the lead time from the design to the final product in the store, which is commonly 6–9 months in the apparel industry, is reduced to 2–5 weeks for Zara, the product variety is about 3–5 times higher for Zara, and the number of mark-down sales is about two times lower for this fast fashion company.

These specific features of the fast fashion concept justify the development of new approaches. Thus, various papers have investigated the new issues raised by the fast fashion strategy, especially from the sales forecasting point of view. These are the main challenges mentioned in these studies related to fast fashion (Choi et al., 2014):

- the high demand uncertainty for fast fashion products because of the ever changing fashion trend and consumer preference,
- the very short lead time to perform the replenishment of stores and, consequently, an even shorter lead time to carry out the sales forecast of a huge number of SKUs,
- the very limited amount of data available for forecasting for a considered SKU, especially because of the short lifespan of products.

Finally, given the aforementioned constraints, the main issue is this: what products should be assorted at each point and when? This issue is much more significant in the fast fashion context since, once again, the selling period is very short, and there is no possibility to make up for a bad assortment.

The volume and date of the replenishment of the store is a crucial task that is often the responsibility of the store manager, who estimates the shipment quantities according to the inventory remaining. Caro and Gallien (2010) have proposed a new process to enhance this practice. This new process is based on a sales forecasting model that relies on past sales data. Considering the enumerated constraints, this sales forecasting technique should be able to perform forecasts in a limited time with limited historical data and to deal with a complex multivariate problem due to the influence of internal and external environments (Kuo and Xue, 1998,1999; Kuo, 2001).

8.3 Sales forecasting methods for fast fashion retailing

In the field of the sales forecasting methods, statistical methods are certainly the most used ones. These statistical techniques include various well-known models that have formal statistical foundations (Chu and Zhang, 2003): exponential smoothing (Brown, 1959), Holt Winters model (Winters, 1960), Box & Jenkins model (Box and Jenkins, 1969), regression models (Papalexopoulos and Hesterberg, 1990), or auto regressive integrated moving average (ARIMA) models.

Despite providing satisfactory results (Kuo and Xue, 1999) in some domains and being popularly used for their simplicity and fast speed, statistical methods suffer a few problems (Liu et al., 2013) when they are implemented for the fashion sector. Indeed, these methods are not suitable for the textile/apparel environment and more generally in any fashion sectors, especially because most time series methods require large historical data sets, a complex optimization of their parameters, and a certain experience of the user, and they are often limited to a linear structure.

Thus, their performances are usually worse when they are compared to more sophisticated methods such as artificial intelligence (AI).

Recently, different reviews on sales forecasting methods for fashion retailing have been proposed in the literature Liu et al. (2013) and Beheshti-Kashi et al. (2015).

In Liu et al. (2013), a study of forecasting techniques over time shows that the implementation of AI methods (pure or hybrid) is growing in the last decade. These methods are able to respond to the main requirements of the fashion sector, mentioned in Section 8.2.

To be specific, fuzzy-based systems are used to deal with long-term forecasts in Thomassey et al. (2002,2005). In Vroman et al. (2001), Wong and Guo (2010), Ni and Fan (2011), ANNs are implemented for the same task. Choi et al. (2011) have proposed a hybrid algorithm that combines the SARIMA method and the wavelet transform method. This hybrid model demonstrates specific capabilities that are particularly required in the fashion industry. Short-term forecasts are carried out with a neuro-fuzzy system in Thomassey et al. (2005). A combination of ART and ANN techniques is also described in Ni and Fan (2011) for two-stage (short- and long-term) forecasting systems.

Some AI models are also developed for the complex issue of sales forecasting of new items without or with a limited amount of historical data. These models are based on clustering and classification techniques using decision trees (Thomassey and Fiordaliso, 2006), ANN (Thomassey and Happiette, 2007), support vendor machine (Teucke et al. 2014), or Grey model combined with ANN (Choi et al., 2012).

In many cases (Ni and Fan, 2011; Wong and Guo, 2010; Sun et al., 2008; Yu et al., 2011; Thomassey et al., 2005; Choi et al., 2011), the best results are obtained with hybrid models, such as by mixing AI with times series techniques or AI techniques together. Indeed, the combination of techniques enables one to overcome the deficiencies of single models and to improve forecasting performance (Khashei and Bijari, 2012).

Finally, more recently, AI techniques have been specifically used for fast fashion forecasting issues. Yesil et al. (2012) have developed a system that combines several methods with a fuzzy model for weekly demand forecasts for a fast fashion apparel company. However, the most used technique in the literature to respond to the fast fashion constraints is obviously the extreme learning machine (ELM).

ELM has demonstrated better learning capabilities, such as processing time and generalization, compared to the ANN with a gradient-based learning algorithm (Zhu et al., 2005). Other benefits are also given to ELM in the literature, especially its ability to avoid many difficulties associated with gradient-based learning methods, such as stopping criteria, learning rate, learning epochs, local minima, and the over-tuned problem (Sun et al., 2008). For these reasons, ELM has been widely used in fashion sales forecasting.

Sun et al. (2008) have proposed a fashion sales forecasting method using ELM. The comparison of monthly data with ANN using two typical back propagation learning algorithms demonstrates the advantages of ELM, especially in terms of training time and generalization performances. Similarly, Wong and Guo (2010) have developed a hybrid model based on ELM for long-term sales forecasting of fashion products. Their system outperforms the models used for comparison, such as ARIMA and Evolutionary Neural Network, on monthly, quarterly, and annual real data.

Du et al. (2015) have implemented an original ANN coupled with a multi-objective evolutionary optimization for a short-term replenishment forecasting problem. If their system is more accurate on real-world sales than ELM models used for comparison, they do not investigate the computing time when a huge number of forecasts has to be performed.

However, if ELM appears as faster and more efficient on limited data than the classical ANN, some drawbacks have been pointed out (Liu et al., 2013). To be specific, in some cases, it is blamed that ELM is unstable or requires more time than expected to achieve a reasonable forecast. However, these problems also arise with other pure statistical and AI methods. Consequently, different enhancements of ELM, often based on hybrid models, have been proposed to provide better forecasting systems to fashion retailers.

For instance, a new ELM has also been successfully implemented in the specific context of fast fashion forecasting (Choi et al., 2014). The proposed hybrid model based on ELM and the Grey model particularly responds to the limited data and time constraints and provides satisfactory forecast results.

As per this literature review, ELM-based models emerge as the most suitable models for fashion sales forecasting, especially to provide short-term forecasts required for the replenishment of the stores of a fast fashion retailer. Indeed, for this issue, the limited amount of data and the huge number of SKUs make ELM particularly suitable for these very specific forecasts. However, as to the best of our knowledge, no investigation has been performed on the implementation of this technique on the whole fast fashion supply process that includes the assortment policy and the two-level forecasting stages: the preseason forecast (long-term forecast) and the ongoing season forecast (short-term forecast). Thus, to quantify the real benefits provided by this method for fast fashion retailing, we propose in the next section to show a simulation of a real case study that includes these features:

- a long-term sales forecasting of several new products (without historical data), carried out with a clustering and classification method introduced in Thomassey and Fiordaliso (2006)
- a short-term sales forecasting based on ELM taking into account the first sales
- a supply of different stores based on the previous forecasts and integrating the assortment strategy described in Caro and Gallien (2010).

8.4 Intelligent system based on sales forecasting and replenishment modules

In fast fashion retailing, the supply of products from warehouse to stores is crucial, especially when the number of references is huge and the network of stores is large. Indeed, the short life span of products does not enable one to deal with a wrong decision. This decision logically depends on the current sales, but it also has to rely on the assortment strategy and the long- and short-term forecasts. Our literature review has demonstrated that many intelligent models exist for each of these issues. However, to accurately evaluate the real benefits of these models, it seems important to combine them to achieve an intelligent, integrated process.

Thus, we propose to simulate a whole supply process from warehouse to stores based on real sales data and the models described earlier.

8.4.1 Long-term forecasting model

In fast fashion retailing, the long-term forecasting (ie, before the beginning of the sales period) is useful for determining the right quantity to purchase, but it could also be used to estimate the first weeks of sales. The main issue is that each product is a new product and consequently has no historical sales. The only way is to take into account the sales of historical products and make relationships between descriptive criteria of these historical products and the new ones.

The aim becomes a classification issue as described, for instance, in Thomassey and Fiordaliso (2006). To deal with this specific forecast, data mining tools have been successfully used, especially the combination of clustering and classification procedures. Thus we decided to implement the system proposed by Thomassey and Fiordaliso (2006), which is based on three stages:

1. A k-means clustering procedure of the historical sales identifies the products with a similar sales curves (called life curves) and extracts the prototypes of sales (mean of the life curve of each cluster).
2. A decision tree is built from the C4.5 algorithm (Quinlan, 1993) and performs the links between prototypes of sales and descriptive criteria of historical items. The optimum number

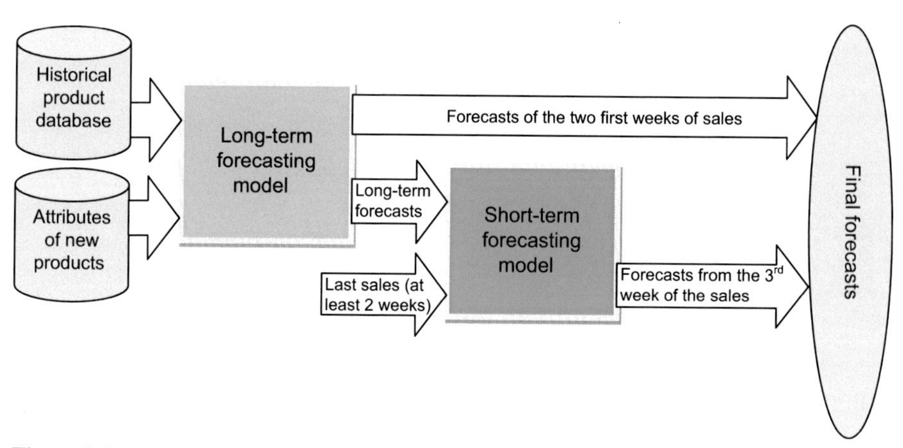

Figure 8.1 The two-stage forecasting model.

of clusters is obtained from a k-fold cross-validation algorithm that minimizes the prediction error of the decision tree.

3. Finally, the classifier, previously trained, assigns each new product to one prototype of sales from its descriptive criteria. This prototype of sales becomes the sales forecast of this new product.

The descriptive criteria used in this simulation are both relevant and available in our database: the price, the starting time of the sales, and the life span of products.

In the proposed system, the long-term forecasts are performed before the sales period. The outputs are used as weekly forecast for the replenishment of stores in the first two weeks of sales and as input of the short-term forecasting model described in next section (Figure 8.1).

8.4.2 Short-term forecasting model

In the proposed system, the short-term forecast is implemented from the third week of sales. The aim is to update the long-term forecast from real sales of the last two weeks. As per the previous section, ELM emerges as a good candidate for this task. Indeed, its learning and generalization capabilities on a limited amount of data in a limited computing time are particularly required in this context.

Thus, we implement an ELM and its learning algorithm proposed by Huang et al. (2012). The inputs of the ELM are (Fig. 8.2) the real sales of the last two weeks and long-term forecasts of the last two weeks and the next week. The activation functions of the hidden neurons are the commonly used sigmoid functions.

To deal with the instability issue of the ELM, we run the ELM n times and define the final forecast as the average of the n obtained outputs. In this work, n is equal to 100. This value is a reasonable compromise between the computation time and the stability of the ELM (Sun et al., 2007; Yu et al., 2011).

To find the best number of hidden neurons, the historical data set is split into a training set and a testing set. On these data, we compute the root mean square error

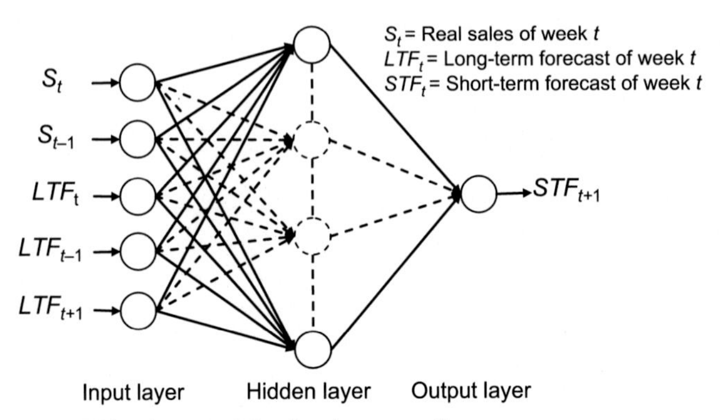

Figure 8.2 The ELM implemented for the short-term forecast.

(RMSE) obtained by ELMs with different number of neurons from 5 to 100 and select the model that provides the minimum RMSE on the testing data.

8.4.3 Store replenishment model

A replenishment model is developed using the discrete event simulation with the Arena Simulation software from Rockwell. Arena Simulation is one of the most popular and powerful simulation software in its domain (Dias et al., 2011). The modules in the software play an important role in modeling automated processes or with an involvement of scheduling, planning, queuing, and waiting. It is integrated with VBA coding, which enhances the possibilities of the simulation model and also could be integrated with other optimization algorithms (Brahmadeep and Thomassey, 2014).

The simulation model is performed on the 12 weeks of the season from the long-term forecasts for the first two weeks and then the short-term forecasts. The model is such that the material flows between five retail stores and one warehouse, and their parameters are designed in the Arena interface, and the interlinked external files are used for the import and export of the data and results, as illustrated in Fig. 8.3.

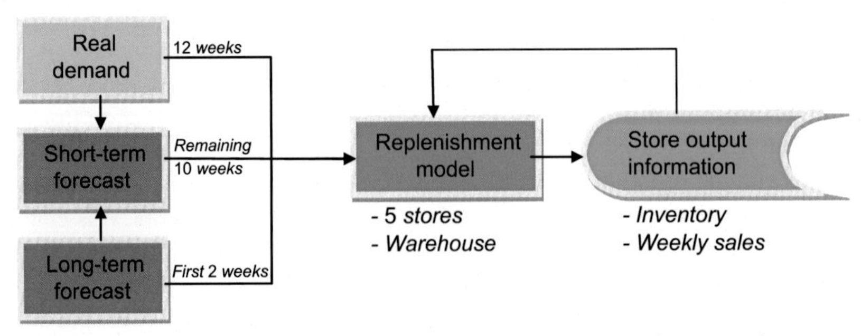

Figure 8.3 Simulation model for replenishment based on long- and short-term forecasts.

8.4.4 Experiment and results

8.4.4.1 Data

The data used for the experimentation of our system are extracted from a real database of a French retailer. These data include the sales and also the descriptive criteria used for the long-term forecast as mentioned in Section 8.4.1. The data used as historical data are composed of 482 products with a sales period of 4 (58%), 6 (28%), and 8 (14%) weeks uniformly distributed during a season of 12 weeks.

For the prediction part, 142 new products are considered with a similar distribution of sales periods: 57% are sold during 4 weeks, 29% during 6 weeks, and 14% during 8 weeks.

8.4.4.2 Models used for comparison

To evaluate the benefits of the two forecasting modules, we simulate the same scenario using a replenishment strategy:

- based only on the long-term forecast, ie, without updating the forecast with the ELM according to the real sales,
- without forecast by completing the inventory each week to reach a constant level in each store. This level is estimated from the average of expected sales. This model is called constant inventory level,
- without forecast by supplying a constant amount of products each week. This supply is estimated from the average of expected sales. This model is called uniform replenishment.

8.4.4.3 Long-term forecasting module

The learning process of the long-term forecasting module composed of a clustering and a classification procedure is based on a tenfold cross-validation process. This process enables the optimization of the number of clusters to reach the minimum forecast error with the classification.

The main shortcoming of this procedure is the computational time involved by the numerous required iterations. However, this process is performed before the start of the season, and consequently, the time constraints are not the same as for the short-term forecast.

For more detail about this process, we recommend reading Thomassey and Fiordaliso (2006).

On our historical database, this process provides an optimal number of clusters of 14. This means that the sales of the 482 products are summarized by 14 prototypes of sales. From these clusters, a decision tree composed of 55 nodes and 28 leaves is built with the C4.5 algorithm to classify the new products from the three descriptive criteria: the price, the starting time of the sales, and the life span of items. The classification stage assigns to a new product a prototype of sales, which becomes the long-term forecast of the product. Examples of long-term sales forecasts are presented in Figs. 8.4 and 8.5.

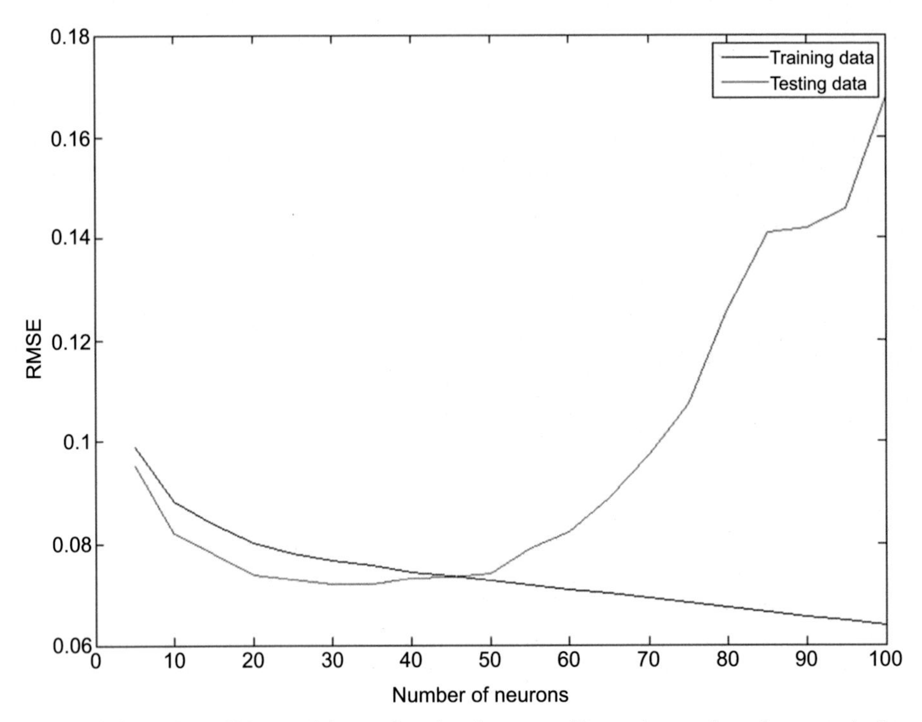

Figure 8.4 ELM RMSE on training and testing data according to the number of neurons in the hidden layer.

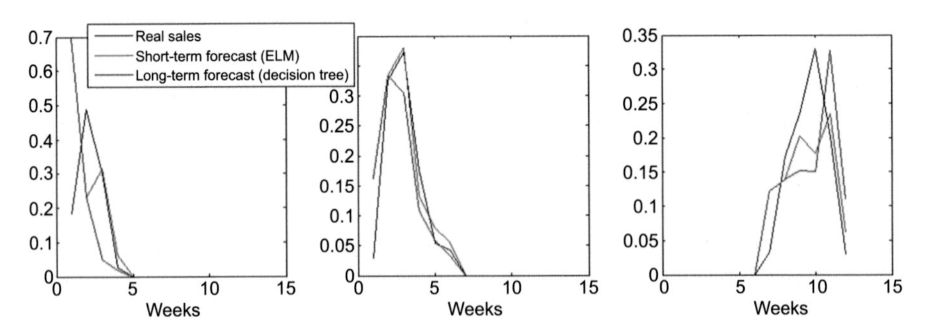

Figure 8.5 Examples of long and short forecasts for products with 4-, 6-, and 8-week periods of sales.

8.4.4.4 Short-term forecasting module

The short-term forecast is performed by the ELM model introduced in Section 8.4.2. As mentioned earlier, the number of hidden neurons has to be optimized on the historical data. Thus, we iteratively train the ELM with an increasing number of hidden neurons from 5 to 100 on a training data set of 362 products and then evaluate it on a testing data set of 120 products. For a given number of hidden neurons, the ELM is run 100 times to enhance the stability of the results. Consequently, this process may be time-consuming. However, it is carried out one time before the beginning of the season, and therefore it does not impact the reactivity of the system during the sales period.

On testing data, the best RMSE is reached with 35 neurons in the hidden layer (Fig. 8.4).

Thus, for each product and store, from the third week of sales an ELM composed of 35 hidden neurons is executed 100 times to provide the final short-term forecast (average of the 100 ELM outputs) of the next week. Examples of short-term sales forecast are illustrated in Fig. 8.5.

Table 8.1 and Fig. 8.6 show that the short-term forecast enables us improve the accuracy of the forecast for more than 80% of the products (114 on 142) and decrease the average RMSE by 13.65%.

Table 8.1 RMSE of long- and short-term forecasts on the 142 new products

	Long-term forecast	Short-term forecast	Improvement
Average RMSE	27.1	23.4	13.65%

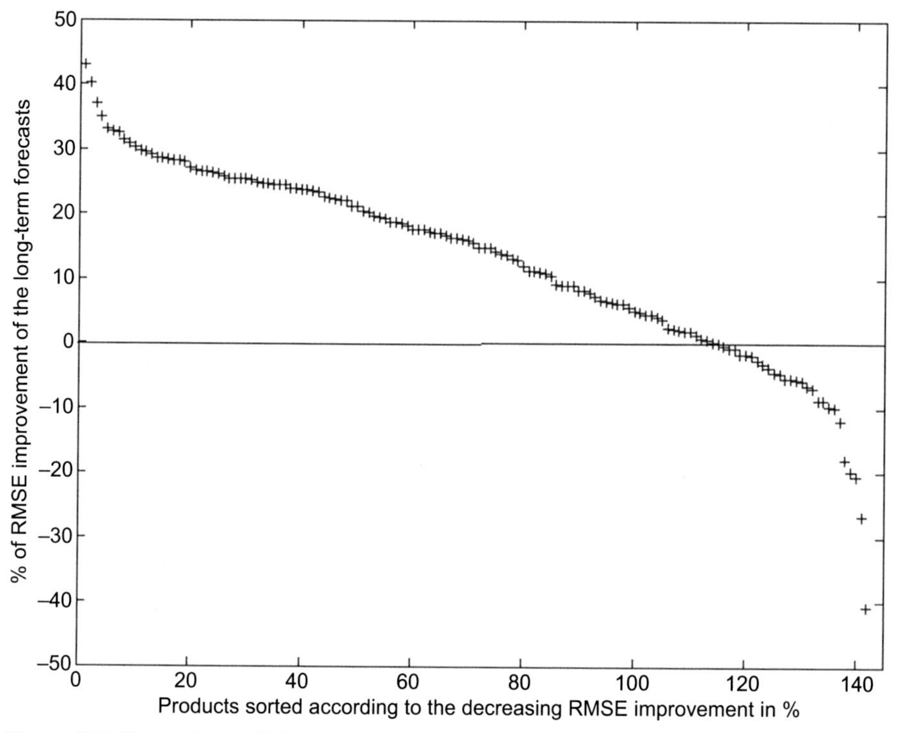

Figure 8.6 Comparison of RMSE between the proposed two-stage forecast and the long-term forecast.

8.4.4.5 Replenishment module

The simulation model is performed on the 12 weeks of the season from the long-term forecasts for the first two weeks and then the short-term forecasts. To simulate the sales demand, we consider the real sales of the 142 new products of the database, and we particularly check the data to ensure that there were no out of stock values. Two indicators are computed for the evaluation of the different simulations: the total sales (satisfied sales) and the residual inventory (in stores and warehouse) at the end of the season. The results obtained with our system and the three used for comparison are illustrated in Fig. 8.7.

It appears that the two forecast-based models, namely the single long-term forecasting model and the two-stage forecasting model (long- and short-term forecast), outperform the other two models, namely the uniform replenishment and the constant

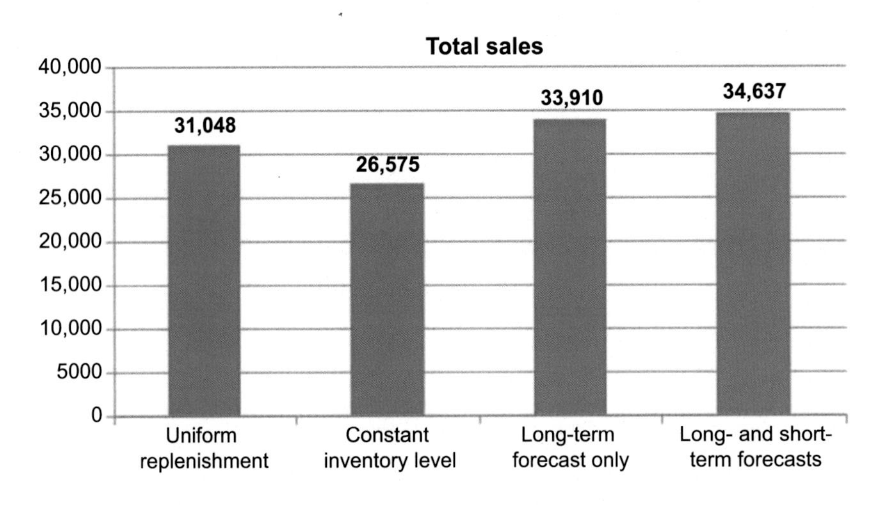

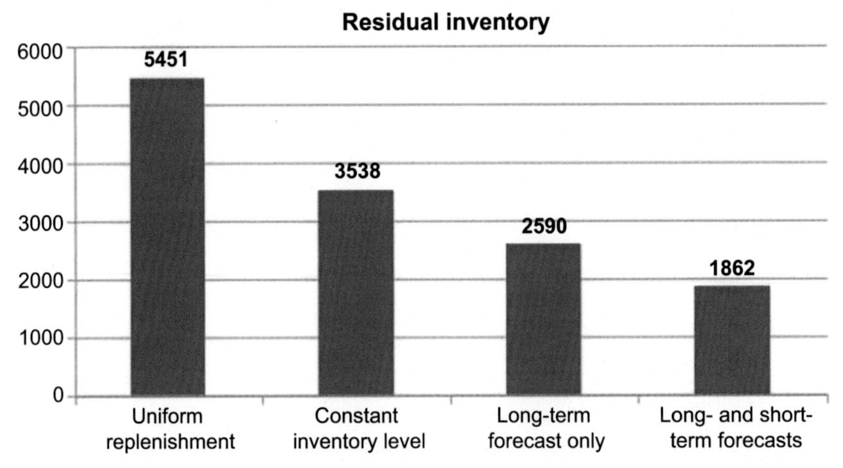

Figure 8.7 Residual inventory and total sales for the four considered models at the end of season.

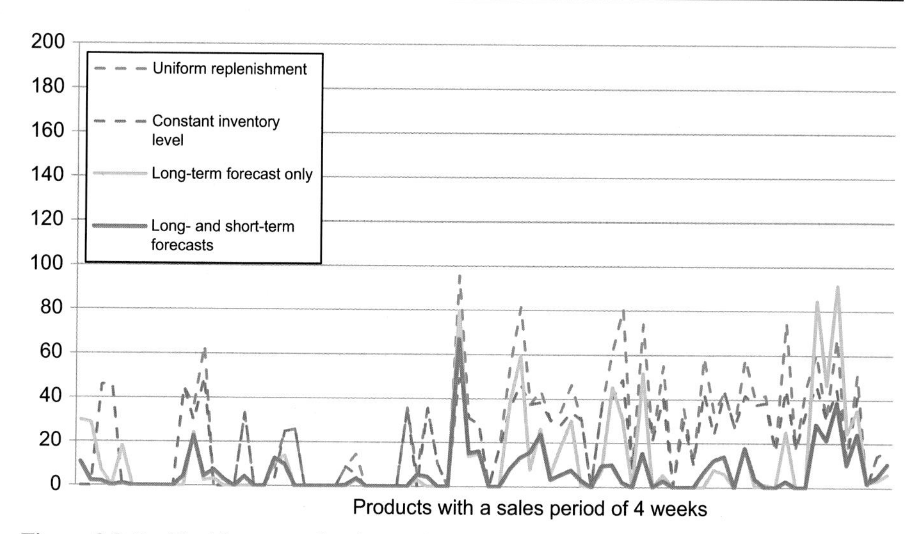

Figure 8.8 Residual inventory for the products with 4 weeks of sales period.

inventory level models (described in Section 8.4.4), on both indicators. If the benefit of the ELM (short-term forecast) compared to the long-term forecast only is relatively low on the total sales (increase of 2%), a real advantage occurs on the residual inventory (reduction of 28%). For the retailer, the profit is multiple:

- fewer end-of-season sales at discounted prices
- larger sales surface available that could be used to sell other references
- fewer purchased products for similar or higher total sales, which means a higher profit margin

Finally, Figs. 8.8 and 8.9 illustrate the residual inventory per product according to the period of sales. This detailed analysis shows that the proposed two-stage

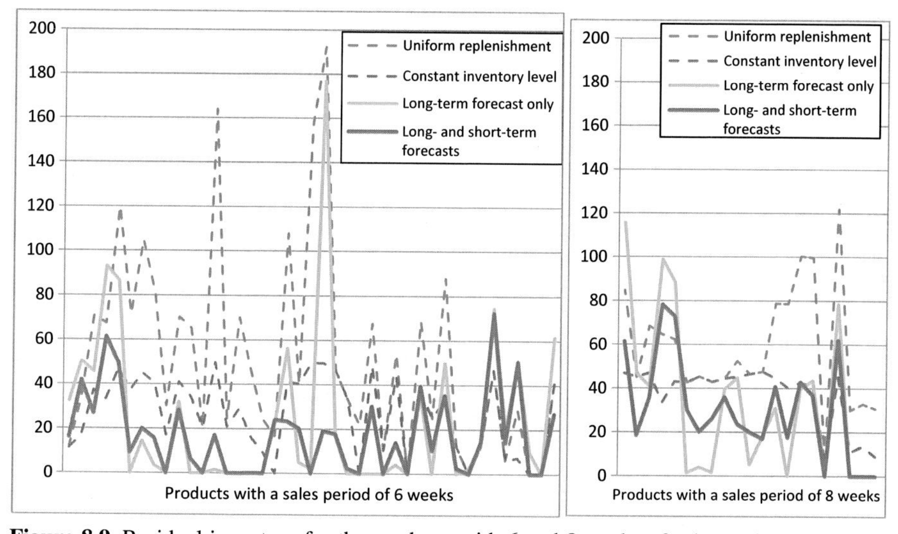

Figure 8.9 Residual inventory for the products with 6 and 8 weeks of sales periods.

forecasting system enables a better inventory management than the other models (single long-term forecasting, uniform replenishment, and constant inventory level models) on almost all of the 482 products regardless of the period of sales (4, 6, or 8 weeks). Furthermore, the comparison between the single long-term and the two-stage (using ELM) forecasting models demonstrates the good performances of the ELM even with a limited number of input data (ie, 2 weeks of real sales) as is commonly the case for fast fashion products.

8.5 Conclusion

Fast fashion requires retailers to develop new strategies to successfully manage the inventory of their stores. Indeed, shortened time between the design and the final product in the store and the limited amount of historical data make the decisions crucial and sometimes uncertain. To enhance the decision process, many models have been proposed in the literature and more especially for sales forecasting. Models based on AI techniques are logically the best candidates, taking into account the specific constraints related to this market. As per the literature review, a two-stage forecasting system combined with an intelligent replenishment model emerges as the more suitable strategy for these reasons:

- A long-term forecast is required to estimate the quantity to purchase and guarantee the first weeks of sales.
- A short-term forecast enables an update of the forecast according to the first sales to quickly react to the demand trend.

The main issue for the long-term sales forecasting model is to perform accurate forecasts without historical sales. In such a situation, a data mining technique such as decision trees or neural networks is often used to establish relationships between the sales and relevant descriptive criteria of products. Thus, the sales forecast of new products can be computed from their descriptive criteria. In this study, we implement the system proposed by Thomassey and Fiordalios (2006) based on clustering and decision trees to achieve the long-term forecast of 142 new products.

For fast fashion products, the short-term forecasts should be both performed in a limited time and on a limited amount of data. These features particularly fit the characteristics of ELM. Indeed, this model has been successfully applied in fast fashion forecasting issues (Choi et al., 2014). Thus, we developed an ELM that updates the long-term forecast from the third week of sales according to the real sales of the last two weeks.

Finally, to really quantify the benefits of these advanced forecasting systems for the retailers, we simulate with a discrete event model the supply of products from the warehouse to the stores during a season of 12 weeks. The replenishment strategy is based on the forecast as recommended in Caro and Gallien (2010).

The simulation demonstrates that the replenishments that rely on suitable forecasts provide both a significant increase of total sales and decrease of residual inventory compared with a common replenishment strategy. The implementation of ELM for short-term forecasts also enables us to substantially reduce the residual inventory by keeping similar total sales to the model based only on long-term forecasts.

However, if this work makes obvious that intelligent sales forecasting systems have a real interest for fast fashion retailers, some limitations or further perspectives should be considered:

- The descriptive criteria used for long-term forecasts are very sensitive. Their availability and reliability are crucial to ensure accurate sales forecasts of new products. Retailers should pay more attention in setting up their information systems to keep and easily extract these data.
- All the results presented in this study concern the products without taking into account the size distribution. This could significantly increase the constraints since the size issue may directly impact the replenishment strategy and the inventory displayed in stores, as mentioned in Caro and Gallien (2010). The integration of size distribution in the forecasts and replenishments should be the aim of future works.
- The replenishment strategies used in the simulation do not integrate the knowledge of the store manager. If the store manager makes decisions that are often considered as too time-consuming and uncertain when the number of references is huge, his expertise could be very efficient in some critical situations. Our discrete event replenishment models do not enable the intervention of the human expert. Therefore, a simulation based on multi-agent systems could be a valuable extension of this work and enables an even more intelligent management of inventories in stores by considering more specifically the environment of the decision.

References

Armstrong, J.S., 2001. Principles of Forecasting – A Handbook for Researchers and Practitioners. Kluwer Academic Publishers, Norwell, MA.

Beheshti-Kashi, S., Karimi, H.,R., Thoben, K.-D., Lütjen, M., Teucke, M., 2015. A survey on retail sales forecasting and prediction in fashion markets. Systems Science & Control Engineering: An Open Access Journal 3 (1), 154–161.

Box, G.E.P., Jenkins, G.M., 1969. Time Series Analysis Forecasting and Control. Prentice Hall, Englewood Cliffs, NJ, USA.

Brahmadeep, Thomassey, S., 2014. A simulation based comparison: manual and automatic distribution setup in a textile yarn rewinding unit of a yarn dyeing factory. Simulation Modelling Practice and Theory 45, 80–90.

Brown, R.G., 1959. Smoothing Forecasting and Prediction of Discrete Time Series. Prentice Hall, Englewood Cliffs.

Cachon, G., Swinney, R., 2011. The value of fast fashion: quick response, enhanced design, and strategic consumer behavior. Management Science 57 (4), 778–795.

Caro, F., Gallien, J., 2007. Dynamic assortment with demand learning for seasonal consumer goods. Management Science 53 (2), 276–292.

Caro, F., Gallien, J., 2010. Inventory management of a fast-fashion retail network. Operations Research 58 (2), 257–273.

Choi, T.M., Yu, Y., Au, K.F., 2011. A hybrid SARIMA wavelet transform method for sales forecasting. Decision Support Systems 51, 130–140.

Choi, T.-M., Hui, C.-L., Ng, S.,F., Yu, Y., 2012. Color trend forecasting of fashionable products with very few historical data. IEEE Transactions on Systems, Man and Cybernetics, Part C 42 (6), 1003–1010.

Choi, T.M., Hui, C.L., Yu, Y., 2014a. Intelligent Fashion Forecasting Systems: Models and Applications. Springer-Verlag Berlin Heidelberg.

Choi, T.M., Hui, C.-L., Liu, N., Ng, S.-F., Yu, Y., 2014b. Fast fashion sales forecasting with limited data and time. Decision Support Systems 59, 84–92.

Choi, T.M., 2007. Pre-season stocking and pricing decisions for fashion retailers with multiple information updating. International Journal of Production Economics 106 (1), 146–170.

Chu, C.W., Peter Zhang, G.Q., 2003. A comparative study of linear and nonlinear models for aggregate retail sales forecasting. International Journal of Production Economics 86, 217–231.

De Toni, A., Meneghetti, A., 2000. The production planning process for a network of firms in the textile-apparel industry. International Journal of Production Economics 65, 17–32.

Dias, L.S., Pereira, G., Vik, P., Oliveira, J.A., 2011. Discrete simulation tools ranking: a commercial software packages comparison based on popularity. In: Proceedings of 9th Annual Industrial Simulation Conference. Industrial Simulation Conference, Venice.

Du, W., Leung, S.Y.S., Kwong, C.K., 2015. A multiobjective optimization-based neural network model for short-term replenishment forecasting in fashion industry. Neurocomputing 151, 342–353.

Ghemawat, P., Nueno, J.L., 2003. ZARA: Fast Fashion. Harvard Business School Case (9-703-497), pp. 1–35.

Huang, G.-B., Zhou, H., Ding, X., Zhang, R., 2012. Extreme learning machine for regression and multiclass classification. IEEE Transactions on Systems, Man, and Cybernetics - Part B: Cybernetics 42 (2), 513–529.

Khashei, M., Bijari, M., 2012. A new class of hybrid models for time series forecasting. Expert Systems with Applications 39, 4344–4357.

Kuo, R.J., Xue, K.C., 1998. A decision support system for sales forecasting through fuzzy neural networks with asymmetric fuzzy weights. Decision Support Systems 24 (2), 105–126.

Kuo, R.J., Xue, K.C., 1999. Fuzzy neural networks with application to sales forecasting. Fuzzy Sets and Systems 108, 123–143.

Kuo, R.J., 2001. A sales forecasting system based on fuzzy neural network with initial weights generated by genetic algorithm. European Journal of Operational Research 129, 496–517.

Liu, N., Ren, S., Choi, T.M., Hui, C.L., Ng, S.F., 2013. Sales forecasting for fashion retailing service industry: a Review. Mathematical Problems in Engineering, 9 pp.

Ni, Y., Fan, F., 2011. A two-stage dynamic sales forecasting model for the fashion retail. Expert Systems with Applications 38 (3), 1529–1536.

Papalexopoulos, A.D., Hesterberg, T.C., 1990. A regression-based approach to short-term system load forecasting. IEEE Transactions Power Systems 5, 1535–1547.

Quinlan, J.R., 1993. C4.5: Programs for Machine Learning. Morgan Kauffman, San Francisco.

Sen, A., 2008. The US fashion industry – a supply chain review. International Journal of Production Economics 114, 571–593.

Sun, Z.L., Au, K.F., Choi, T.M., 2007. A neuro-fuzzy inference system through integration of fuzzy logic and extreme learning Machines. IEEE Transactions on Systems, Man, and Cybernetics, Part B 37 (5), 1321–1331.

Sun, Z.-L., Choi, T.M., Au, K.-F., Yu, Y., 2008. Sales forecasting using extreme learning machine with applications in fashion retailing. Decision Support Systems 46 (1), 411–419.

Teucke, M., Ait-Alla, A., El-Berishy, N., Beheshti-Kashi, S., Lütjen, M., 2014. Forecasting of seasonal apparel products. In: Kotzab, H., Pannek, J., Thoben, K.D. (Eds.), Dynamics in Logistics. Fourth International Conference, LDIC 2014 Bremen, Germany, February 2014 Proceedings. Springer.

Thomassey, S., Fiordaliso, A., 2006. A hybrid sales forecasting system based on clustering and decision trees. Decision Support Systems 42 (1), 408–421.

Thomassey, S., Happiette, M., 2007. A neural clustering and classification system for sales forecasting of new apparel items. Applied Soft Computing Journal 7 (4), 1177–1187.

Thomassey, S., Happiette, M., Castelain, J.M., 2002. An automatic textile sales forecast using fuzzy treatment of explanatory variable. Journal of Textile and Apparel, Technology and Management 2 (4), 1–12.

Thomassey, S., Happiette, M., Castelain, J.M., 2005. A short and mean-term automatic forecasting system - application to textile logistics. European Journal of Operational Research 161 (1), 275–284.

Thomassey, S., 2010. Sales forecasts in clothing industry: the key success factor of the supply chain management. International Journal of Production Economics 128, 470–483.

Thomassey, S., 2014. Sales forecasting in apparel and fashion Industry: a review. In: Choi, T.M., Hui, C.L., Yu, Y. (Eds.), Intelligent Fashion Forecasting Systems: Models and Applications. Springer-Verlag Berlin Heidelberg.

Vroman, P., Happiette, M., Vasseur, C., 2001. A hybrid neural model for mean-term sales forecasting of textile items. Studies in Informatics and Control 10 (2), 149–168.

Winters, P.R., 1960. Forecasting sales by exponential weighed moving averages. Management Science 6, 324–342.

Wong, W.K., Guo, Z.X., 2010. A hybrid intelligent model for medium-term sales forecasting in fashion retail supply chains using extreme learning machine and harmony search algorithm. International Journal of Production Economics 128 (2), 614–624.

Yesil, E., Kaya, M., Siradag, S., 2012. Fuzzy forecast combiner design for fast fashion demand forecasting. In: Proceedings of the IEEE International Symposium in Innovations in Intelligent Systems and Applications (INISTA '12), pp. 1–5.

Yu, Y., Choi, T.M., Hui, C.L., 2011. An intelligent fast sales forecasting model for fashion products. Expert Systems with Applications 38, 7373–7379.

Zhu, Q.-Y., Qin, A.K., Suganthan, P.N., Huang, G.B., 2005. Evolutionary extreme learning machine. Pattern Recognition 38 (10), 1759–1763.

Fashion design using evolutionary algorithms and fuzzy set theory – a case to realize skirt design customizations

9

P.Y. Mok[1], J. Xu[1,2], Y.Y. Wu[1,3]
[1]The Hong Kong Polytechnic University, Hung Hom, Hong Kong; [2]Wuhan Textile University, Wuhan, China; [3]Cornell University, Ithaca, NY, United States

9.1 Introduction

The fashion and textiles industries have gone through many changes over the years; fast fashion, supply chain management, e-tailing, green technology, and smart materials are some buzzwords of the decade. To cope with all these changes, information technology (IT) tools and computer systems have been adopted to increase overall operational effectiveness and efficiency along the supply chain. For example, enterprise resources planning (ERP) systems were used to manage the workflow of entire business transactions (Monk and Wager, 2009). Product lifecycle management (PLM) systems were introduced to help the fashion industry to better manage detailed product data and information (Grieve, 2006). We have observed a large amount of artificial intelligence research work that supports various manufacturing operations (Kwong et al., 2006; Mok et al., 2007; Leung et al., 2008; Guo et al., 2009) along the textile and clothing supply chain (Dong and Leung, 2009; Pan et al., 2009). There are also some research studies focused on downstream retail operations, eg, in sales forecasting (Thomassey et al., 2004; Thomassey and Fiordaliso, 2006; Sun et al., 2008) and cross-selling recommendations (Wong et al., 2009a,b).

Today's consumers always look for affordable products that are customized to their unique needs. Yet it is still challenging for fashion companies to accommodate both the consumer's desire for high-quality customized garments and the industrial need for low-cost production and fast product development. The difficulties lie in the product development process, in particular the areas of sketch design and pattern design.

9.1.1 Sketch design and its related work

In the fashion industry, there are different kinds of sketch. First, it is the routine job of every designer to keep a close eye on trends and the runway shows. Whenever

Information Systems for the Fashion and Apparel Industry. http://dx.doi.org/10.1016/B978-0-08-100571-2.00009-9

anything interesting is found, designers will record their creative ideas on paper with a pen, as hand sketches. Later, these hand sketches will be organized and redrawn for different purposes: some for presentation purposes, eg, fashion illustration (see Fig. 9.1(a)) and others for facilitating communication between designers and the production teams, eg, technical sketches (see Fig. 9.1(b)).

Among the different types of sketch, the technical sketch is the most important in the product development process. Technical sketches, also called "*flats*" or technical drawings, are drawn to scale and include sewing and construction information. Flats are specific, precise, and literal, and they serve as a blueprint to explain the design to the production team for pattern making and prototype production. Flats usually involve both a front view and a back view, and are often developed with the help of croquis – template human figures (see Fig. 9.1(c)).

Today, designers mainly use commercial software, such as CorelDraw and Adobe Illustrator, to develop technical drawings from hand sketches stroke by stroke, which is time-consuming and skill-dependent. In recent years, digital fashion libraries like SnapFashun™ have been provided for designers to compose their sketches more efficiently. Other researchers investigated creating designs using artificial intelligence computational techniques. For example, Ogata and Onisawa (2007) proposed a clothes design system based on an interactive genetic algorithm (IGA), but unreasonable (infeasible) designs were generated. Mok et al. (2013) proposed a design knowledge

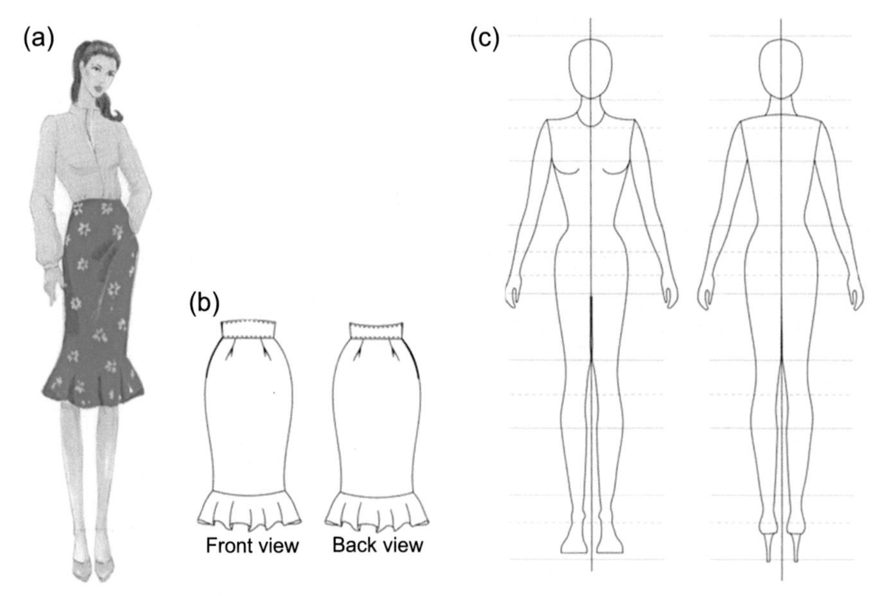

Figure 9.1 Illustration of sketches. (a) Fashion illustrations; (b) technical sketches (flats); (c) croquis (human figure template).

model to create realistic and practical fashion designs using IGA. However, they did not link up with downstream patternmaking.

9.1.2 Pattern design and its related work

Draping and flat patternmaking are two traditional approaches to patternmaking. Draping, also called 3D modeling, is a three-dimensional method that creates a style on a dress form by manipulating a piece of fabric, and the 3D shape developed is later flattened to obtain relevant clothing patterns. Flat patternmaking is a two-dimensional method that manipulates a set of basic blocks on the flat to create styles, which are recorded in the form of technical sketches. Flat patternmaking is the dominant approach to patternmaking, using which over 80% of ready-to-wear products are developed. The obvious advantages of flat patternmaking include that it can develop a large variety of styles in a relatively short period of time and in a less costly manner.

In flat patternmaking, the very first step is style analysis (also called design analysis). Upon receiving a flat, the patternmaker will carefully analyze the sketches in order to gather information like (1) the type of garment, eg, skirt, pants, jacket, etc.; (2) the physical dimensions and shape of the garment; (3) the number of garment sections involved; (4) the exact type of style features found in each section; and (5) the type of garment construction needed (Joseph-Armstrong, 2000). Designs are categorized into the appropriate garment structures and their respective details in design analysis (see Fig. 9.2).

Based on the design analysis, patternmakers will manipulate blocks to create patterns for the particular styles. Patternmakers develop their own sets of rules in pattern manipulation by trial and error. The actual practice of flat patternmaking thus varies from patternmaker to patternmaker, yet the general rules or principles of their practices are more or less the same. The traditional patternmaking process is highly manual. Patternmakers take years of training to master the techniques and skills for making

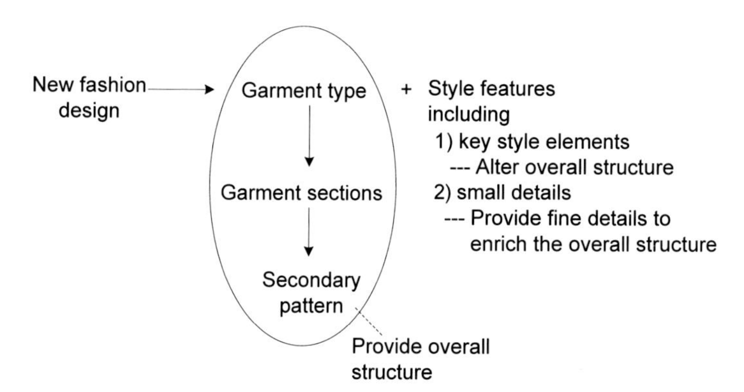

Figure 9.2 Traditional process of analyzing a fashion design.

patterns for a specific type of garment, such as a dress or jacket. Patternmaking is complex considering the diversity of styles, so sampling or prototyping is mandatory to ensure the correct fit and to confirm the design, as shown in Fig. 9.3.

Computer technology has been used to improve the patternmaking process. For example, made-to-measure (MTM) patternmaking was once regarded as a technology for mass customization that can produce apparel with an accurate fit that satisfies customers' individual needs (Istook, 2002). However, it has been implemented so far only for specific types of clothing products, eg, uniform and suits, because MTM requires a lengthy set-up and it does not support easy style alteration.

For the application of artificial intelligence techniques, Hu et al. (2008) proposed an interactive co-evolutionary CAD system for the parametric pattern design of a leisure shirt. However, only one style of pattern was designed. To improve the fit performance of 2D parametric patternmaking, some researchers have used artificial intelligence techniques like fuzzy logic to optimize garment patternmaking (Chen et al., 2009), while others used 3D techniques to verify and improve the fit of the clothing patterns. Lu et al. (2010) introduced an expert knowledge base to assist customized patternmaking. However, the accuracy and performance of this method have not been systematically verified.

9.1.3 Cutting-edge CAD developments

In support of fashion product development, a large number of clothing CAD-related research studies were reported in the past decade, either following a 2D-to-3D approach or a 3D-to-2D approach. The 2D-to-3D approach simulates the sampling and fitting process by virtually stitching 2D pattern pieces around a 3D mannequin (Sul and Kang, 2006; Volino and Magnenat-Thalmann, 2005; Meng et al., 2010, 2012; Kim and Park, 2006). In the 2D-to-3D approach, digital 2D garment patterns are first prepared by either digitizing paper patterns or drafting patterns on computers. Later, the digital 2D patterns are virtually stitched around a 3D human model to simulate the try-on effect for fit evaluation in 3D space (Cho et al., 2005). The 3D human models can be either scanned models or obtained by deformation (Zhu et al., 2013). It is noted that the 2D-to-3D approach integrates traditional 2D patternmaking with virtual sewing simulations, and thus indeed mimics the traditional "sampling" process. Patternmakers can alter the 2D patterns when needed, and they re-run the simulation to check the fit again; the process is repeated until a satisfactory look is achieved. This

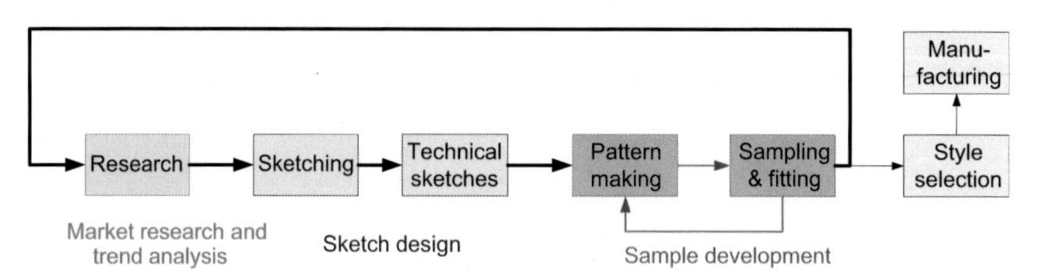

Figure 9.3 Fashion product development cycle.

reduces the number of sampling cycles for design confirmation and fit optimization, and speeds up the process of apparel product development. Well-known commercial clothing CAD systems like OptiTex, Browzwear, and Modrias 3D Fit all follow the 2D-to-3D approach.

By contrast, the 3D-to-2D approach creates 3D garments directly with reference to digital human models and later flattens 3D garment surfaces into 2D patterns (Wang, 2005; Cho et al., 2006; Huang, 2011). 3D garment modeling is the core part of the 3D-to-2D garment design and patternmaking approach. It can be either geometrical-based (Fontana et al., 2006) or physical-based (Choi and Ko, 2005). After the 3D modeling stage, the 3D garments are flattened into 2D patterns. This process is considered surface flattening. This means that, given a 3D freeform surface and material properties, finding its counterpart pattern in the plane and a mapping relationship between the two, when the 2D pattern is reversely folded back to a 3D surface, the amount of distortion – wrinkles and stretches – is minimized (McCartney et al., 1999).

9.1.4 A new approach to customized fashion product design

By reviewing the related literature in the past decade, we can find that most studies focused on the sample development stage of the product development cycle or the downstream manufacturing, as shown in Fig. 9.3. Sketch design is a less explored area of research in fashion product development. As discussed, today's consumers are increasingly complex and more powerful. Empowering customers in the product development process not only helps identify product requirements and features that are important to customers, but also improves customer enjoyment and satisfaction (Piller and Tseng, 2003). It is appealing to develop CAD systems so as to empower general customer participation in the design process, even though customers do not have drawing skills and professional knowledge of design.

Mok et al. (2013) developed a computer system that generates realistic and practical fashion designs by IGA. Extending the work of Mok et al. (2013), we improve the design knowledge model by modeling the aesthetics and harmony of different combinations of design features using fuzzy numbers in this chapter. The output sketches governed by such a design knowledge model are a feasible design with all digital design details, including the silhouette, the design elements involved with location and other fine details, eg, pleats, number of pleats, location of pleats, and size of pleats, etc. It, in fact, provides all the detailed information in a traditional design analysis. With such precise detailed information on the styles, we develop an automatic pattern design module that automatically generates the patterns of the styles.

Fig. 9.4 depicts a new approach of computer-aided fashion design. It involves customers as co-designers in the sketch design process, and it streamlines the patternmaking process by computer automation. To the best of our knowledge, it is the first published work on mass customization of fashion products from sketch design to pattern development.

As shown in Fig. 9.4, the system development process starts with a systematic style classification, resulting in a style feature database, which is detailed in Section 9.2.

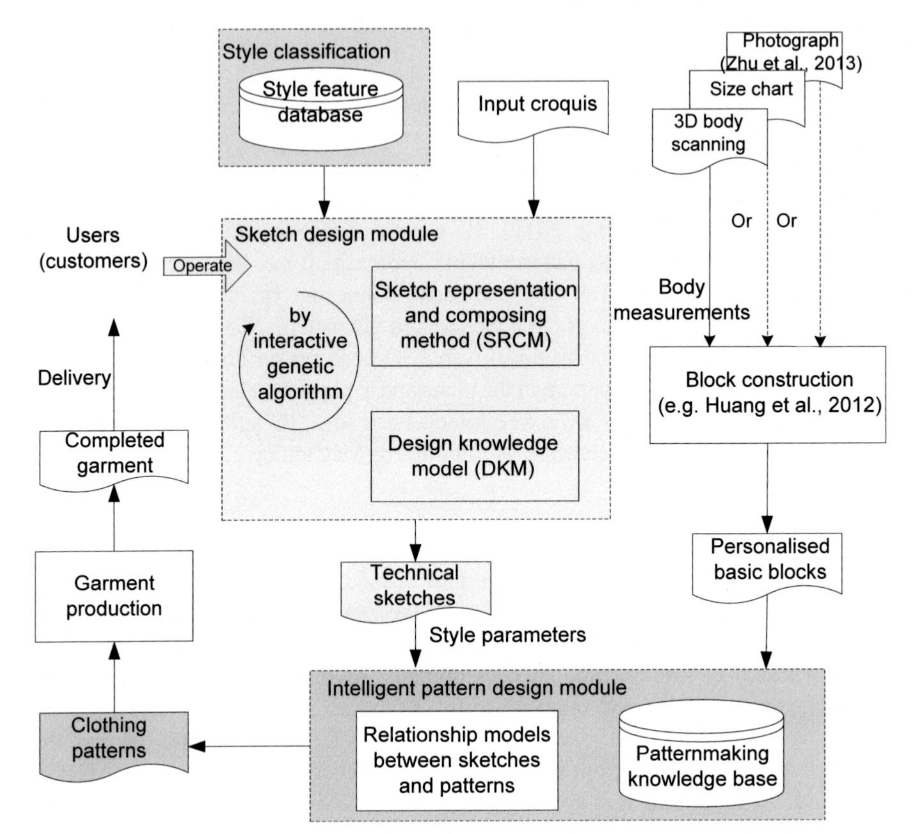

Figure 9.4 System overview: a new approach for customized fashion product design.

Following the style classification, sketch design is conducted, where customers select from a pool of computer-generated designs their preferred ones. Iteratively, customers can confirm their preferred sketches. In sketch design, a sketch representation and composing method (SRCM) is developed to generalize different style features as parametric models and composed as vector graphs. Interactive genetic algorithms (IGAs) are used to hide the tedious parameters definition of SRCM from users and allow them to realize their design ideas by simple computer operations. A design knowledge model is proposed using fuzzy set theory that bridges SRCM and IGAs and ensures that all the generated designs are practical and aesthetically appealing. Sketch design is detailed in Section 9.3.

The output from the sketch design process is technical sketches (flats) with explicit information on the designs. An intelligent pattern design module is developed, which formulates mathematical relationships between flats and patterns, automating the relevant patternmaking in optimized sequential computer operations.

Skirts are used to illustrate the proposed method for sketch design and automatic patternmaking. Nevertheless, other fashion products, such as trousers and blouses, can also be created by the same method.

9.2 Style classification and style feature database

Based on a comprehensive review of the ready-to-wear skirt styles, a three-level design model (see Fig. 9.5) is proposed to define the skirt style from the perspective of patternmaking.

The design model classifies the skirt design elements into three levels. To be more specific, the first level classifies the skirt silhouette, including the waist and hem information plus the shape category. Each skirt design can be assigned to one of the shape categories, such as tapered, straight, round, and so forth. The second level defines key design elements which can contribute to the overall skirt shape. It defines the parameters of each design element which indicate the corresponding shape, size, and position. The third level classifies additional details of the design, including pockets and openings. Following this three-level classification system, no matter how complicated the skirt styles are, they can be decomposed and represented as different combinations of three-level design elements. A skirt style feature library (see Figs. 9.6–9.8) is thus constructed by reviewing the design literature and recording the design elements of the ready-to-wear styles available in the online resources.

9.3 Sketch design using fuzzy numbers and IGA

9.3.1 Sketch representation and composing method (SRCM)

As shown in Fig. 9.4, the sketch design module includes a sketch representation and composing method (SRCM), a design knowledge model, and a design process modeled by an interactive genetic algorithm. Technical sketches are developed with

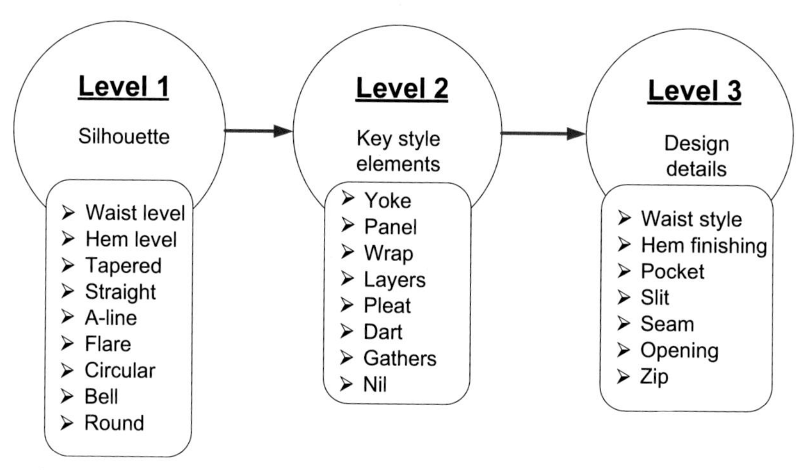

Figure 9.5 The three-level design model of skirt.
Source: Mok, P.Y., Xu, J., Wang, X.X., Fan, J.T., Kwok, Y.L., Xin, J.H., 2013. An IGA-based design support system for realistic and practical fashion designs. Computer-Aided Design 45(11), 1442–1458.

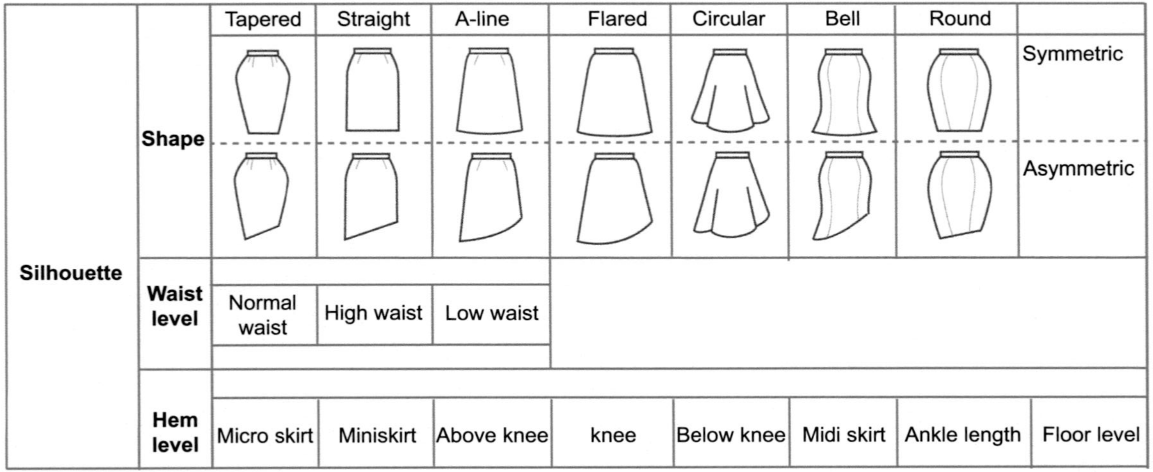

Silhouette									
	Shape	Tapered	Straight	A-line	Flared	Circular	Bell	Round	
									Symmetric
									Asymmetric
	Waist level	Normal waist	High waist	Low waist					
	Hem level	Micro skirt	Miniskirt	Above knee	knee	Below knee	Midi skirt	Ankle length	Floor level

Figure 9.6 Silhouette feature.
Source: Mok, P.Y., Xu, J., Wang, X.X., Fan, J.T., Kwok, Y.L., Xin, J.H., 2013. An IGA-based design support system for realistic and practical fashion designs. Computer-Aided Design 45(11), 1442–1458.

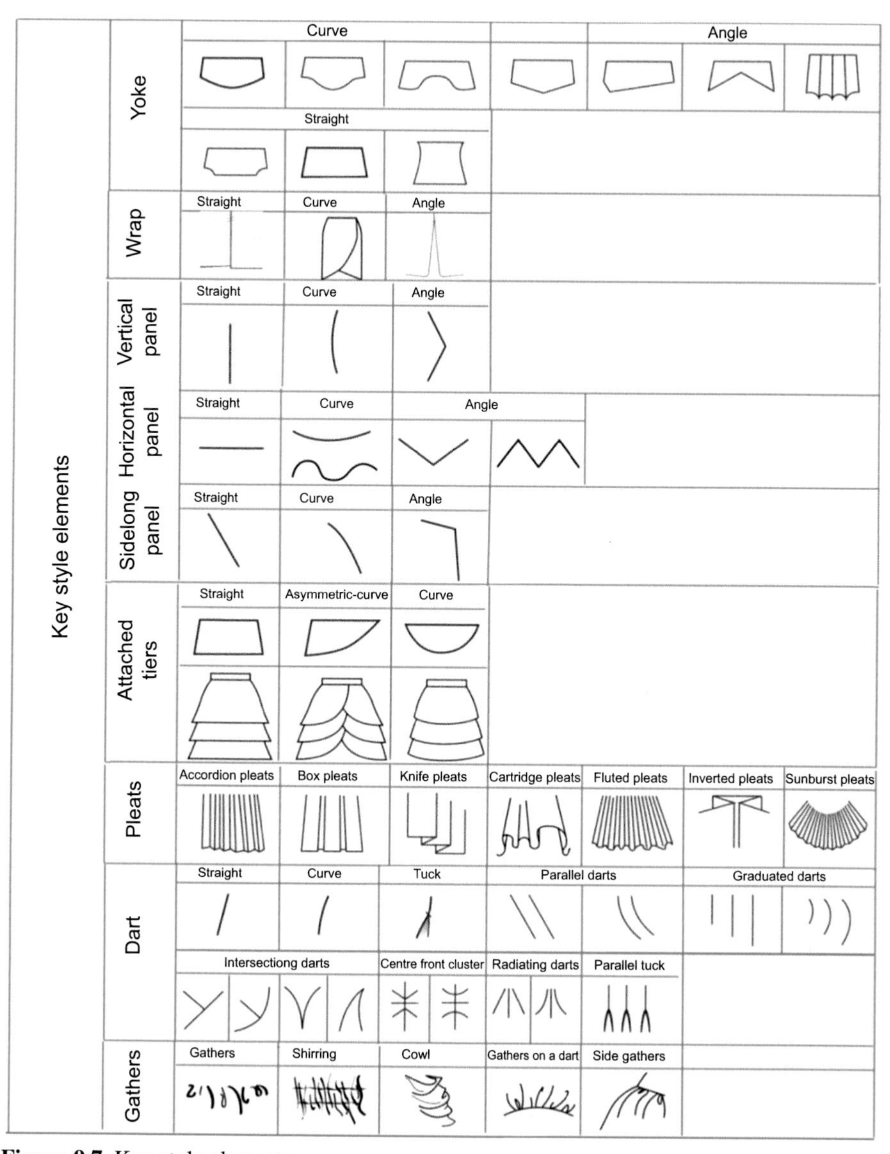

Figure 9.7 Key style elements.
Source: Mok, P.Y., Xu, J., Wang, X.X., Fan, J.T., Kwok, Y.L., Xin, J.H., 2013. An IGA-based design support system for realistic and practical fashion designs. Computer-Aided Design 45(11), 1442–1458.

the aid of croquis; each fashion company has its own set of croquis (see Fig. 9.9) and they are prepared based on target customers' body proportions and shapes.

In order to create sketches with compatible design elements that can automatically "fit" different input croquis, parametric models are defined for each level of design element, according to those recorded in the style feature library.

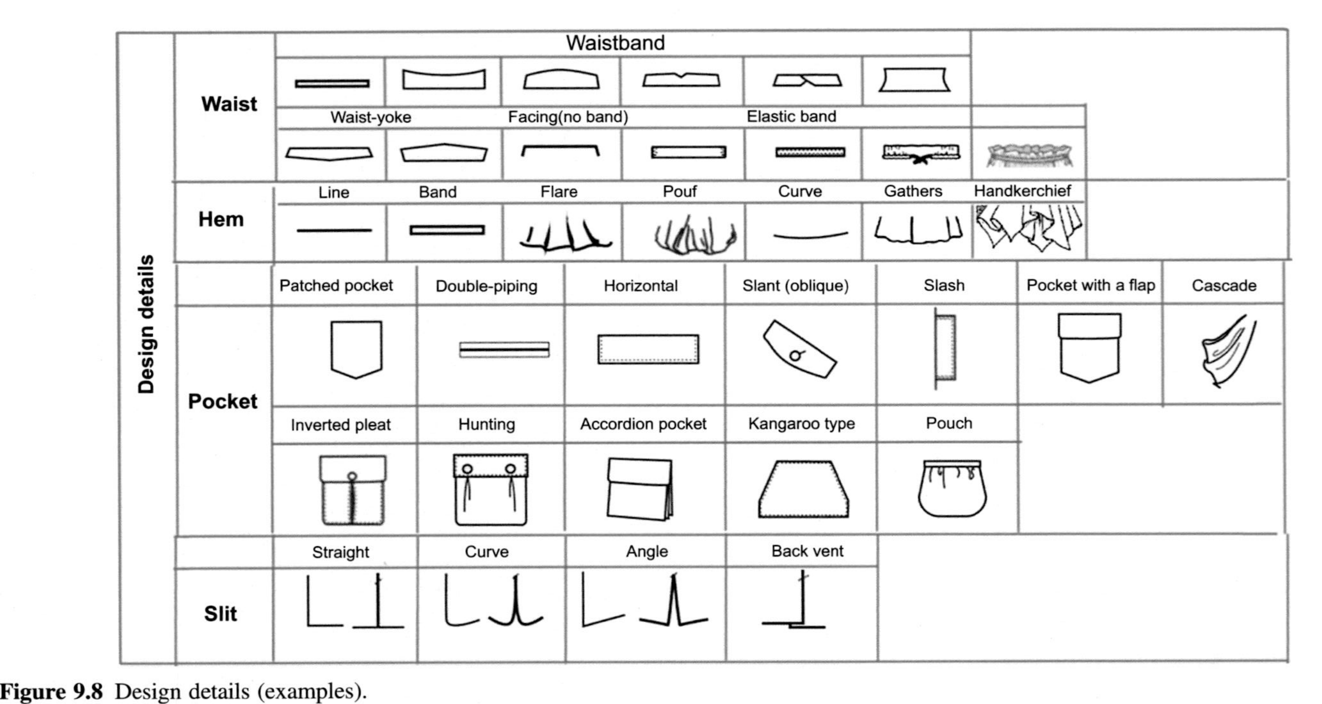

Figure 9.8 Design details (examples).
Source: Mok, P.Y., Xu, J., Wang, X.X., Fan, J.T., Kwok, Y.L., Xin, J.H., 2013. An IGA-based design support system for realistic and practical fashion designs. Computer-Aided Design 45(11), 1442–1458.

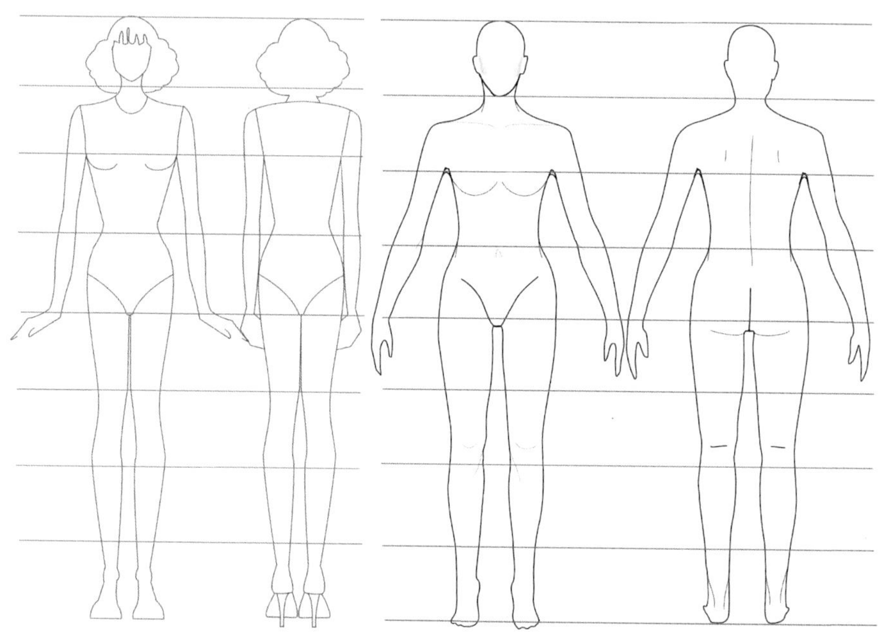

Figure 9.9 Croquis examples.

For example, the first-level design element usually defines the general shape/silhouette of a fashion product. A total of eight parameters $\boldsymbol{S} = [S_1 \ S_2 \ \ldots \ S_8]$ are used to define the skirt silhouette, as shown in Eq. [9.1] and Table 9.1:

$$\text{Silhouette}(\text{Shape}_1, \text{Shape}_2, \ldots, \text{Shape}_N) = f^{(S)}(S_1, S_2, \ldots, S_8) \quad [9.1]$$

where N is the number of shape groups.

Table 9.1 Silhouette parameters

Parameters	Description	With respect to human template
S_1	Symmetry information	
S_2	Waist level parameter	Vertical position
S_3	Hem level parameters	Vertical position
S_4	Hip width parameter	Horizontal position
S_5	Hem width parameter	Horizontal position
S_6	Complex shape indicator	
S_7	Style line parameter	Vertical position
S_8	Style line width parameter	Horizontal position

Source: Mok, P.Y., Xu, J., Wang, X.X., Fan, J.T., Kwok, Y.L. Xin, J.H., 2013. An IGA-based design support system for realistic and practical fashion designs. Computer-Aided Design 45(11), 1442–1458.

These parameters ***S*** are defined as ratio parameters to some standard measurements, for instance, the body waist width, of the given template model or croquis. Based on fashion design knowledge, the value ranges of these parameters are defined, as shown in Table 9.2. By defining different values of these silhouette parameters, various shapes can be obtained, such as tapered, straight, A-line, round, and bell. Fig. 9.10 shows some example silhouettes and relevant silhouette parameters ***S***. The appropriateness of the value ranges has been verified by professional designers.

With the silhouette parameters, a set of shape classifiers $\boldsymbol{C} = [C_1\ C_2\ \ldots\ C_7]$ is calculated by Eq. [9.2].

$$\text{ShapeClassifier}(C_1,\ C_2, \ldots,\ C_7) = f^{(C)}(S_1,\ S_2, \ldots,\ S_8) \quad [9.2]$$

These shape classifiers define the characteristics of the shapes (see Table 9.3), which are also the knowledge/information obtained from a design analysis. Patternmakers

Table 9.2 The value ranges of parameters *S*

Parameters	Value ranges	Remarks	Types
S_1	1, 0	1 – Symmetric silhouette and design details; 0 – Asymmetric silhouette and design details	Symmetry
S_2	[0.5, 1.5]	1 Normal waist; (1, 1.5) high waist; (0.5, 1) low waist	Waist level
S_3	[0.17, 2.5]	The greater the value, the longer the skirt	Hem level
S_4	[0.5, 0.95]	The greater the value, the wider the hip width	Shape parameters
S_5	[0.7, 2]	The greater the value, the wider the hem width	
S_6	1, 0	1 – Additional bit to define *style point* for bell and round shape; 0 – Shape with straight side seam from hip to hem	
S_7	[0.2, 0.9]	*Style point* level ratio. If $S_6 = 1$, S_7 is valid. The greater the value, the lower the *style point*	
S_8	[0.7, 1.6]	*Style point* width ratio (*styw*). If $S_6 = 1$, S_8 is valid; if $styw < 1$, bell shape; $styw = 1$ straight; $styw > 1$, round	

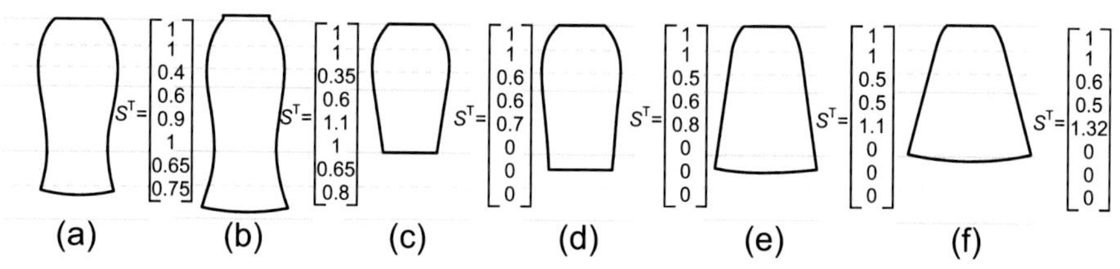

Figure 9.10 Example silhouettes and parameters ***S***.

Table 9.3 Shape classifying parameters

Parameters	Description
C_1	Proportion parameter
C_2	Design ease at hip
C_3	Hip curve shape classifier
C_4	Above hip shape classifier
C_5	Below hip shape classifier
C_6	Design ease at style line level
C_7	Below style line level shape classifier

Source: Mok, P.Y., Xu, J., Wang, X.X., Fan, J.T., Kwok, Y.L., Xin, J.H., 2013. An IGA-based design support system for realistic and practical fashion designs. Computer-Aided Design 45(11), 1442–1458.

then apply such knowledge/information and manipulate the basic blocks to achieve the desired shape by the incorporation of key style elements.

The allowable key style elements for a given skirt silhouette are governed by the shape classifiers ***C***:

$$\text{KeyStyleElements}\left(\text{DF}^{(1)}, \ldots, \text{DF}^{(K)}, \ldots, \text{DF}^{(H)}\right) = f_2^{(\text{DF})}(C_3, C_5/C_7) \quad [9.3a]$$

where $\text{DF}^{(K)}$ ($K = 1, 2, \ldots, H$) are H different categories of key style elements. For example, $\text{DF}^{(K)}$ represents the key style element category of the yoke when $K = 1$. For skirts, other common categories of key style element include panel ($K = 2$), wrap ($K = 3$), layers ($K = 4$), pleats ($K = 5$), dart ($K = 6$), and gathers ($K = 7$). Each category of key style element $\text{DF}^{(K)}$ includes a number of M feature subgroups, denoted as $\text{DF}_i^{(K)}$, where $i = 1, \ldots, M$. For example, curved yoke and angle yoke are two style subgroups of yoke. For each $\text{DF}_i^{(K)}$, n parameters, denoted as $\text{DFT}_{ij}^{(K)}$ ($j = 1, \ldots, n$), are used to describe the characteristics/feature details of the style feature subgroup, and are defined by Eq. [9.3b].

$$\text{DesignFeatureDetails}(\text{DFT}_{i1}^{(K)}, \text{DFT}_{i2}^{(K)}, \ldots, \text{DFT}_{in}^{(K)}) = f^{(\text{DFT})}\left(\text{DF}_i^{(K)}\right) \quad [9.3b]$$

The third-level design element – design details, which will not alter the silhouette but provide decorative details or additional functions for a style – can also be defined in a similar manner, as shown in Eqs. [9.4]–[9.8]. Parts (a) of the equations depict which higher-level design elements are used in the parametric definition design details, and such formulation ensures the mutual compatibility of design elements. The various characteristics of the design details are as defined in parts (b) of the equations.

$$\text{WaistStyle}(W_1, W_2, \ldots) = f^{(W)}(S_2,\ \text{DF}) \qquad [9.4a]$$

$$\text{WaistStyleDetails}(\text{WT}_{i1}, \text{WT}_{i2}, \ldots) = f^{(\text{WT})}(W_i) \qquad [9.4b]$$

$$\text{HemFinishing}(\text{HF}_1, \text{HF}_2, \ldots) = f^{(\text{HF})}(\text{DF}, C_5/C_7) \qquad [9.5a]$$

$$\text{HemFinishingDetails}(\text{HFT}_{i1}, \text{HFT}_{i2}, \ldots) = f^{(\text{HFT})}(\text{HF}_i) \qquad [9.5b]$$

$$\text{Slits}(\text{SL}_1, \text{SL}_2, \ldots) = f^{(\text{SL})}(S_3, C_5, \text{HF}_i) \qquad [9.6a]$$

$$\text{SlitsDetails}(\text{SLT}_{i1}, \text{SLT}_{i2}, \ldots) = f^{(\text{SLT})}(\text{SL}_i) \qquad [9.6b]$$

$$\text{Pockets}(P_1, P_2, \ldots) = f^{(P)}(\text{DF}) \qquad [9.7a]$$

$$\text{PocketsDetails}(\text{PT}_{i1}, \text{PT}_{i2}, \ldots) = f^{(\text{PT})}(P_i) \qquad [9.7b]$$

$$\text{Opening}(O_1, O_2, \ldots) = f^{(O)}(\text{DF}, W) \qquad [9.8a]$$

$$\text{OpeningDetails}(\text{OT}_{i1}, \text{OT}_{i2}, \ldots) = f^{(\text{OT})}(O_i) \qquad [9.8b]$$

Upon defining the parametric design model, a sketch-composing engine is developed to construct technical sketches, which are indeed different combinations of the three-level parametric design elements. In this study, a pilot system was developed in C# language based on the Teigha.NET development kit, which outputs sketches in DXF file format. DXF file format is supported by most commercial CAD systems.

The sketch-composing engine consists of two types of algorithm, including silhouette-drafting algorithms (SDA) and adaptation and attachment algorithms (AAA). The composing engine first reads a human template model/croquis; some key landmarks (Hwang, 2004), as well as the body outline curves necessary for sketch design, are defined. Next, the skirt silhouette shape is sketched, based on the body information and a set of silhouette parameters $\boldsymbol{S}$ using SDA. Later, level 2 key design elements and level 3 waist styles and hem finishing are sketched in an optimized sequence, based on the parametric models. Lastly, other level 3 design elements are sketched. AAA is used in this process.

9.3.2 Design knowledge model by fuzzy numbers

Design is different from art, and design is more than purely looking for aesthetic appeal. Fashion designers often use their knowledge of material properties and pattern-making in the creation process.

According to the parametric design model, any combination of the three-level design elements can compose a style. However, the generated parametric styles may not be practical or aesthetically attractive ones. Due to the textile material properties, a certain silhouette must be achieved by incorporating certain style features in pattern-making. For example, a trumpet (bell-shape) skirt with panels (Fig. 9.11(b)) can be realized in fabrics, but darts cannot realize the same shape of skirt. Therefore, the sketch in Fig. 9.11(a) is not a feasible design.

In this section, a design knowledge model is developed to govern the creation of sketches and to overcome impractical designs. Fuzzy sets are used to model the design knowledge base. The design knowledge model is represented by relationship matrices, indicating the harmony and attractiveness of the relevant combinations of design elements. For example, the second-level design model (Fig. 9.5) defines the key style elements, which are introduced to realize the silhouette defined in the first level of the design model. The design knowledge is represented by a relationship matrix $\boldsymbol{F}^{(K)}$

$$\boldsymbol{F}^{(K)} = \begin{bmatrix} f_{1,1}^{(K)} & \cdots & f_{1,j}^{(K)} & \cdots & f_{1,N}^{(K)} \\ \vdots & \ddots & \vdots & \ddots & \vdots \\ f_{i,1}^{(K)} & \cdots & f_{i,j}^{(K)} & \cdots & f_{i,N}^{(K)} \\ \vdots & \ddots & \vdots & \ddots & \vdots \\ f_{M,1}^{(K)} & \cdots & f_{M,j}^{(K)} & \cdots & f_{M,N}^{(K)} \end{bmatrix}_{M\times N} = \begin{bmatrix} F_1^{(K)} \\ \vdots \\ F_i^{(K)} \\ \vdots \\ F_M^{(K)} \end{bmatrix}_{M\times N} \qquad [9.9]$$

and $F_i^{(K)} = \begin{bmatrix} f_{i,1}^{(K)} & \cdots & f_{i,j}^{(K)} & \cdots & f_{i,N}^{(K)} \end{bmatrix}_{1\times N}$

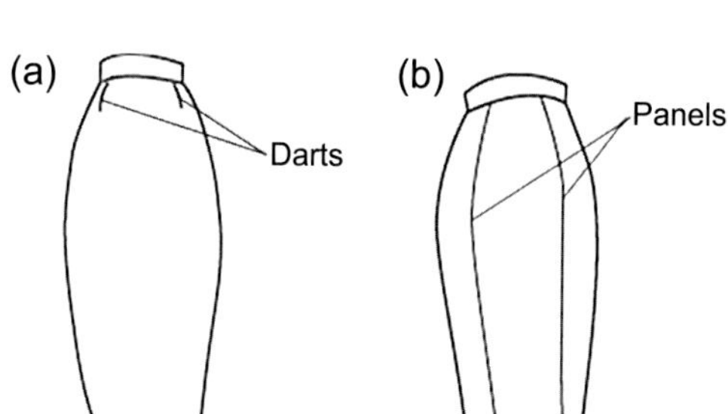

Figure 9.11 Bell-shape trumpet skirt: (a) impractical design and (b) practical design.
Source: Mok, P.Y., Xu, J., Wang, X.X., Fan, J.T., Kwok, Y.L., Xin, J.H., 2013. An IGA-based design support system for realistic and practical fashion designs. Computer-Aided Design 45(11), 1442–1458.

Table 9.4 Fuzzy number definition for design assessment

	Linguistic description	Fuzzy numbers
Harmony $\widetilde{h}$	NOT	0
	Fair	(0, 0.2, 0.4)
	Average	(0.2, 0.4, 0.6)
	Good	(0.4, 0.6, 0.8)
	Excellent	(0.6, 0.8, 1.0)
Attractiveness $\widetilde{a}$	Poor	(−0.2, 0, 0.2)
	Fair	(0, 0.2,0.4)
	Average	(0.2, 0.4, 0.6)
	Good	(0.4, 0.6, 0.8)
	Excellent	(0.6, 0.8, 1.0)

where $K = 1, 2, \ldots, H$ represents different categories of key style element. Assuming there are M feature subgroups of a specific category of key style element K in the style database, the relationship matrix $\boldsymbol{F}$ is thus an $M \times N$ matrix, indicating the relationship between the M feature subgroups and N shape groups.

Fuzzy set theory (Zadeh, 1965) is an attractive framework to deal with "fuzzy" concepts like the harmony and attractiveness of a design. The harmony and aesthetic characteristics are defined by fuzzy numbers in this study, as shown in Table 9.4.

The NOT fuzzy set is used if the two design elements are not compatible; see Fig. 9.11(a) as an example. A group of professional designers are invited to evaluate the harmony and aesthetics of any combination of two design elements. Fuzzy arithmetic operations of addition and scalar multiplication (Kaufmann and Gupta, 1991) are used to average the assessment of different designers so as to obtain the fuzzy numbers presenting the respective harmony and attractiveness. Multiplication of fuzzy numbers is used to obtain the combined effect of harmony and attractiveness. The expected value of the resulting fuzzy number is then calculated, which gives the respective entry value $f_{i,j}^{(K)}$ of $\boldsymbol{F}^{(K)}$.

The expected value of resulting fuzzy number $\widetilde{C}$ is calculated by Yager's (1981) index of linear weighting (see Fig. 9.12).

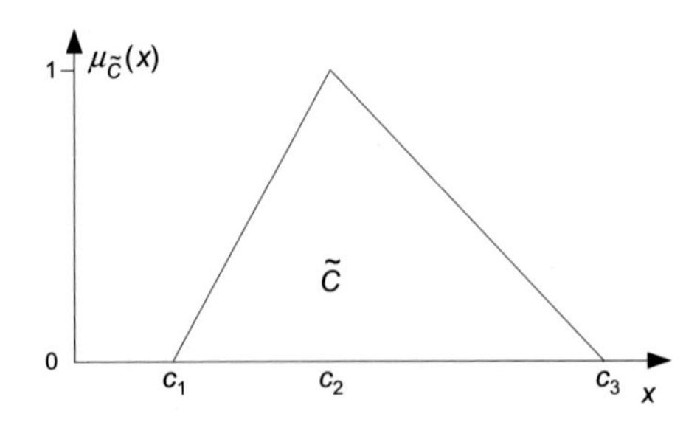

Figure 9.12 Expected value of fuzzy number (c_1, c_2, c_3).

$$f_{i,j}^{(K)} = m(\widetilde{C}) = \frac{\int x\mu_{\widetilde{C}}(x)dx}{\int \mu_{\widetilde{C}}(x)dx} \quad [9.10]$$

With known $\boldsymbol{F}^{(K)}$, a row vector $\boldsymbol{FR}^{(K)} = \left[fr_1^{(K)} \ \cdots \ fr_j^{(K)} \ \cdots \ fr_N^{(K)} \right]_{1\times N}$ is produced by summing the relationship matrix $\boldsymbol{F}^{(K)}$ across columns, where each entry is calculated by:

$$fr_j^{(K)} = \sum_{i=1}^{M} f_{i,j}^{(K)} \quad \text{for} \quad j = 1, 2, \ldots, N. \quad [9.11]$$

$fr_j^{(K)}$ denotes whether key style element K, among all M feature subgroups, is compatible with different shapes j ($j = 1, \ldots, N$). If the incorporation of such a key style element can realize the j-th shape, then $fr_j^{(K)} > 0$; otherwise, $fr_j^{(K)} = 0$.

Another matrix $\boldsymbol{T}$ is defined based on the row vectors $FR^{(K)}$ for the H different categories of key style elements as follows:

$$\boldsymbol{T} = \begin{bmatrix} t_{1,1} & \cdots & t_{1,j} & \cdots & t_{1,N} \\ \vdots & \ddots & \vdots & \ddots & \vdots \\ t_{K,1} & \cdots & t_{K,j} & \cdots & t_{K,N} \\ \vdots & \ddots & \vdots & \ddots & \vdots \\ t_{H,1} & \cdots & t_{H,j} & \cdots & t_{H,N} \end{bmatrix}_{H\times N} \quad [9.12]$$

where the entry value of $\boldsymbol{T}$ is $t_{K,j} = fr_j^{(K)}$.

In addition to key style elements, the knowledge on the use of various design details can be defined in a similar way. In patternmaking, the possible design details depend on the silhouette as well as the selected key style elements. As shown in Eqs. [9.4]–[9.8], different design details are functions of the silhouette parameters S_i, shape classifying parameters C_i, and selected features. This is consistent with the three-level design model, where level 2 design elements (key style elements) are dependent on level 1 elements (the silhouettes), while level 3 elements (design details) are dependent on the elements of both level 1 and level 2.

Let z be a third-level design element; assuming that there are P style subgroups of z in the style database, the relationships between these P subgroups and various key style elements (classified in H categories) are defined by matrix $\boldsymbol{U}$:

$$\boldsymbol{U} = \begin{bmatrix} u_{1,1} & \cdots & u_{1,K} & \cdots & u_{1,H} \\ \vdots & \ddots & \vdots & \ddots & \vdots \\ u_{i,1} & \cdots & u_{i,K} & \cdots & u_{i,H} \\ \vdots & \ddots & \vdots & \ddots & \vdots \\ u_{P,1} & \cdots & u_{P,K} & \cdots & u_{P,H} \end{bmatrix}_{P\times H} = \begin{bmatrix} U_1 \\ \vdots \\ U_i \\ \vdots \\ U_P \end{bmatrix}_{P\times H} \quad \text{and} \quad [9.13]$$

$$U_i = \left[u_{i,1} \ \cdots \ u_{i,K} \ \cdots \ u_{i,H} \right]$$

The design element z also depends on a certain silhouette parameter S_i or C_i, which classifies the design characteristics into Q subgroups. The relationships between the P style groups of z and the Q silhouette subgroups are defined by matrix $\boldsymbol{V}$:

$$\boldsymbol{V} = \begin{bmatrix} v_{1,1} & \cdots & v_{1,j} & \cdots & v_{1,Q} \\ \vdots & \ddots & \vdots & \ddots & \vdots \\ v_{i,1} & \cdots & v_{i,j} & \cdots & v_{i,Q} \\ \vdots & \ddots & \vdots & \ddots & \vdots \\ v_{P,1} & \cdots & v_{P,j} & \cdots & v_{P,Q} \end{bmatrix}_{P\times Q} = \begin{bmatrix} V_1 \\ \vdots \\ V_i \\ \vdots \\ V_P \end{bmatrix}_{P\times Q} \quad \text{and} \qquad [9.14]$$

$$V_i = \begin{bmatrix} v_{i,1} & \cdots & v_{i,j} & \cdots & v_{i,Q} \end{bmatrix}$$

The entry values of $\boldsymbol{U}$ and $\boldsymbol{V}$ are defined in a similar way to those for $\boldsymbol{F}$ by fuzzy numbers.

The Kronecker product, $\boldsymbol{Z}$, is calculated for the row vectors of matrices $\boldsymbol{U}$ and $\boldsymbol{V}$ by Eq. [9.15], to define design element z's harmony and attractiveness in different combinations of H categories of key feature and Q subgroups of the silhouette:

$$\boldsymbol{Z} = \begin{bmatrix} U_1 \otimes V_1 \\ \vdots \\ U_i \otimes V_i \\ \vdots \\ U_P \otimes V_P \end{bmatrix}_{P\times HQ} \qquad [9.15]$$

9.3.3 Sketch design by interactive genetic algorithm

In this section, interactive genetic algorithms are used to construct attractive and practical flats. As discussed, a design can be obtained by defining relevant parameters of the parametric design model. Tedious parameter definition is hidden from users by formulating design parameters as individual "chromosomes" in IGA. The decoding of individual chromosomes is governed by the design knowledge model, which ensures that the output sketches are practical.

9.3.3.1 Individual representation and its decoding

As shown in Fig. 9.13, the individual representation (the string) is composed of three segments. The first segment of the string indicates the design silhouette, corresponding to silhouette parameters $\boldsymbol{S}$ (defined in Table 9.1). The second segment of the string comprises bits for selecting the key style elements category K (see the second part of Fig. 9.9), the style subgroup of the selected key style element DF (defined by Eq. [9.3a]) and the detailed characteristics of the selected key feature style DFT_i

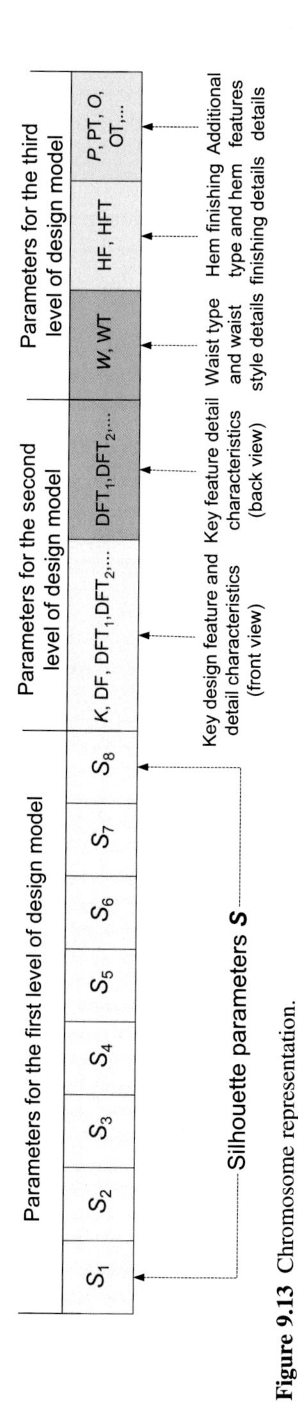

Figure 9.13 Chromosome representation.
Source: Mok, P.Y., Xu, J., Wang, X.X., Fan, J.T., Kwok, Y.L., Xin, J.H., 2013. An IGA-based design support system for realistic and practical fashion designs. Computer-Aided Design 45(11), 1442–1458.

(defined by Eq. [9.3b]). Both the front-view and back-view details are defined. The third segment of the string defines the parameters for design details (see the third part of Fig. 9.9) to increase the complexity of the design, including waist style, hem finishing, slits, pockets, and opening (defined by Eqs. [9.4]–[9.8]). The formulation of the individual chromosome is a string of real numbers, each in the range of [0.0, 1.0].

The decoding of the string with bit values in the range of [0.0, 1.0] is done by linear interpolation to different value ranges for individual parameters. The decoding is done in segments, ie, the first segment of the string is decoded first, then the second segment, and last the third segment. The decoding of the latter segments depends on the decoded value of the previous segments.

The first segment of the individual is a string of eight real numbers with the value range [0.0, 1.0], and the decoding is to convert the bit value to a specific parameter value range $[S_{i\_lower}, S_{i\_upper}]$ by linear interpolation. These parameter value ranges $[S_{i\_lower}, S_{i\_upper}]$ are defined in Table 9.2. The shape-classifying parameters $\boldsymbol{C}$ are then calculated from the decoded $\boldsymbol{S}$ using Eq. [9.2]. The $\boldsymbol{S}$ and $\boldsymbol{C}$ define to which of the N shape groups the current silhouette belongs, eg, the j-th group.

The second segment of the string defines the relative parameter values of the key design features, corresponding to the second-level elements of the design model. The first bit of the second string segment defines which of the H key style element categories is selected for the given j-th shape group. This can be done by converting the bit value by proportion, in a value range of $\left[0, \sum_{K=1}^{H} t_{K,j}\right]$. The upper limit of the parameter value range $\sum_{K=1}^{H} t_{K,j}$ is the summation of the j-th column of matrix $\boldsymbol{T}$ in Eq. [9.15]. The decoded value α indicates which one of the allowable key style element categories is selected for the given j-th silhouette shape. Next, the second bit of the second segment defines which feature subgroup of key style element K is selected from the style database for the given j-th shape group. The other bits of the second segment are decoded to define the required parameters for detailed characteristics of key style element $\mathrm{DFT}_i^{(K)}$ for both front-view and back-view sketches.

The third segment of the string defines the relative parameter values of the design details, which correspond to the third-level elements of the design model. In this segment, the decoding of bit values is done in a similar way to that of the feature subgroup of key style element K, with reference to matrix $\boldsymbol{Z}$ based on the selected key feature $\mathrm{DF}^{(K)}$ and other dependent silhouette parameters.

9.3.3.2 *Other design decisions and genetic operators*

Population size is one of the design decisions in evolutionary applications. On the one hand, it is not recommended to overload users with too many designs to choose, as only a limited amount of information can be accessed by individuals. On the other hand, it is also necessary to ensure good coverage in the design search space. A popular size of nine is used, taking into account the user interface design and the above two concerns.

To ensure a wider coverage of the search space and accelerate the design process, a special initialization process is introduced such that different shapes/silhouettes of sketches are included in the initial population. Other than that, users are allowed to provide design preferences, eg, in terms of skirt length and waist level, so the system can generate designs closer to the users' preferences for a faster search.

Crossover and mutation operations are typical genetic operations to introduce variations to the population. For crossover, a multipoint discrete crossover operator is applied to recombine the individuals. In this sketch design application, two types of crossover operator are proposed. The first type limits the crossover operation to take place only after the first segment, whereas the second type does not have such a positional constraint. The first type avoids changing the product silhouette shape but alters the key features and design details, which can be used in the later stage of genetic evolution.

Both Gaussian mutation and uniform mutation operators are introduced. The Gaussian mutation operator is employed on the first segment of the string, except for the silhouette parameters S_1 and S_6. The uniform mutation operator is employed on the second and the third segments of the string.

To avoiding losing good designs, an elitism operation is used to retain individuals in the evolution process. If the user is satisfied with the major features of some individuals, then these individuals (design sketches) can be "locked" by users. The locked individuals will be copied directly to the next generation, so that preferred designs or style features are preserved. This improves the search speed while maintaining design variability.

In most IGA-based applications, users are required to rank and grade every individual in the population (Meghna and Barbara, 2010; Bush and Sayama, 2011). In this specific application, fitness evaluation is combined with a selection operator: users are required to select from a pool of nine designs a few individuals (design sketches) that appeal to them the most. Combining fitness evaluation with selection reduces user fatigue without compromising the diversity of individuals.

9.3.3.3 Evolution process

An open-ended system design is adopted such that users are given more control over the design process. Apart from selecting their preferred design, users can determine the genetic operations to perform in each design cycle, ie, which crossover operation and whether crossover and mutation operations are employed in each cycle. In addition to selecting the genetic operations to be included in each design cycle, users can also select an arbitrary number of individuals for crossover and mutation operations. The crossover rate and the mutation rates are not fixed throughout the experiment. Users are allowed to define the mutation and crossover probabilities in every design cycle on the graphic user interface (Fig. 9.14). Fig. 9.15 outlines the steps of our IGA-based sketch design process.

9.4 Intelligent pattern designs

An intelligent pattern design module is developed to generate clothing patterns based on flats obtained from the sketch design module. Traditional flat patternmaking techniques are used in the construction of the pattern design module.

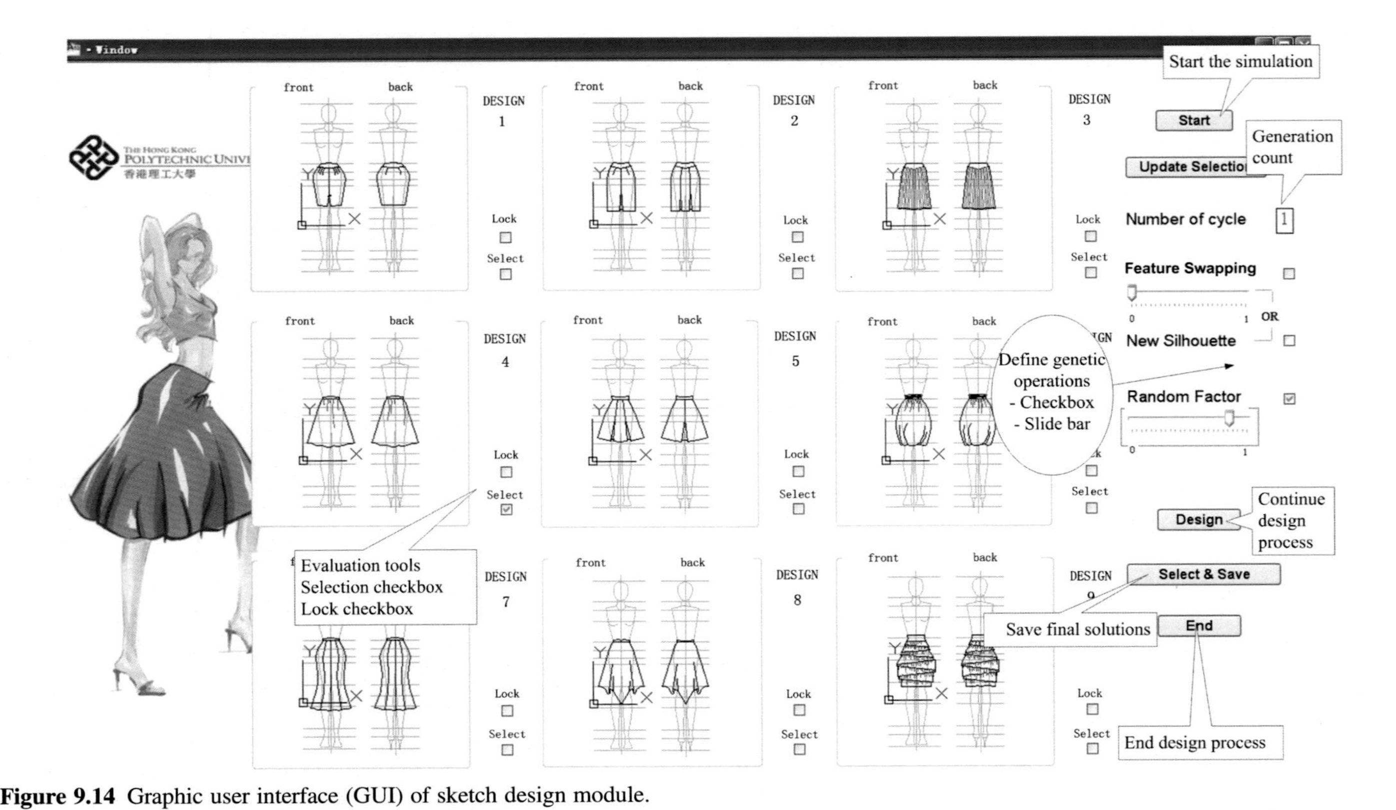

Figure 9.14 Graphic user interface (GUI) of sketch design module.
Source: Mok, P.Y., Xu, J., Wang, X.X., Fan, J.T., Kwok, Y.L., Xin, J.H., 2013. An IGA-based design support system for realistic and practical fashion designs. Computer-Aided Design 45(11), 1442–1458.

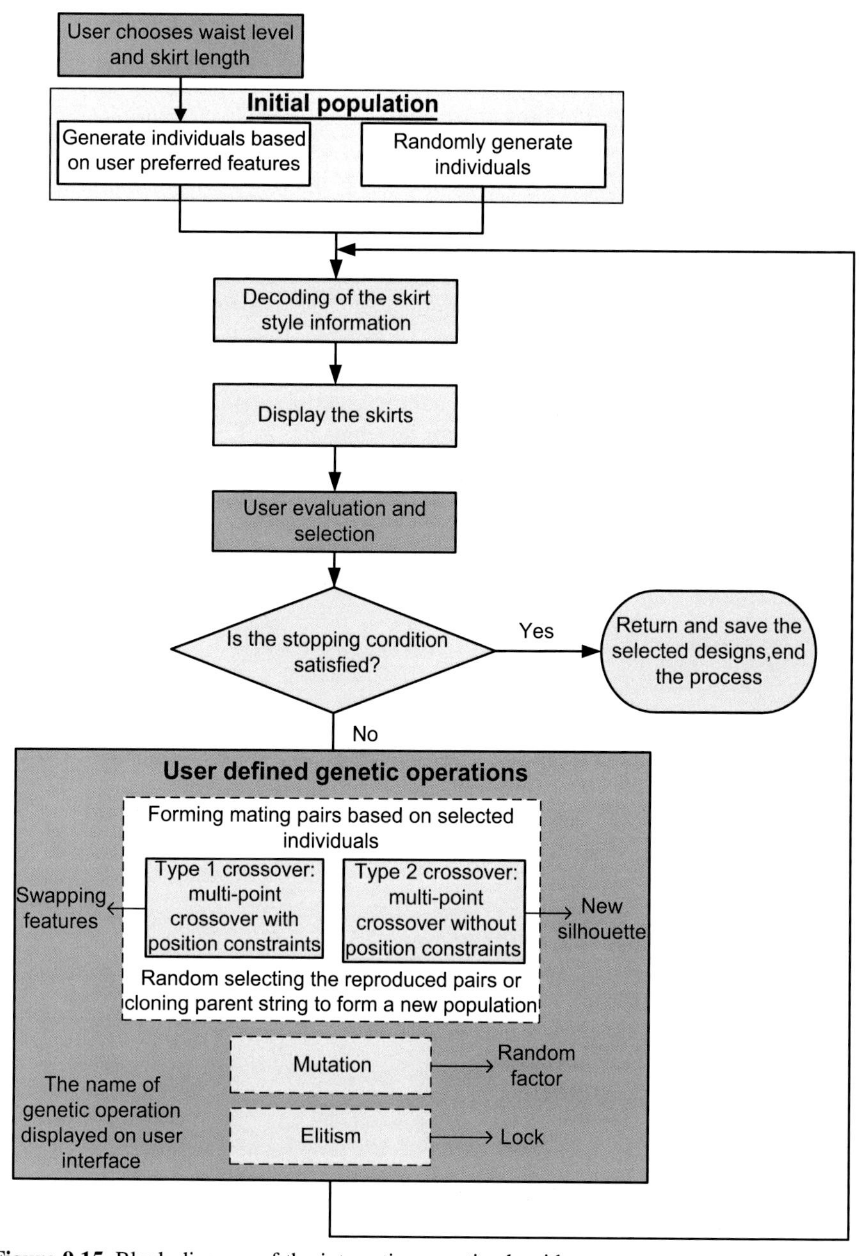

Figure 9.15 Block diagram of the interactive genetic algorithm.
Source: Mok, P.Y., Xu, J., Wang, X.X., Fan, J.T., Kwok, Y.L., Xin, J.H., 2013. An IGA-based design support system for realistic and practical fashion designs. Computer-Aided Design 45(11), 1442–1458.

As shown in Fig. 9.16(a), flat patternmaking generally involves three steps: basic block construction, fashion silhouette modification, and fashion feature creation (Liu, 2005; Aldrich, 2008; Zamkoff and Price, 2009). First of all, a set of basic blocks is constructed representing the respective shape of an individual using his/her body measurements. *Wearing ease*, which refers to the necessary room between the human body and the garment for basic body movements, is introduced in the stage of basic block construction. Next, the basic blocks are modified to achieve a desired silhouette, as depicted in flats, by incorporating *design ease*. Design ease is closely related to the design idea, while wearing ease is related to the customer's body shape. Secondary patterns are obtained after silhouette modification, and are further manipulated to incorporate fashion features, such as yokes, gathers, and pockets, to achieve a final look. The process involves sequential operations of modifying the overall shape and size of the pattern pieces, making small adjustments to certain areas of the pattern pieces, and designing additional pieces for attachments, and so forth. After all the necessary pattern operations, information like grain, notches, and seam allowances is added on the final patterns, which are then ready for garment production.

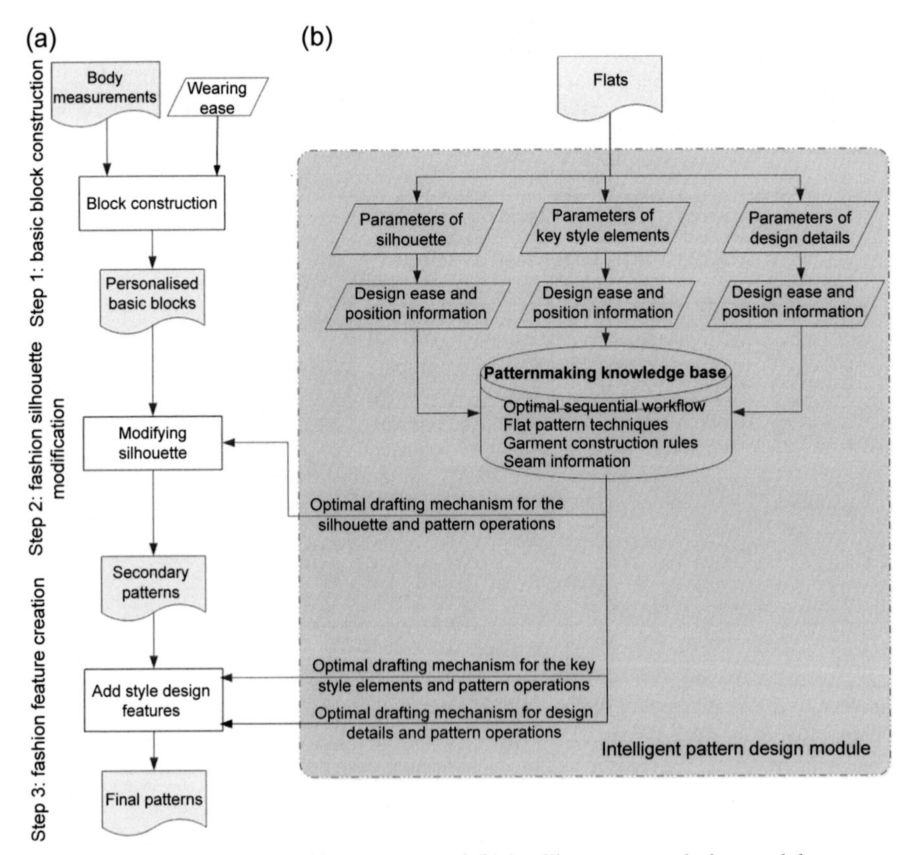

Figure 9.16 (a) Flat patternmaking process and (b) intelligent pattern design module.

The literature of clothing pattern design is focused on block construction – the first step of the above three-step process (Aldrich, 2008; Joseph-Armstrong, 2000; Liu, 2005; Miyoshi, 2001). Only limited work has been done on the subsequent steps of silhouette modification and fashion feature creation. The pattern design module automates these pattern operations using personalized basic blocks (Huang et al., 2012) and flats from the sketch design module as inputs.

Two key developments are involved in the pattern design module (as shown in Fig. 9.4): (1) mathematical relationship models between the sketches (flats) and patterns and (2) a patternmaking knowledge base. The mathematical relationship models are formulated in terms of design ease. The patternmaking knowledge base houses a suite of computer algorithms that automate the pattern operations on the input blocks in a systematic sequential manner.

9.4.1 *Mathematical relationship between sketches and patterns*

In the pattern design literature (Joseph-Armstrong, 2000; Aldrich, 2008; Lu et al., 2010), a fashion silhouette is obtained by adding appropriate amounts of wearing ease (WE) and design ease (DE) to the body measurements (BM):

$$\text{BM} + \text{WE} + \text{DE} = \text{Fashion Silhouette} \qquad [9.16]$$

Wearing ease was introduced in the basic block development, and design ease was introduced in subsequent pattern manipulations. As shown in Eq. [9.16], body measurements are used in pattern construction. As discussed in Section 9.3, sketches are composed of design elements, which are defined by parameters, eg, ratios or proportions to some standard sizes. To realize the correct shapes in pattern design, the design parameters expressed in the sketches must be converted into actual pattern measurements. It is thus required to establish mathematical relationship models between sketches (flats) and clothing patterns. These relationship models are defined with reference to the basic skirt shape.

Take the bell-shape skirt in Fig. 9.17 as an example. To achieve a bell-shape silhouette, the amounts of design ease at hip level, turn level, and hem level should be defined.

In Fig. 9.17, the dashed line represents the shape of the basic skirt (from basic blocks), and the solid line represents the desired bell-shape silhouette. As shown, no extra design ease is added to the hip level ($\text{DE}_h = 0$), because the solid line is of the same width as that of the basic blocks (dashed line) at hip level. L_{AB} is the parameter defining the ease at turn level (S8 in Eq. [9.1] and Table 9.1) on the sketch, which is represented as the ratio of distance AB to half the skirt width (distance from the dashed line to the center line). By proportion, the design ease at turn level (DE_t) can be calculated by

$$\text{DE}_t = \frac{4 \times L_{\text{AB}}}{\text{SWE}_h + \text{SH}} \qquad [9.17]$$

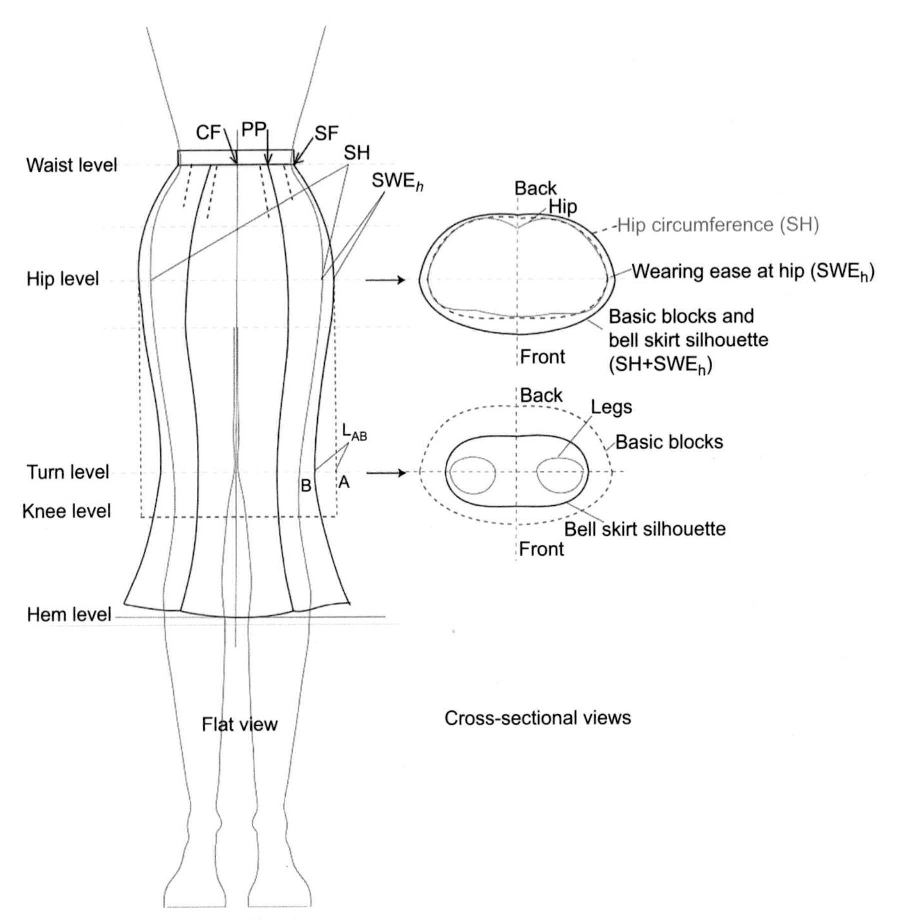

Figure 9.17 Bell-shape skirt sketch.

where SH is the standard body measurement, namely the hip measurement (SH); standard wearing ease (SWE$_h$) is included in the construction of basic blocks. Similarly, the design ease at hem level can also be calculated. With these design eases, the pattern measurements can be calculated according to Eq. [9.16].

9.4.2 Patternmaking knowledge base

A patternmaking knowledge base is developed to automate the patternmaking process, in which pattern operations are generalized and coded as computer algorithms using the Teigha.NET development kit. The rationale behind this is that complex pattern design can be achieved by sequential geometrical operations/manipulation of the basic blocks. The knowledge base contains four types of patternmaking knowledge: sequential workflow, flat pattern techniques, garment construction rules, and seam information. The optimal workflow specifies the required sequence of generic pattern operations for each style element. Flat pattern techniques are detailed pattern manipulation techniques, such as pivoting, slash, and spread, adding fullness. Garment construction rules are for avoiding conflicts among different design elements, ensuring pattern accuracy and compatibility (ie, curve truing and right angles at joining points), and being ready for manufacturing and virtual simulation (Meng et al., 2010, 2012). Seam information guides the finalizing process, in which seam allowances, annotations, and other manufacturing instructions such as notches and holes are added to the pattern pieces.

9.5 Results and discussions

9.5.1 Experimental results

Figs. 9.18–9.21 show four style results generated by the system, including the technical sketches, the corresponding clothing patterns, and trial fitting photos of sample skirts.

9.5.2 System characteristics

The sketch design process is an iterative process, where users assess the generated flats and select from them the ones they like until their preferred designs (flats) are created. It usually takes about three to four generations before users confirm their preferred designs. The system uses 1–6 s to generate a flat, depending on the complexity of the style. The sketch design module has the following characteristics/advantages.

First, output flats have fully compatible design elements and can adapt to various input croquis. The parametric sketch model has a hierarchical structure, which defines style elements/design details using information from the upper level. Therefore, the interdependence among the features/details and silhouette/shape is well defined and ensures the mutual compatibility of all design elements. Moreover, the output flats fit various input croquis, allowing the automation of sketch design by selecting design elements of the relevant parameters. The full compatibility of the output sketches is illustrated in Fig. 9.22.

Second, the design process is formulated by interactive genetic algorithms, which hide the tedious definition of parameters to generate parametric sketches. This allows general users to create styles by simple computer operations – selecting their preferred styles. Furthermore, fuzzy set theory was used to model design knowledge, taking into account the design harmony and aesthetic appeal. It also ensures that the output sketches are feasible designs that can be produced as real garments, considering the properties of different textile materials.

Third, the output sketches are parameterized, with explicit and precise design information. This minimizes style misinterpretation and eases the communication among different parties along the fashion supply chain. The precise and explicit design information provides all the necessary information in a design analysis, which can be readily used as input for the downstream pattern development process.

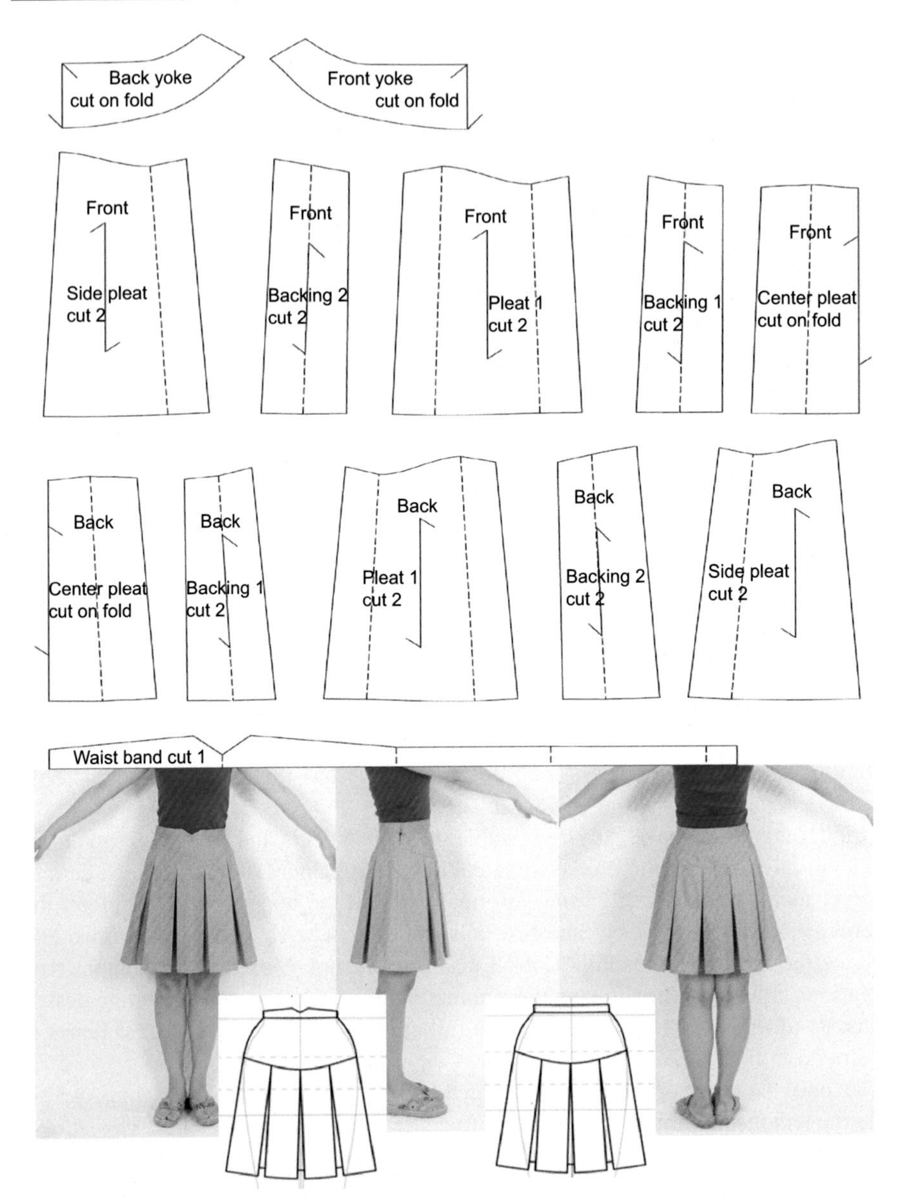

Figure 9.18 Style 1 result – output sketches, output patterns, and trial fitting photos.

The pattern design process takes 1 or 2 minutes to generate all the clothing patterns for a particular input sketch. The output clothing patterns are ready for 3D virtual simulation and downstream garment manufacturing. In terms of pattern design, the key contributions are as follows.

First, a new theoretical model is established for automatic patternmaking, which realizes customized fit and achieves diversity of styles. Although "fit" is a complex

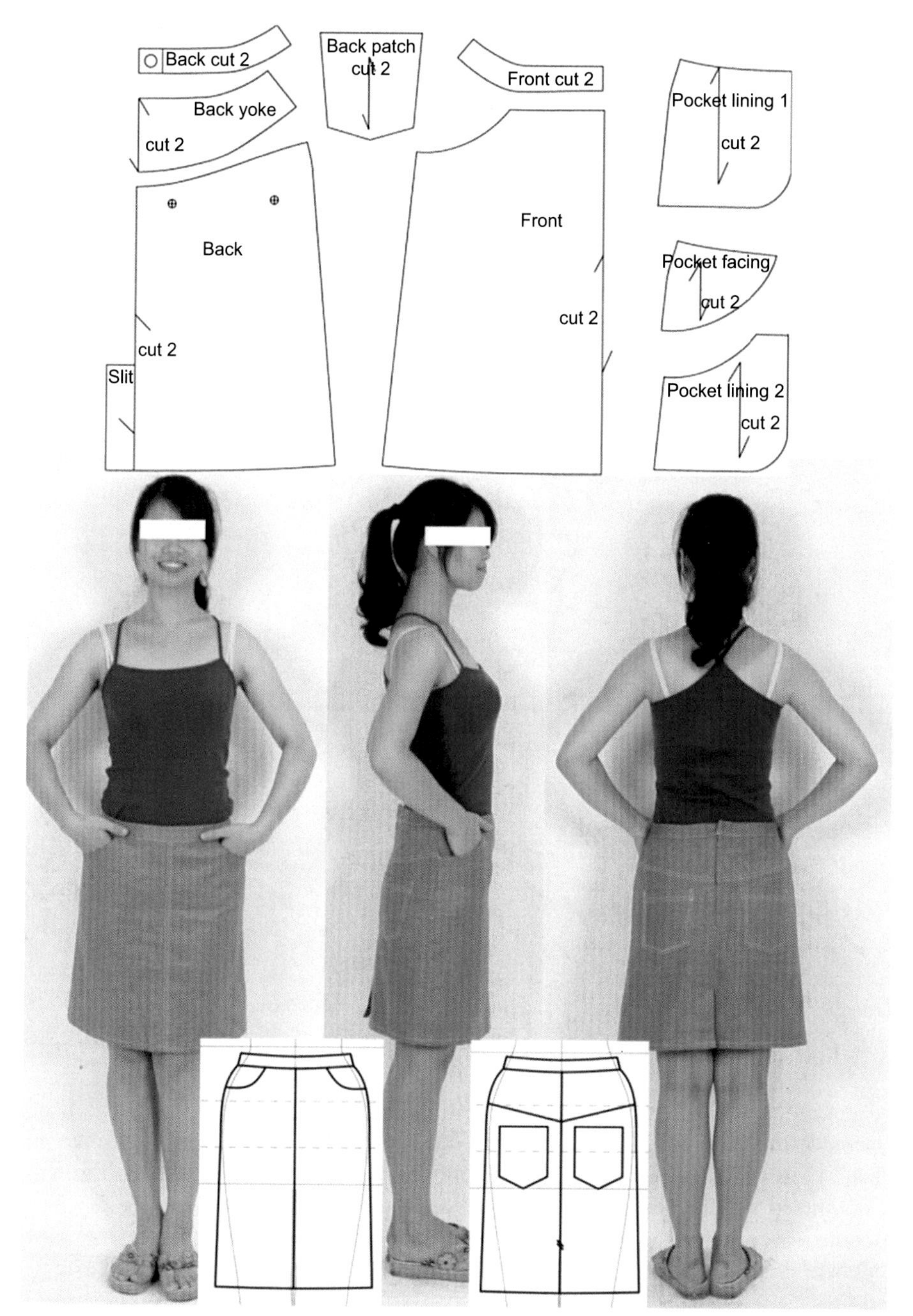

Figure 9.19 Style 2 result – output sketches, output patterns, and trial fitting photos.

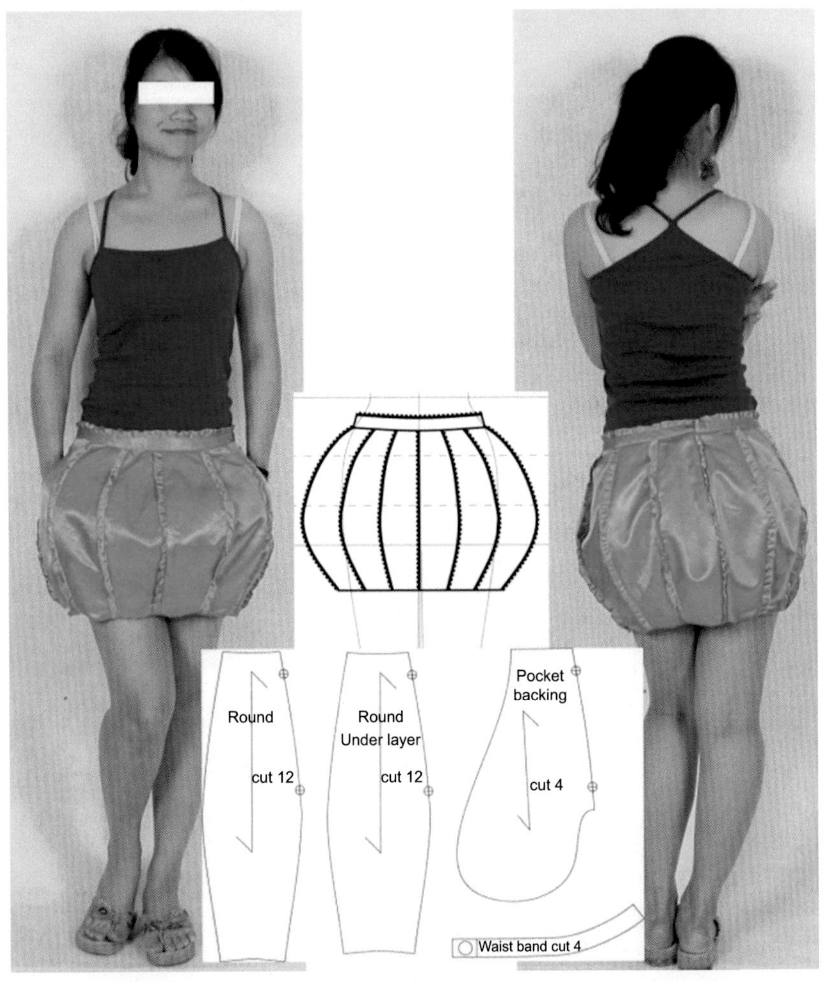

Figure 9.20 Style 3 result — output sketches, output patterns, and trial fitting photos.

concept in fashion design and has ambiguous criteria, iterative sampling is often used to achieve the ideal fit. The method provides a theoretical framework to realize an "ideal" customized fit. This is done by applying personalized basic blocks through an optimized pattern design process. The personalized blocks incorporate the best amount of wearing ease, and the intelligent patternmaking process derives the best amount of design ease to include. This is the first theoretical model to realize a customized fit.

Second, a pattern knowledge base is developed. This is a pioneer study of generalizing and optimizing flat patternmaking workflow and techniques for the purpose of automation. By this novel work, automatic pattern design for ready-to-wear styles can be realized.

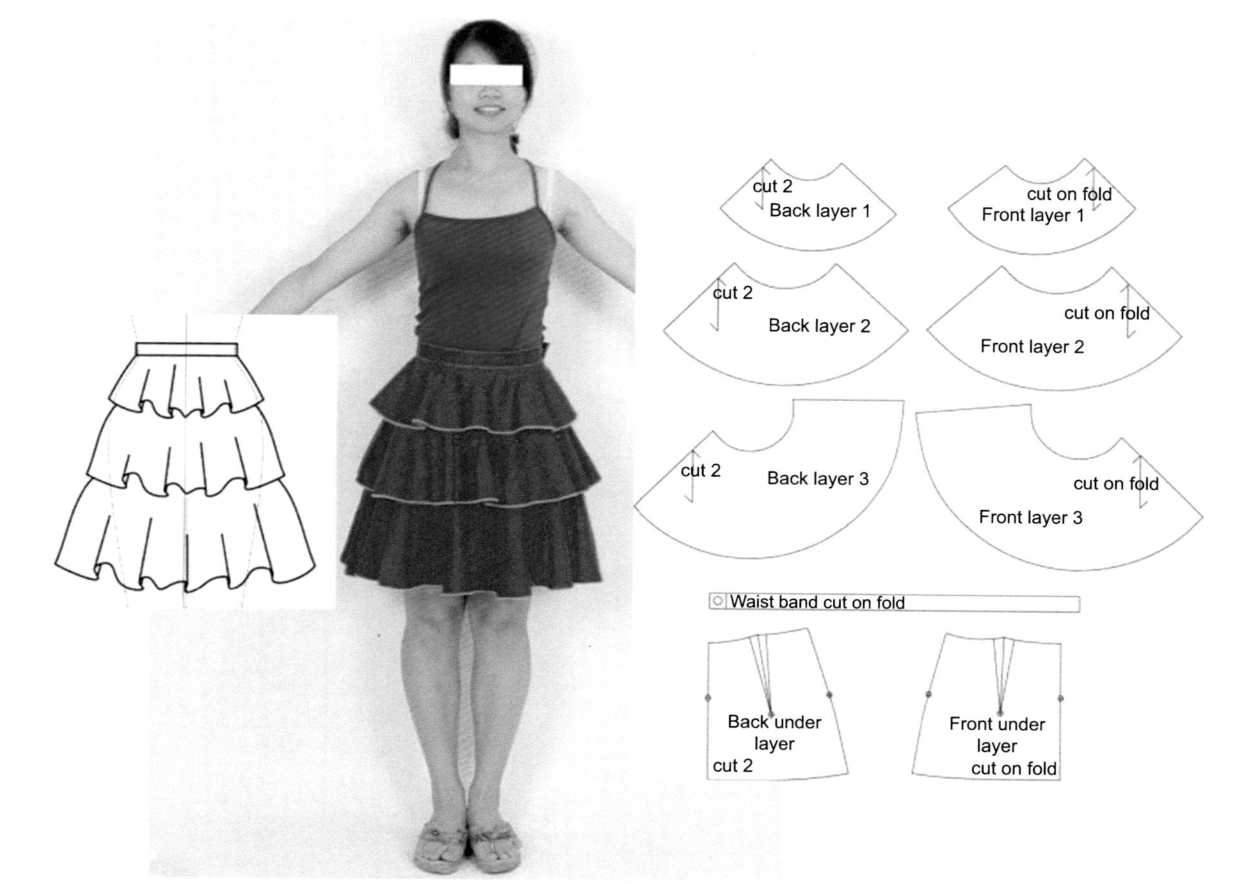

Figure 9.21 Style 4 result – output sketches, output patterns, and trial fitting photos.

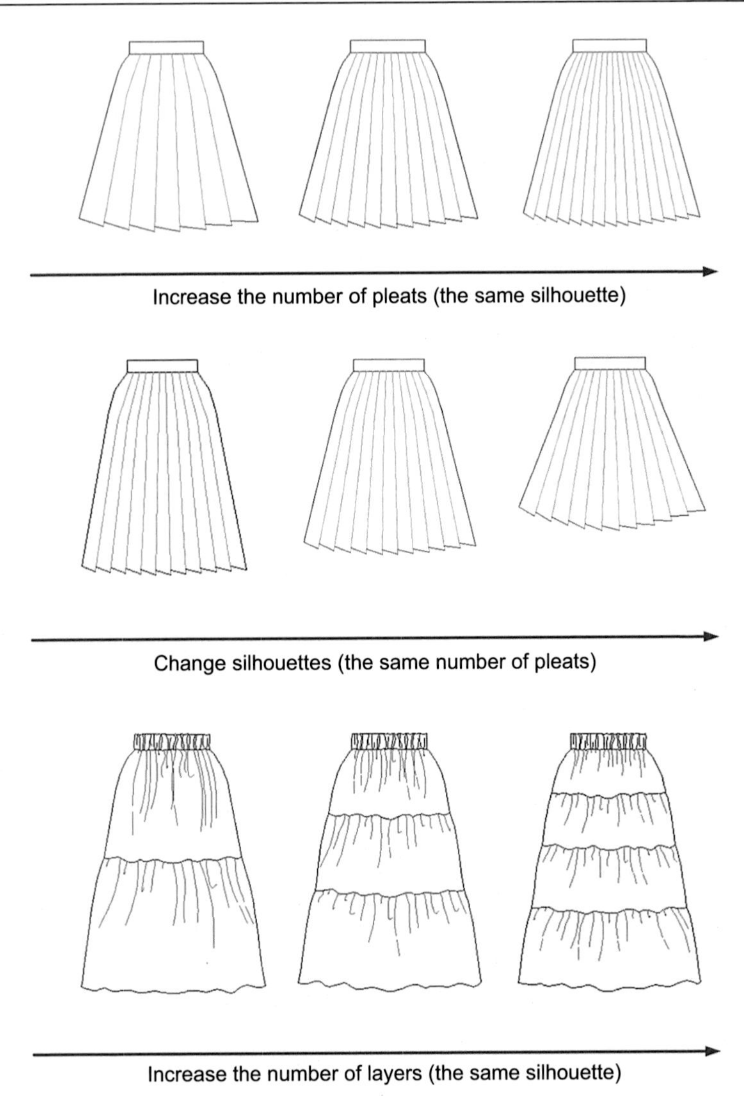

Figure 9.22 Examples illustrating that design elements are fully compatible.

Both the output sketches and output patterns are in the file format of DXF, which can be imported and further processed by most commercial CAD. The sketch design and pattern design modules are interoperable and fully compatible. The output sketches from the sketch design module can be "understood and interpreted" by the pattern design module to automatically generate relevant clothing patterns. This allows users without professional knowledge of design and patternmaking to create their own designs. It realizes a truly customized design from sketch to patternmaking.

9.6 Conclusions and future research

In this study, a customized fashion design system has been proposed for nonprofessional users (general customers) to create fashion designs and make corresponding patterns in a user-friendly way. First, a sketch representation and composing method is proposed. This method defines a new sketch representation and output sketch by integrating various compatible silhouette and design elements. Second, a design knowledge model has been developed to regulate feasible sketch composition, and to ensure aesthetically appealing and practically feasible styles. Third, an IGA-based design model has been designed to hide tedious design parameter definitions from customers, while still constructing effective design sketches based on customer preferences. Finally, an intelligent pattern design module is developed to improve the efficiency of the process from the generated sketch to final patterns. All these components are integrated in a user-friendly design support system, by which customers can easily create their preferred designs and corresponding patterns.

The current work can be enhanced further in the following three areas.

First, the style database should be expanded in the future. We also intend to transplant the system to the Web platform to maximize the efficiency of using and database update. In the Web structure, updating the database at the server does not influence the usage at the front end. Users only need to refresh their computer browsers instead of downloading, so the database can be frequently updated.

Second, the current system generates technical sketches that include only lines and curves without any texture or color information. It is appealing to add color or texture information to the technical sketches. Because the system generates sketches by combining different levels of design elements in a systematic and mutually compatible manner, filling color and texture mapping can be added easily.

Third, although it is generally agreed that fabric properties have a significant impact on patternmaking (Fan et al., 2004; Aldrich, 2008), this has not been carefully studied. A systematic and comprehensive study should be carried out in the future to quantify such impacts in a mathematical model, probably also in terms of design ease, between fabric properties and patterns.

Acknowledgments

The work described in this paper was financially supported by projects of the Hong Kong Polytechnic University (Project code: RT7B and A-PL78). It was also partially supported by a grant from the Research Grants Council of the Hong Kong Special Administrative Region, China (Project No. PolyU 5218/13E).

References

Aldrich, W., 2008. Metric Pattern Cutting for Women's Wear. Blackwell Science Publications, Oxford.

Bush, B.J., Sayama, H., 2011. Hyperinteractive evolutionary computation. IEEE Transactions on Evolutionary Computation 15 (3), 424–433.

Chen, Y., Zeng, X., Happiette, M., Bruniaux, P., Ng, R., Yu, W., 2009. Optimisation of garment design using fuzzy logic and sensory evaluation techniques. Engineering Applications of Artificial Intelligence 22 (2), 272–282.

Cho, Y., Komatsu, T., Inui, S., Takatera, M., Shimizu, Y., Park, H., 2006. Individual pattern making using computerized draping method for clothing. Textile Research Journal 76 (8), 646–654.

Cho, Y., Okada, N., Park, H., Takatera, M., Inui, S., Shimizu, Y., 2005. An interactive body model for individual pattern making. International Journal of Clothing Science and Technology 17 (2), 91–99.

Choi, K.J., Ko, H.S., 2005. Research problems in clothing simulation. Computer-Aided Design 37 (6), 585–592.

Dong, A., Leung, S.Y.S., 2009. A simulation-based replenishment model for the textile industry. Textile Research Journal 79, 1188–1201.

Fan, J., Yu, W., Hunter, L., 2004. Clothing Appearance and Fit: Science and Technology. Woodhead Publishing Ltd., Cambridge.

Fontana, M., Rizzi, C., Cugini, U., 2006. A CAD-oriented cloth simulation system with stable and efficient ODE solution. Computer & Graphics 30 (6), 391–406.

Grieves, M., 2006. Product Lifecycle Management: Driving the Next Generation of Lean Thinking. McGraw-Hill Companies, Inc, USA.

Guo, Z.X., Wong, W.K., Leung, S.Y.S., Fan, J.T., 2009. Intelligent production control decision support system for flexible assembly lines. Expert Systems with Application 36, 4268–4277.

Hu, Z., Ding, Y., Zhang, W., Yan, Q., 2008. An interactive co-evolutionary CAD system for garment pattern design. Computer-Aided Design 40 (12), 1094–1104.

Huang, H.Q., 2011. Development of 2D Block Patterns from Fit Feature-aligned Flattenable 3D Garments [Doctor of philosophy dissertation]. The Hong Kong Polytechnic University, Hong Kong.

Huang, H.Q., Mok, P.Y., Kwok, Y.L., Au, J.S., 2012. Block pattern generation: from parameterizing human bodies to fit feature-aligned and flattenable 3D garments. Computers in Industry 63, 680–691.

Hwang, S.J., 2004. Standardization and Integration of Body Scan Data for Use in the Apparel Industry (Ph.D. Thesis). North Carolina State University, USA.

Istook, C.L., 2002. Enabling mass customization: computer-driven alteration methods. International Journal of Clothing Science and Technology 14 (1), 61–76.

Joseph-Armstrong, H., 2000. Patternmaking for Fashion Design. Prentice-Hall, Inc., Upper Saddle River, New Jersey.

Kaufmann, A., Gupta, M.M., 1991. Introduction to Fuzzy Arithmetic. Van Nostrand Reinhold, New York.

Kim, S., Park, C.K., 2006. Development of a platform for realistic garment drape simulation. Fibers and Polymers 7 (4), 436–441.

Kwong, C.K., Mok, P.Y., Wong, W.K., 2006. Determination of fault-tolerant fabric-cutting schedules in a just-in-time apparel manufacturing environment. International Journal of Production Research 44, 4465–4490.

Leung, S.Y.S., Wong, W.K., Mok, P.Y., 2008. Multiple-objective genetic optimization of the spatial design for packing and distribution carton boxes. Computers & Industrial Engineering 54, 889–902.

Liu, R.P., 2005. Pattern Making Theory & Technology. China Textile & Apparel Press, Beijing, China.

Lu, J.M., Wang, M.J.J., Chen, C.W., Wu, J.H., 2010. The development of an intelligent system for customized clothing making. Expert Systems with Applications 37 (1), 799–803.

McCartney, J., Hinds, B.K., Seow, B.L., 1999. The flattening of triangulated surfaces incorporating darts and gussets. Computer Aided Design 31 (4), 249–260.

Meghna, B.S., Barbara, M., 2010. A case-based micro interactive genetic algorithm (CBMIGA) for interactive learning and search: methodology and application to groundwater monitoring design. Environmental Modelling and Software 25 (10), 1176–1187.
Meng, Y., Mok, P.Y., Jin, X., 2010. Interactive virtual try-on clothing design systems. Computer-Aided Design 42 (4), 310–321.
Meng, Y., Mok, P.Y., Jin, X., 2012. Computer aided clothing pattern design with 3D editing and pattern alteration. Computer-Aided Design 44 (8), 721–734.
Miyoshi, M., 2001. Clothing Construction. Bunka Publishing Bureau, Tokyo, Japan.
Mok, P.Y., Kwong, C.K., Wong, W.K., 2007. Optimisation of fault-tolerant fabric-cutting schedules using genetic algorithms and fuzzy set theory. European Journal of Operational Research 177 (3), 1876–1893.
Mok, P.Y., Xu, J., Wang, X.X., Fan, J.T., Kwok, Y.L., Xin, J.H., 2013. An IGA-based design support system for realistic and practical fashion designs. Computer-Aided Design 45 (11), 1442–1458.
Monk, E.F., Wagner, J.W., 2009. Concepts in Enterprise Resources Planning. Course Technology Cengage Learning, Boston, USA.
Ogata, Y., Onisawa, T., 2007. Interactive clothes design support system. In: Proceedings of the 14th International Conference on Neural Information Processing, Part II, pp. 657–665. Kitakyushu, Japan.
Pan, A., Leung, S.Y.S., Moon, K., Yeung, K., 2009. Optimal reorder decision-making in the agent-based apparel supply chain. Expert Systems with Applications 36, 8571–8581.
Piller, F.T., Tseng, M.M., 2003. New Directions for Mass Customization. Springer, Berlin.
Sul, I.H., Kang, T.J., 2006. Interactive garment pattern design using virtual scissoring method. International Journal of Clothing Science and Technology 18 (1), 31–42.
Sun, Z., Choi, T., Au, K., Yu, Y., 2008. Sales forecasting using extreme learning machine with applications in fashion retailing. Decision Support System 46, 411–419.
Thomassey, S., Happiette, M., Castelain, J., 2004. A short and mean-term automatic forecasting system – application to textile logistics. European Journal of Operational Research 161, 275–284.
Thomassey, S., Fiordaliso, A., 2006. A hybrid sales forecasting system based on clustering and decision trees. Decision Support Systems 42, 408–421.
Volino, P., Cordier, F., Magnenat-Thalmann, N., 2005. From early virtual garment simulation to interactive fashion design. Computer-Aided Design 37 (6), 593–608.
Wang, C.C.L., 2005. Parameterization and parametric design of mannequins. Computer-Aided Design 37 (1), 83–98.
Wong, W.K., Zeng, X., Au, W., 2009a. A decision support tool for apparel coordination through integrating the knowledge based attribute evaluation expert system and the T-S fuzzy neural network. Expert Systems with Applications 36, 2377–2390.
Wong, W.K., Zeng, X., Au, W., Mok, P.Y., Leung, S.Y.S., 2009b. A fashion mix-and-match expert system for fashion retailers using fuzzy screening approach. Expert Systems with Applications 36, 1750–1764.
Yager, R.R., 1981. A procedure for ordering fuzzy subsets of the unit interval. Information Sciences 24, 143–161.
Zadeh, L.A., 1965. Fuzzy sets. Information and Control 8, 338–353.
Zamkoff, B., Price, J., 2009. Basic Pattern Skills for Fashion Design. Fairchild Pulications, New York, NY.
Zhu, S., Mok, P.Y., Kwok, Y.L., 2013. An efficient human model customization method based on orthogonal-view monocular photos. Computer-Aided Design 45 (11), 1314–1332.

Intelligent systems for managing returns in apparel supply chains

Y. Li, F. Xu, X. Li
Nankai University, Tianjin, China

10.1 Introduction

The fashion and apparel industry is a major sector of developed and developing countries, contributing to economic development and social employment. As a relatively traditional industry, the apparel industry is gradually being combined with more popular elements. Specifically, fast fashion, which aims for production and selling in a timely and cost-efficient manner, has become a classic business model. For example, Zara achieves an impressive "15 days' magic," in which the whole process of product design, production, packaging, and sale takes a mere 15 days. At present, a high return rate is a new challenge faced by modern apparel enterprises based on the new features of the apparel industry. The impulsive purchase of consumers is an important factor of the high return rate in the apparel industry. For fast fashion products, stock-out can sometimes stimulate a more "fanatical pursuit" of fast fashion products. This fast fashion retailing strategy enhances consumers' impulse purchases to some extent. In the apparel industry, product return rates are reported to range from 10% to 20% for casual apparel and as high as 35–40% for high fashion. Chinese Industry Research Network indicates that the average apparel return rate is 25% during "Double 11" in 2013, and this rate even reaches up to 40% in some enterprises.

Therefore, enterprises in the apparel industry should focus more on maximizing the efficiency of return process management and minimizing the cost associated with product returns. However, returns management is not easily performed because of its unpredictability, low visibility, and low traceability. In addition, the operational and strategic decision-making issues in a fast fashion supply chain are different from those in a conventional supply chain system. Moreover, returns management can be regarded as a complex system engineering, which requires intelligent selections and optimal decisions, especially in apparel supply chains. As a modern management tool, information systems have been gaining academic as well as practitioner attention in the apparel industry. For instance, several apparel retailers in the United States, including Guess, Sports Authority, and the Limited, have addressed the return problem by using a shopper tracking technology called Verify-1 to detect consumers who are inclined to return items and to control the quantity of product returns to some extent (Campanelli, 2005). In addition, enterprise resource planning (ERP) systems have been widely applied in the fashion industry, and relevant literature shows that ERP system implementation can bring significant benefits. An efficient intelligent system (IS) allows supply chains to achieve agility; reduce uncertainty, cycle-time, and inventory; and collaborate with

Information Systems for the Fashion and Apparel Industry. http://dx.doi.org/10.1016/B978-0-08-100571-2.00010-5

networked members closely and maximize efficiency of returns management. Consequently, building an efficient IS for managing returns in apparel supply chains can raise the utilization ratio of returned clothes, accelerate the process of returns, improve brand image, and promote sustainable development of companies.

This chapter focuses on returns management with ISs in the apparel industry, summarizes the extant research in this field, and aims to provide recent research and relevant results for decision makers in business practice. We discuss this problem in the following aspects. In Section 10.2, we provide an overview of recent research on ISs for managing returns in apparel supply chains. In Section 10.3, we analyze the unique characteristics of product returns in apparel supply chains, which are critical to the design and implementation of ISs of returns management that satisfy the specific requirements of apparel supply chains. After the general features are realized, a quantity model can help to analyze the research problem in detail before an IS is designed and implemented. Therefore, based on different resources of returns and selling models, several mathematical models are established to analyze the apparel supply chain in Section 10.4, and Section 10.5 provides a systematic perspective to IS of returns management in the apparel industry based on differences in properties between the supply chains that carry fast fashion products and conventional newsvendor-type productions introduced in Sections 10.3 and 10.4. Last, we summarize the whole chapter and discuss the future research in Section 10.6.

10.2 Literature review

10.2.1 Returns management in the apparel supply chain

A "returns policy" is one of the various policies implemented for supply chain management that has been widely applied in the real world. And the return policy has become a very important tool for manufacturers and retailers. Enterprises (manufacturers and retailers) and consumers make corresponding optimal decisions based on the market environment and their own preference (Mukhopadhyay and Setaputra, 2007; Su, 2009; Hsieh and Lu, 2010; Li et al., 2013). According to the return policy research in traditional supply chains, online selling has become a new research topic in returns management (Mukhopadhyay and Setoputro, 2004; Chen and Bell, 2011; Zhai and Li, 2011; Li et al., 2013; Nakhata and Magi, 2015).

Different from conventional newsvendor-type production, fast fashion brands plan to reach stock-out because it is a feature of fast fashion that obtains certain benefits. In addition, a returns management policy and mechanism must be established in industry practice. Based on fast fashion features, Li et al. (2014a) build an analytical MV optimization model for a two-echelon fast fashion supply chain with a returns policy. Yao et al. (2005) investigate the role of the returns policy in the coordination of fashion supply chains. In this model, a manufacturer provides a returns policy for unsold goods to two competing retailers who face uncertain demands. Considering the features of fashion products, Hu et al. (2014) develop a system of a rent-based, closed-loop supply chain to address returns from consumers. In this research, the supply chain processes, operations management issues, and sustainability promotion aspects are investigated. Li et al. (2014a) study the returns policy and the advance-selling strategy of a retailer who sells fashionable products in light of potential opportunistic returns of consumers. In this model, the consumers face valuation uncertainty and know their valuation realization only after product acquisition. Choi (2013) analytically examines the optimal return service charge policy under mass customization (MC) service in the fashion industry. Liu et al. (2012) study the optimal policy by constructing an analytical model with both demand and return uncertainties under MC service.

10.2.2 Intelligent systems for managing returns in the textile and apparel supply chain

The textile and apparel industry is characterized by unpredictable demand, short product life cycles, quick response time, large product variety, and a volatile, inflexible, and complex supply chain structure (Fischer, 1997). ISs can improve efficiency (Srinivasan et al., 1994; Gunasekaran and Ngai, 2004) and effectiveness (Levary, 2001; Ngai et al., 2014) in supply chain management, especially in the competitive market environment of the textile and apparel industry. Several kinds of ISs are commonly used in the apparel industry, including expert systems (Ford and Rager, 1995; Metaxiotis, 2004; Wong et al., 2009; Shahrabi et al., 2013), genetic algorithms (Lin, 2007, 2009; Hsu et al., 2009; Mok, 2011; Wong et al., 2013), artificial neural networks (Thomassey and Happiette, 2007; Hui et al., 2007; Sun et al., 2008; Vigneswaran et al., 2012), and decision support systems (Wong et al., 2009, Rong-Chang et al., 2006; Choy et al., 2013; Chen et al., 2014; Ruiz-Torres et al., 2013).

Product returns, especially in the apparel industry, are characterized by uncertainty and a need for timing and processing. ISs use data and mathematical models that possess the characteristics of flexibility, adaptability, memory, comprehension, and the ability to manage uncertain and constantly changing information. Therefore, ISs significantly help to improve efficiency and effectiveness of reverse supply chain management, particularly within a competitive environment such as the textile and apparel industry.

The application of ISs on returns management has been studied by many scholars. Temur et al. (2014) develop a fuzzy expert system to design a robust forecast of returns quantity to handle uncertainties from the returns process in a reverse logistics (RL) network. Guo et al. (2015) propose a multi-period and dynamic joint construction model to build the RL network and verify the feasibility of the model by adopting a particle swarm optimization algorithm and genetic algorithm (GA) with a case study of the apparel e-commerce enterprises of Shanghai. Turki and Mounir (2014) propose an integrated decision support system (DSS) that uses specifically treated issues to make a complete RL conceptual framework to develop better decision models. The DSS aims to help managers structure and organize their returned products. Improved communication among all elements in an RL system leads to improved control of returned products, which is crucial in correct and efficient decision-making.

Specific to apparel supply chains, relevant research on ISs to manage returns in the apparel industry seems to be insufficient. For example, Hwang (2008) explores the application of RFID technology on centralized reverse supply chain management in the apparel industry, and relevant results show that RFID-based ISs improve operational, tactical, and strategic levels, such as operational efficiency in return management and return forecasting accuracy; reduce return inventory and labor; and strengthen partnerships in the reverse supply chain. Madaan et al. (2012) propose a flexible reverse enterprise system (RES) model that can handle returned products according to different features. From an overall perspective, the extant research and studies in the field of ISs for returns management in apparel supply chains focus more on the basic guidelines and technical framework of ISs in returns management, lacking a profound study on specific algorithms and decision support mechanisms.

10.3 Critical factors of returns management in apparel supply chains

The fashion and apparel industry has its own market characteristics, such as short product life cycle, high volatility, low predictability, and a high level of impulsive purchase. A successful IS for managing returns in apparel supply chains must be designed with consideration of these specific features and corresponding management strategies. Consequently, identifying the characteristics of returns management in apparel supply chains is critical in designing and implementing ISs for managing apparel returns, not only for scholars but also for enterprise managers.

10.3.1 Common types of product returns in the apparel industry

In the apparel industry, product returns are caused by different reasons. Hwang (2008) summarizes four main kinds of product returns: non-defective return, defective return, overstock return (excess return), and bad stock return. Among them, consumers generate the first two type of product returns. And the last two types come from retailers.

1. Nondefective return
   A nondefective return is always caused by consumer dissatisfaction of a product without a quality problem. In the apparel industry, certain retailers guarantee consumers that they can return the sold product within the warranty period (7 or 15 days).
2. Defective return
   Defective return is always caused by quality problems of products. As a more common reason, poor product quality may lead to product returns. If a product is found by consumers to have quality problems, then it is returned to retailers or manufacturers.
3. Overstock return
   Overstock return refers to first-quality products that the retailer has in excess, and it is sometimes caused by demand overestimation of retailers. Retailers must return the unsold products if they ordered significantly more than the actual sales. In addition, overstock return can also result from an overzealous manufacturer. This phenomenon may be caused by manufactures' inaccurate forecasts or inappropriate production quantity.
4. Bad stock return
   Bad stock is another type of product returns in the apparel industry. It means returning the bad stock that has remained in stock for an extended period. In these cases, the retailer always finds selling out the products to be considerably difficult, and the stock cost is relatively high. Therefore, the bad stock will be returned to manufacturers.

10.3.2 Characteristics of product returns in the apparel industry

At present, enterprises gradually realize the importance of returns management and increasingly focus on optimizing the return process based on the specific features of products and the industry. In the apparel industry, managing returns is a more serious problem. Compared with other industries, returns management in apparel supply chains has its own characteristics (Hwang, 2008).

1. High return rate
   The return rate in the apparel industry is relatively high (approximately 19.4%). Such a value is roughly three times of the return rates of other industries. This high return rate is caused by the characteristics of apparel products and the population characteristics of consumers. Apparel products are highly customer-sensitive products. Therefore, after being purchased, these products should meet the necessary conditions for customer satisfaction such as size, color, state after washing, and view of others.
2. High marginal value of time
   Apparel products have high marginal value of time (MVT), which indicates the degree of value of the products after launching. The flow of returned products represents a sizeable asset stream for many companies, but a significant amount of this asset value is lost in the reverse supply chain as time progresses (Blackburn et al., 2004). MVT is represented by the slopes of the lines in Fig. 10.1. Time-sensitive products (high MVT) such as apparel products always lose their value at rates in excess of 1% per week, as shown in Fig. 10.1.
3. Season feature
   Third, the value of apparel products tends to vary with the seasons. To be specific, apparel products' value decreases during the season and degrades dramatically after the season ends (Fig. 10.1). In other words, after the end of a selling period, returns have to be sold at a lower price in the secondary market or simply disposed. Therefore, in the apparel industry, returned products without quality defect should be resold quickly because of high MVT. Hence, the efficiency of returns handling is very important to the apparel industry.

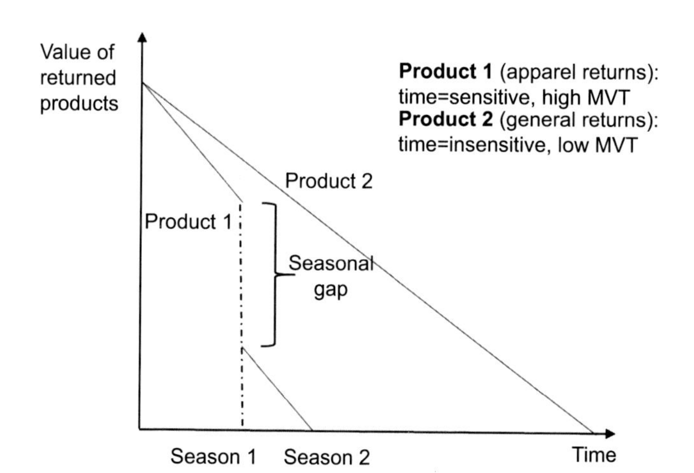

Figure 10.1 Marginal value of time of apparel returns and general returns.
Compiled from Hwang, Y.M., 2008. The Impact of RFID on the Centralized Reverse Supply Chain Management (RSCM) in Apparel Industry. Information and Communications University.

4. High return cost
High cost is another feature of returns management in the apparel industry. Compared to other industries, the stock keeping units are usually higher. For apparel manufacturers and retailers, the combination of different sizes, colors, styles, fabrics, price lines, and consumer groups means that a retailer must carry an enormous range of different returned products. Therefore, returns handling processes, including picking, sorting, and inspection, are always difficult and time-consuming. In addition, because of the characteristics of this labor-intensive industry, the handling cost of apparel returns is relatively high.

10.3.3 Strategies of returns management in the apparel industry

Based on the characteristics of returns management in apparel supply chains, the corresponding return policies should be decided based on the unique features of the apparel industry. Lund (1984) proposes two different strategies for a flow system: to be efficient and to be flexible. These strategies can be extended further to the product recovery system to develop an **efficient** RES designed to recover products at low cost and a **flexible** RES designed for the speed of response of the recovery option. Fig. 10.2 shows the relationship between flexibility and efficiency for product returns.

Based on the characteristics of product returns in the apparel industry, including high return rate and high MVT, **a flexible DIS (Decision and Information Synergy)**

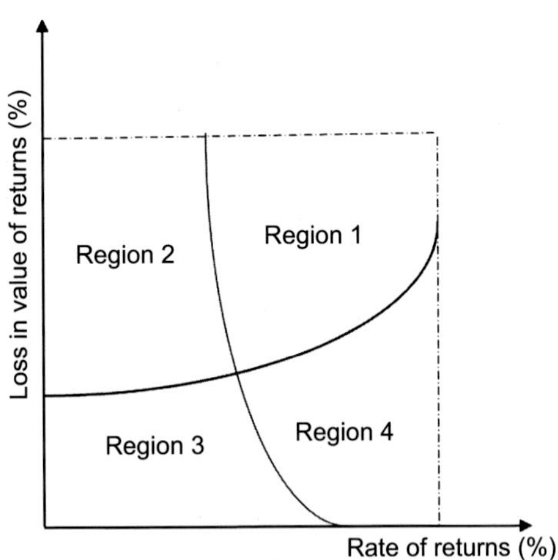

Figure 10.2 Trade-off between flexibility and efficiency in building DIS (Decision and Information Synergy) for RES.
Compiled from Madaan, J., Kumar, P., Chan, F.T.S., 2012. Decision and information interoperability for improving performance of product recovery systems. Decision Support Systems 53(3), 448–457.

strategy (Region 1) that can handle more returns with less value loss during returns processing is more appropriate for apparel returns. In other words, decreasing delays is critical to returns management in apparel supply chains.

10.3.4 Identifying return-inclined consumers in the apparel industry

For manufacturers and retailers in apparel supply chains, identifying consumers who are more likely to return apparel products is highly critical in their daily operations and strategy formulation. It is a prerequisite of an efficient and professional IS.

Kang and Johnson (2009) conducted an empirical study to investigate the relationships between apparel return behavior and its potential influencing factors (fashion innovativeness, buying impulsiveness, and consideration of return policies).

Via data collection from questionnaires and statistical analysis, four proposed hypotheses were tested and rectified based on confirmed reliabilities. The ultimate statistical results showed that the frequent apparel return behavior of participants was positively related to **buying impulsiveness** and **consideration of the return policies of a store** and was predicted by these two variables. However, apparel return behavior was not related to fashion innovativeness. Moreover, the innovative and impulsive purchase behaviors of the participants had no significant relationships with their consideration of return policies. These findings have implications for both retailing practices and consumer research.

10.4 Quantity model for managing returns in apparel supply chains

In addition to qualitative analysis, mathematical modeling is another way of realizing the research problem in depth. Understanding basic operational rules is a prerequisite to establish and implement ISs. Therefore, quantity model analysis is indispensable in studying ISs to manage returns in apparel supply chains. To explore the operational mechanism of returns policy in the apparel industry, many scholars propose mathematical models based on the features of apparel supply chains. This section reviews typical literature in this research field. For apparel returns from retailers, Li et al. (2014a) build an analytical MV optimization model for a two-echelon fast fashion supply chain. Furthermore, for apparel returns from consumers, optimal return policies are discussed in different sales models (advanced selling and mass customization (MC)). The relevant conclusions provide a theoretical foundation for IS implementation in apparel supply chains.

10.4.1 Optimal returns policy for retailers in apparel supply chains

In apparel supply chains, retailers and suppliers are important members in the sale stage. And returns from retailers are an indispensable part in reverse flow of apparel supply chains. Aiming at this phenomenon, Li et al. (2014a) established an optimization model to explore the research problem in depth.

A two-echelon make-to-order (MTO) apparel supply chain is considered, with one supplier and one retailer. The retailer sells a fast fashion product and faces an *i.i.d* stochastic market demand, X, with a pdf f and a cdf F, in a short selling period. The supplier incurs a unit production cost, c, and supplies the product to the retailer at a unit wholesale price, w. The retailer sells the product in the market at a unit price, p. The order quantity of the retailer is denoted by, q, which is also the production quantity of the supplier because the supply chain is MTO. During the selling season, owing to the characteristics of the fast fashion market, any stock-out will incur a unit additional value π to the retailer. At the end of the selling season, any unsold quantity from the retailers can be returned to the supplier at a standard returns refund, b. The supplier will salvage the returned quantities in a secondary market at a unit value, v.

For notational convenience, subscripts R, S, and SC represent the retailer, supplier, and supply chain, respectively. Define $A \wedge B = \min(A, B)$. The expressions of the profits of all the parties are shown as follows:

$$P_{SC}(q) = p(q \wedge x) + v(q - x)^{+} + \pi(x - q)^{+} - cq,$$

$$P_{R}(q) = p(q \wedge x) + b(q - x)^{+} + \pi(x - q)^{+} - wq,$$

$$P_{S}(q) = (w - c)q - (b - v)(q - x)^{+}.$$

The expressions of expected profit (EP) and variance of profit (VP) are also given based on these basic profit functions.

Based on quantitative analysis of this basic model, comparisons of structural properties of EPs and VPs between fast fashion products and newsvendor products are made, and some important propositions are identified. According to different risk types of the retailer, the basic model is analyzed in the following two cases: (1) risk-neutral retailer and (2) risk-averse retailer. The coordination mechanisms are also explored under the two cases. Based on the basic model and its corresponding analysis, the extended model, which consists of a single-supplier and multi-retailers, is also explored in this paper.

10.4.2 Optimal returns policy for consumers in apparel supply chains

The proportion of returns from consumers in an apparel supply chain is relatively high. And the quality and value of returned apparel products from consumers vary wildly. Therefore, it is urgent for apparel enterprises to manage consumer returns scientifically and efficiently. Motivated by contemporary industrial practice, Li et al. (2014b) and Choi (2013) analytically examine the optimal return service charge policy under two sales models: **advanced selling (AS)** and **MC**. AS is a commonly observed industrial practice in which a retailer allows consumers to pre-book a fashionable product before the real selling season starts. MC service is a pertinent industrial practice in the fashion apparel industry. It refers to a service offered by a seller to consumers who wish to purchase customized products or services. Implementation models for optimal returns management have to be decided based on the characteristics and sales models in apparel supply chains.

10.4.2.1 Advanced selling

The following model studies the AS strategy for a retailer who sells a newsvendor-type of fashionable product in light of potential opportunistic returns of consumers. In the AS season that ends before the normal selling season, the retailer announces an AS price, ξp, in the advance-selling season, and a selling price, p, in the normal selling season. The retailer then offers a refund, r, to allow consumers to return the product when they are not satisfied with it. At the beginning of the normal selling season, based on the demand realization, x, in the AS season, the retailer can update the demand forecast, Y, in the normal selling season and determine the order quantity, $Q+x$, from the manufacturer. Then, the retailer delivers the preordered product, x, and takes back the returns of the consumers. However, the retailer starts to sell the product at a price, p, and the leftover unit has a salvage value, s.

In this model, the consumers face valuation uncertainty and realize their valuation only after product acquisition. In the AS season, the consumers' valuation $V_i \sim U[1,h]$ is identically and independently drawn from the distribution $G_i(\cdot)$. According to the concern of consumers on AS, consumers are classified into two groups: informed and uninformed. Informed and uninformed consumers are myopic in purchasing products. Thus, informed consumers cannot transit to the normal selling season. Furthermore, informed consumers can also be classified into two groups: one has high valuation of the product, and the other has lower valuation. Specifically, the high-valuation consumers group is marked as $V_H = V_i$ with pdf $g_H(\cdot)$ and cdf $G_H(\cdot)$, and the low-valuation consumers group is marked as $V_L = \varphi V_i (0 < \varphi \leq 1)$ with pdf $g_L(\cdot)$ and cdf $G_L(\cdot)$, and their proportions are α and $(1-\alpha)$ $(0 \leq \alpha \leq 1)$, respectively. The trial value of consumers is $\beta V_i (\beta \geq 0)$. Uninformed consumers can only buy during the selling season, and their valuation on the product is V_u.

In the selling season, the retailer faces a standard newsvendor problem and determines the optimal order quantity, $Q+x$, given that the demand of the AS period is realized as x. Meanwhile, consumers acquire the product booked in advance and determine whether to return or hold it. If the consumers return the product, they can be paid with a full or partial refund. Thereafter, the retailer decides the AS discount rate, ξ, to attract all informed consumers to order the product early. The demand of the uninformed customer in the AS case is $D_{AS} = \sum_{i=1}^{y} 1(v_i \geq p)$. Given the demand realization, x, in the AS season, the retailer can update their demand forecast, Y, in the selling season. The new mean and standard deviation of the demand in the selling season can be easily proposed.

The model is analyzed in two cases: (1) AS with full refund (AS with AFP) and (2) AS with partial refund (AS with APP).

Case I: AS with AFP

In the AFP case, the retailer offers a full refund, that is, consumers can return the product with a full refund. The total EP of the retailer is shown as follows:

$$\pi_{\text{FRN}} = \max_{Q \geq 0} E_{D_{\text{AS}}|X=x}[pE\min\{Q, D_{\text{AS}}\} + s[Q - E\min\{Q, D_{\text{AS}}\}] - wQ].$$

The first term is the AS revenue. The second term represents the salvage value of the returned product from the consumers, and the last term is the procurement cost for the product in the AS season. The optimal order quantity and the AS discount rate are derived through mathematical analysis.

Case II: AS with APP

In the APP case, the informed consumers make two decisions sequentially. In the AS season, they decide on whether to preorder the product. In the selling season, after the products are delivered, they decide on whether to keep the product after observing it with their own valuation privately. For consumers who do not preorder in the AS season, only part of them whose valuations are at least as high as the selling price $v_i \geq p$ will choose to purchase in the selling season. First, the decisions of the consumers in the AS season are considered. In the AS season, the expected consumption utilities of groups H and L are these:

$$\text{EU}_H = \int_{\frac{r}{1-\beta}}^{+\infty} (v_H - \xi_{\text{PR}}p)g_H(v_H)dv_H + \int_0^{\frac{r}{1-\beta}} (r - \xi_{\text{PR}}p)g_H(v_H)dv_H,$$

$$\text{EU}_L = \int_{\frac{r}{1-\beta}}^{+\infty} (v_L - \xi_{\text{PR}}p)g_L(v_L)dv_L + \int_0^{\frac{r}{1-\beta}} (r - \xi_{\text{PR}}p)g_L(v_L)dv_L.$$

In total, the expected consumption utility of each consumer is shown:

$$\text{EU} = \alpha\text{EU}_H + (1 - \alpha)\text{EU}_L.$$

Then, the APP case is investigated under different selling pricing policies (penetration pricing policy and skimming pricing policy), and the optimal refund is derived under the two policies. Interesting conclusions can be derived from the comparison of optimal pricing policies and profits in these two cases.

10.4.2.2 Mass customization

Based on the industrial practice, Choi (2013) uses a mathematical modeling approach to examine the optimal returns service policy in two cases: the risk-neutral and the risk-averse MC service provider.

Following industrial practice in the fashion MC program with a consumer returns policy, the service provider offers a full refund minus the return service charge, $r = p - l$. Thus, the demand function and return function are proposed as a function of the return service charge, l, in the following:

$$D(l) = \alpha - \beta p + \gamma(p - 1) + \delta m + \varepsilon_D,$$

$$R(l) = \phi + \psi(p - 1) + \varepsilon_R,$$

where p is the retail selling price, l is the return service charge, m is the level of modularity that reflects the degree of customization, $\alpha > 0$ is the base demand, $\beta > 0$ is the price−demand sensitivity coefficient, $\gamma > 0$ is the refund−demand sensitivity coefficient, $\delta > 0$ is the modularity−demand sensitivity coefficient, ϕ is the base−return quantity, ψ is the refund return−quantity sensitivity coefficient, and both ε_D and ε_R are continuous random variables.

The profit function can be easily derived as follows:

$$\pi(l) = (p - w)(\alpha - \beta p + \gamma(p - l) + \delta m) - (\phi + \psi(p - l))((p - l) - vm) + (p - w)\varepsilon_D - ((p - l) - wm)\varepsilon_R.$$

Thereafter, given expectation and variance, the EP $E[\pi(l)]$ and the variance of the profit $V[\pi(l)]$ can be expressed.

This model is considered in two cases: the risk-neutral and the risk-averse MC service provider.

Case 1: risk-neutral MC fashion service provider

For the risk-neutral case, the optimal return service charge of the MC service provider, denoted by l_{RN*}, is the solution to the following optimization problem:

$${l_{RN}}^* = \arg\max_{l \geq 0}\{E[\pi(l)]\}.$$

Therefore, a closed-form expression of the optimal return service charge is derived,

$${l_{RN}}^* = p - \frac{vm}{2} + \frac{\phi - \gamma(p - w)}{2\psi}.$$

Case 2: risk-averse MC fashion service provider

According to relevant literature, the mean−variance objective of capturing the decision-making problem for the risk-averse MC service provider is defined as follows:

$$U(l) = E[\pi(l)] - kV[\pi(l)],$$

where $k > 0$ is the risk sensitivity parameter.

For the risk-averse case, the optimal return service charge of the MC service provider, denoted by $l_{RA}{}^*$, is the solution to the following optimization problem (with a trade-off between EP and the VP):

$$l_{RA}{}^* = \arg\max_{l \geq 0}\{U[\pi(l)]\}.$$

The analytical expression of the optimal return service charge when the MC service provider is risk-averse is given:

$$l_{RA}{}^* = [k\sigma_R A(\sigma_R, \sigma_D, \rho) + B]/\left(d\sigma_R^2 + \psi\right),$$

$$A(\sigma_R, \sigma_D, \rho) = (p - vm)\sigma_R - \rho\sigma_D(p - w),$$

$$B = [\phi + \psi(2p - vm) - \gamma(p - w)]/2.$$

Finally, the optimal conditions in offering a zero return service charge are also analyzed.

10.4.3 Summary

In Section 10.4, we review several analytical models about returns management in apparel supply chains. To improve the systematism and comprehensibility of the aforementioned quantity models, we summarize critical features of the models in Table 10.1.

Table 10.1 Summary of critical features of three quantity models in this section

	Business model	Optimization objective	Members of supply chain	Source of returns	Returns policy
Model 1 (Section 10.4.1)	–	MV optimization	Retailer, supplier	Retailer	Return refund
Model 2 (Section 10.4.2.1)	Advanced selling	Profit maximization	Retailer, consumer	Consumer	Full refund/ partial refund
Model 3 (Section 10.4.2.2)	Mass customization	MV optimization	MC service provider, consumer	Consumer	Full refund minus service charge

Based on the mathematical analysis of these models, some important and interesting conclusions can be summarized:

- Considering the existence of a returns policy, the supply chain optimization mechanism for fast fashion supply chains is different from that for supply chains carrying the conventional newsvendor-type products.
- A simple returns policy can be applied to coordinate the fast fashion supply chain with retailers in different risk types, even in the presence of multiple retailers.
- Full refund does not always lead to higher consumer returns compared to partial refund. The partial refund policy can reduce the moral-hazard risk through a lower refund amount, and the partial refund policy with skimming pricing can thoroughly eliminate the effect of the moral hazard on the profit of the retailer. However, this pricing policy cuts off the revenue from low-valuation consumers.
- MC service providers should offer differentiated return services to different market segments. To be specific, free return service can be offered by the company in the following situations: (1) the revenue that the company makes from consumers is high enough, (2) "reusable value" of returned products is sufficiently high, and (3) return uncertainty is sufficiently low.

10.5 Intelligent system implementation for managing returns in apparel supply chains

An increasing number of enterprises have gradually realized the importance of returns management and now pay more attention to optimizing the return process based on specific features of the products and the industry. ISs, with their dynamic development and wide range of applications, provide new solutions to complicated problems. Based on the general characteristics and operational rules of returns management in apparel supply chains introduced in Sections 10.3 and 10.4, the application of an IS on returns management in apparel supply chains is discussed in this section. First, an expert system for forecasting returns quantity is introduced. Second, relevant research in DSS for managing returns in apparel supply chains is summarized.

10.5.1 Expert system for forecasting returns quantity in apparel supply chains

Temur et al. (2014) develop a fuzzy expert system to design a robust forecast of returns quantity to handle uncertainties in the returns process of a RL network.

The factors that influence product returns quantity are first defined, and then dimension redundancy analysis is implemented to eliminate collinearity with other factors. By training the data of selected factors with the fuzzy expert system, the return amounts of alternative cities are forecasted. The inputs (which can be named as "characteristic features" or "predictors") of the model are obtained from the environment. In increasing the precision of forecasting, ineffective predictors are eliminated by

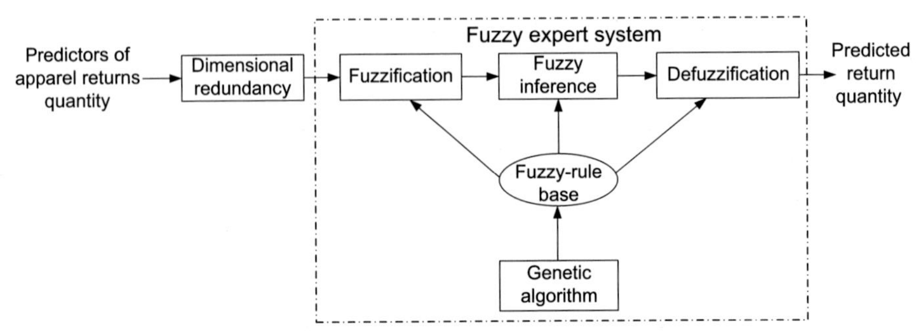

Figure 10.3 Framework of expert system for forecasting apparel returns quantity. Compiled from Sun, Z.L., Choi, T.M., Au, K.F., Yu, Y., 2008. Sales forecasting using extreme learning machine with applications in fashion retailing. Decision Support Systems 46(1), 411–419 and Temur, G.T., Balcilar, M., Bolat B., 2014. A fuzzy expert system design for forecasting return quantity in reverse logistics network. Journal of Enterprise Information Management 27(3), 316–328.

dimensional redundancy, and "dimensionality reduction" algorithms are used. The critical part of the decision model is the fuzzy expert system. Fuzzy expert systems, which work based on the fuzzy-logic approach, can model the rules obtained from fuzzy preferences of experts and can provide outputs by using these rules. The main elements of a fuzzy expert system are fuzzy logic, fuzzy base rule, fuzzy inference, and learning method (Siler and Buckley, 2005).

Through a case study, the ultimate result indicates that fuzzy expert systems can be used as a supportive tool for forecasting the returns quantity of alternative areas. Several factors are found as important in affecting product returns quantity. Accordingly, through the collection of data on these factors, the returns quantity can be forecasted.

Although the case study is based on an e-recycling facility and the forecasting object is the quantity of e-waste recycling in this research, the principle and model of forecasting is also applicable to the forecasting of returns quantity in the apparel industry. The inputs of fuzzy expert systems should be adjusted according to the features of product returns in apparel supply chains. As a result, the quantity of product returns can be predicted similarly. Specific to the apparel industry, the framework of an expert system for forecasting returns quantity is shown in Fig. 10.3.

10.5.2 DSS for returns management in apparel supply chains

The RES supplies the means of efficient planning, managing, and controlling relevant information from the consumption point to the starting point, to counter the overall production cost of the supply chain (Guide, 2000). RES can be now realized as a specialized area of an enterprise system that involves handling individual incoming returns; opening and inspecting products; communicating with internal departments, customers, and vendors; and then directing products into disposal channels that will provide the highest value. On the one hand, a good RES supports the decisions on returns management. On the other hand, it improves the performance of the whole supply chain, including the forward and reverse supply chains.

Scholars from the University of Bordeaux in France propose a model for interoperable decisions and information in RES. The model consists of the information system and the decision system. The information system collects, processes, and eventually provides data and valid information for the decision system, which supports the operation decision in daily returns management. The model depicts when and how decisions are made under various levels of information, and the decision system is divided into two subsystems: the operational system and physical system. The operational system provides operation management strategies, and the physical system supports decisions on material flow (products, sales, return, and controlled disposal).

Using an RES perspective, Madaan et al. (2012) propose a flexible DIS model that can handle products with various options and greater return volumes and variability. The model synergizes the role of product availability and quality information in improving the effectiveness of decisions in RES. And Experts Decision System and Knowledge for Recovery Decision support decisions in the returns management process in this model. Based on features of returns in the apparel industry, we modify the proposed generic DIS model for a product returns system (Madaan et al., 2012), as shown in Fig. 10.4.

As a critical factor of returns management in the apparel industry (described in Section 10.3), the feature of high MVT leads to the decision objective of an IS: flexibility, to minimize delays in the process of returns of apparel products. Based on this rule, Madaan et al. (2012) present an analytical model that computes DIS delays in RES and analyzes the influence of DIS delays on various levels of flexibility in RES. In this model, the expected utility (*EU*) of a DIS delay, d_j, is given by the expression:

$$\mathrm{EU}_E = \max_{d_j} \sum_{i=1}^{n} p(h_i|E)u(h_i, d_j),$$

where $p(h_i|E)$ represents the probability of the state, h_i, of the product, given the set of information, E, about the product, and $u(h_i,d_j)$ represents the utility of choosing option d_j when the state of the product is h_i. With the expected benefit of reducing delays and selecting the best recovery decision on time, EBI_E can be expressed as the difference between the expected benefits that can be obtained when the decision-maker chooses the best recovery option without delay (because its state is known with complete certainty) and the expected benefits of making the decision with delays (because its state is not known with complete certainty).

$$\mathrm{EBI}_E = \mathrm{EU}_E - \mathrm{EU}_E(d).$$

The maximum expected benefit that can be obtained from reducing DIS delays related with a product recovery is given by the expression:

$$\mathrm{EBI}_{\max} = -\frac{\varepsilon_1 \varepsilon_2}{\varepsilon_1 + \varepsilon_2},$$

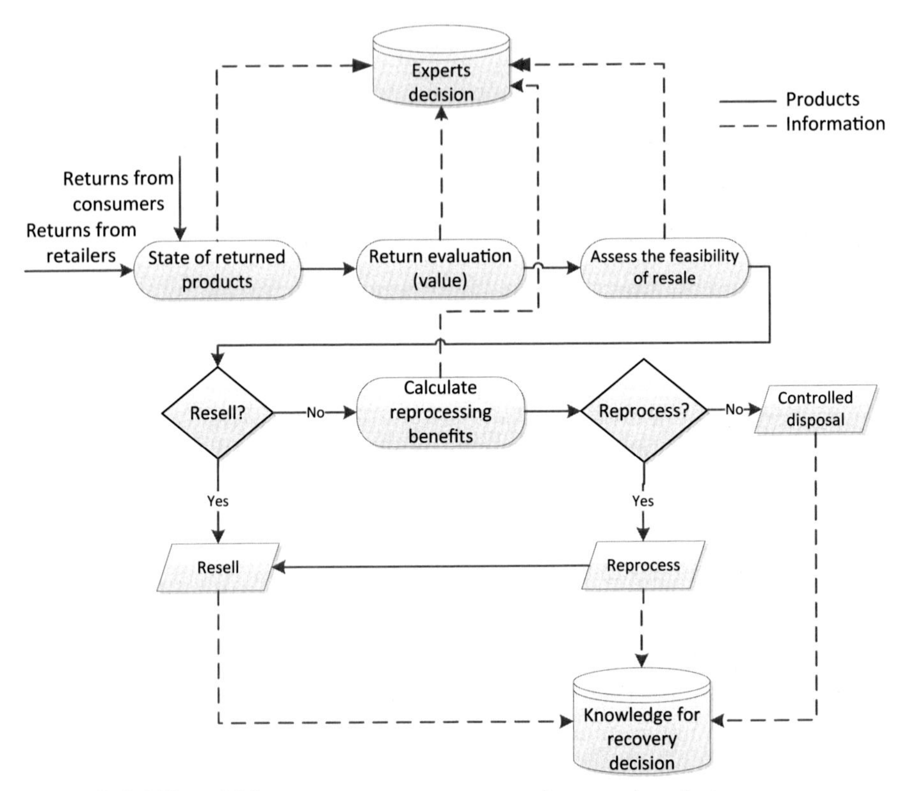

Figure 10.4 DIS model for return management system for apparel products.
Compiled from Madaan, J., Kumar, P., Chan, F.T.S., 2012. Decision and information interoperability for improving performance of product recovery systems. Decision Support Systems 53(3), 448–457.

where $\varepsilon_1\varepsilon_2$ denote the DIS delay penalties incurred from erroneous decisions. In combining the aforementioned equations, the optimal DIS delay decision, d_e, based on the available information, E, is given by the expression:

$$d_e = \begin{cases} d_1, & \text{if } p\left(h_1|E \geq \dfrac{\varepsilon_2}{\varepsilon_1 + \varepsilon_2}\right) \\ d_2, & \text{if } p\left(h_1|E \leq \dfrac{\varepsilon_2}{\varepsilon_1 + \varepsilon_2}\right) \end{cases}.$$

Madaan et al. (2012) further verify the aforementioned results through simulation and find that different nodes act relatively autonomously to decide on the type and level of flexibility to use, what and when the information is required to be shared for improving decision performance, and the benefits obtained from the return process.

Table 10.2 can be adjusted as follows

S/D		
S/A	**High**	**Low**
High	Type A	Type B
Low	Type D	Type C

S/D, sales/demand prediction; S/A, sales/attractiveness.

As introduced in Section 10.3, part of the returns in apparel supply chains is caused by stock factors, including overstock and bad stock. Viewing from the perspective of retailers, Hwang (2008) analyzes the decision model under overstock and bad stock cases and supports the return decisions.

1. Overstock decision model
   Retailers tend to keep stock in their backroom until sale season is over. Although high costs of stock handling can occur, retailers keep their unnecessary stock to avoid losing a sales opportunity. Based on the dilemma, an overstock return decision model is proposed, and products are divided into four types. Each type of product has its appropriate returns policy. The decision model is shown in Table 10.2.
   The "attractiveness" can be described by the formula:

   $$f1\text{:}\ \ A = c1 * \text{lifting count} + c2 * \text{fitting count (where } c1 \text{ and } c2 \text{ are weighting variables).}$$

   They also put forward that the product should be considered as overstock for returns of type *C*. The product ought to be returned or transshipped quickly because customers do not frequently look for this type of product, and most of them do not have any intention to purchase the product.

2. Bad stock decision model
   Bad stock is the stock that has not been sold and has been in storage for a long period. It requires a more efficient returns management and decision-making process. Consequently, disclosing bad stock in the backroom with high efficiency is very important, and the following expression represents the amount of bad stock and supports the return decisions:

   $$f2\text{: Bad-stock} = (\text{Current date} - \text{Inbound date in back room}) > \text{Max(item keeping days)}.$$

Consequently, retailers can make optimal decisions more efficiently based on the estimated amount of bad stock.

The aforementioned studies focus on the background of ISs of returns management. Considering the development phase, Turki and Mounir (2014) propose a web-based DSS that can ease product and shipping information tracking, storing, referencing, and reporting for managers (decision makers).

10.6 Conclusions and future direction

Recent legislation on customer protection by governments worldwide, together with the high expectation of customers toward product quality and good customer service, has resulted in an increase in return rates of merchandizes. The topic of this chapter relates to ISs in managing returns in the apparel industry. Based on the review of relevant literature, the following three classes of problems are discussed in this chapter.

First, the common types of unique characteristics of returns in apparel supply chains are analyzed. Successful ISs in managing returns in apparel supply chains have to been designed by considering these specific features: high return rate, high MVT, decreased value during the season, various reasons of returns, and high return cost. The corresponding management strategy is issued based on the aforementioned features. Second, several mathematical models are proposed and analyzed based on the features of apparel supply chains. The optimal returns policies and operational strategies are obtained by theoretical deduction, and theoretical basis is provided for the establishment and implementation of ISs in the apparel industry. Third, some preliminary applications of ISs on returns management in the apparel industry are discussed. The summary of extant research achievement provides a theoretical basis and research direction for ISs in the future.

Future research endeavors indicate various promising areas to explore. Possible future directions include the returns management problem with a more complicated supply chain structure considering the feature of fast fashion in the apparel industry adequately, especially in the current accelerated society. In addition, on the one hand, the extant literature on ISs for returns management in the apparel industry focuses more on basic guidelines and technical framework of ISs in returns management, as opposed to specific algorithms and decision support mechanisms. On the other hand, research on the application of ISs on forward supply chains is more adequate and mature compared to those on reverse supply chains in the apparel industry. Consequently, another promising research opportunity may be the combination of returns management with information technology and computer algorithms, such as RFID, expert systems, GA, artificial neural networks, knowledge-based systems, DSS, and fuzzy-logic systems, for effective reverse supply chain management in the apparel industry. In conclusion, the most important aspect of the application of ISs in returns management in the apparel industry is the improvement of the operational efficiency and effect in decision-making on returns management.

Acknowledgments

We acknowledge the support of grants from The Major Program of National Social Science Fund of China (Grant No. 13 & ZD147), and National Natural Science Foundation of China (NSFC), Research Fund Nos. 71372100 and 71372002.

References

Blackburn, J.D., Guide, V.D.R., Souza, G.C., et al., 2004. Reverse supply chains for commercial returns. California management review 46 (2), 6–22.

Campanelli, M., April 2005. Happy Returns? E-tailers Are Rethinking Their Policies to Curb Return Fraud. Should You? Entrepreneur.

Chen, J., Bell, P.C., 2011. Coordinating a decentralized supply chain with customer returns and price-dependent stochastic demand using a buyback policy. European Journal of Operational Research 212 (2), 293–300.

Chen, M.K., Wang, Y.-H., Hung, T.-Y., 2014. Establishing an order allocation decision support system via learning curve model for apparel logistics. Journal of Industrial & Production Engineering 31 (5), 274–285.

Choi, T.M., 2013. Optimal return service charging policy for a fashion mass customization program. Service Science 5 (1), 56–68.

Choy, K.L., Lam, H.Y., Lin, C., et al., 2013. A hybrid decision support system for storage location assignment in the fast-fashion industry. Technology Management in the IT-Driven Services (PICMET). In: 2013 Proceedings of PICMET '13. IEEE, pp. 468–473.

Fischer, M.L., 1997. What is the right supply chain for your product? Harvard Business Review 75, 105–116.

Ford, F.N., Rager, J., 1995. Expert system support in the textile industry: end product production planning decisions. Expert Systems with Applications 9 (2), 237–246.

Gunasekaran, A., Ngai, E.W.T., 2004. Information systems in supply chain integration and management. European Journal of Operational Research 159 (2), 269–295.

Guide, V.D.R., 2000. Production planning and control for remanufacturing: industry practice and research needs. Journal of Operations Management 18 (4), 467–483.

Guo, J., Liu, X., Jo, J., 2015. Dynamic joint construction and optimal operation strategy of multi-period reverse logistics network: a case study of Shanghai apparel E-commerce enterprises. Journal of Intelligent Manufacturing 1–13.

Hsieh, C.C., Lu, Y.T., 2010. Manufacturer's return policy in a two-stage supply chain with two risk-averse retailers and random demand. European Journal of Operational Research 207, 514–523.

Hsu, H.M., Hsiung, Y., Chen, Y.Z., Wu, M.C., 2009. A GA methodology for the scheduling of yarn-dyed textile production. Expert Systems with Applications 36 (10), 12095–12103.

Hu, Z.H., Li, Q., Chen, X.J., et al., 2014. Sustainable rent-based closed-loop supply chain for fashion products. Sustainability 6 (10), 7063–7088.

Hui, C.L., Chan, C.C., Yeung, K.W., Ng, S.F., 2007. Application of artificial neural networks to the prediction of sewing performance of fabrics. International Journal of Clothing Science and Technology 19 (5), 291–318.

Hwang, Y.M., 2008. The Impact of RFID on the Centralized Reverse Supply Chain Management (RSCM) in Apparel Industry. Information and Communications University.

Kang, M., Johnson, K., 2009. Identifying characteristics of consumers who frequently return apparel. Journal of Fashion Marketing & Management 13 (1), 37–48 (12).

Levary, R.R., 2001. Computer integrated supply chain. International Journal of Materials and Product Technology 16 (6–7), 463–483.

Li, J., Choi, T.M., Cheng, T.C.E., 2014a. Mean variance analysis of fast fashion supply chains with returns policy. IEEE Transactions on Systems Man & Cybernetics Systems 44 (4), 422–434.

Li, Y.J., Xu, L., Choi, T.M., 2014b. Optimal advance-selling strategy for fashionable products with opportunistic consumers returns. IEEE Transactions on Systems Man & Cybernetics Systems 44 (7), 938–952.

Li, Y.J., Xu, L., Li, D.H., 2013. Examining relationships between the return policy, product quality, and pricing strategy in online direct selling. International Journal of Production Economics 144 (2), 451–460.

Lin, J.J., 2007. Intelligent decision making based on GA for creative apparel styling. Journal of Information Science & Engineering 23 (6), 1923–1937.

Lin, M.T., 2009. The single-row machine layout problem in apparel manufacturing by hierarchical order-based genetic algorithm. International Journal of Clothing Science and Technology 21 (1), 31–43.

Liu, N., Choi, T.M., Yuen, C.M., et al., 2012. Optimal pricing, modularity, and return policy under mass customization. IEEE Transactions on Systems, Man and Cybernetics, Part A: Systems and Humans 42 (3), 604–614.

Lund, R.T., 1984. Remanufacturing. Technology review 87 (2), 18–23.

Madaan, J., Kumar, P., Chan, F.T.S., 2012. Decision and information interoperability for improving performance of product recovery systems. Decision Support Systems 53 (3), 448–457.

Metaxiotis, K., 2004. RECOT: an expert system for the reduction of environmental cost in the textile industry. Information Management & Computer Security 12 (3), 218–227.

Mok, P.Y., 2011. Intelligent apparel production planning for optimizing manual operations using fuzzy set theory and evolutionary algorithms. In: 2011 IEEE 5th International Workshop on Genetic and Evolutionary Fuzzy Systems (GEFS). IEEE, pp. 103–110.

Mukhopadhyay, S.K., Setaputra, R., 2007. A dynamic model for optimal design quality and return policies. European Journal of Operational Research 180 (3), 1144–1154.

Mukhopadhyay, S.K., Setoputro, R., 2004. Reverse logistics in e-business: optimal price and return policy. International Journal of Physical Distribution & Logistics Management 34 (1), 70–89.

Nakhata, C., Magi, A., 2015. Does a multi-channel return policy affect online purchase intention? In: Proceedings of the Academy of Marketing Science.

Ngai, E.W.T., Peng, S., Alexander, P., et al., 2014. Decision support and intelligent systems in the textile and apparel supply chain: an academic review of research articles. Expert Systems with Applications 41 (1), 81–91.

Rong-Chang, C., Chih-Chang, L., Shiue-Shiun, L., 2006. An automatic decision support system based on genetic algorithm for global apparel manufacturing. International Journal of Soft Computing 1 (1), 17–21.

Ruiz-Torres, A.J., Ablanedo-Rosas, J.H., Mukhopadhyay, S., 2013. Supplier allocation model for textile recycling operations. International Journal of Logistics Systems and Management 15 (1), 108–124.

Shahrabi, J., Hadavandi, E., Salehi Esfandarani, M., 2013. Developing a hybrid intelligent model for constructing a size recommendation expert system in textile industries. International Journal of Clothing Science and Technology 25 (5), 338–349.

Siler, W., Buckley, J.J., 2005. Fuzzy Expert Systems and Fuzzy Reasoning. Wiley.

Srinivasan, K., Kekre, S., Mukhopadhyay, T., 1994. Impact of electronic data interchange technology on JIT shipments. Management Science 40 (10), 1291–1304.

Su, X., 2009. Consumer returns policies and supply chain performance. Manufacturing & Service Operations Management 11 (4), 595–612.

Sun, Z.L., Choi, T.M., Au, K.F., Yu, Y., 2008. Sales forecasting using extreme learning machine with applications in fashion retailing. Decision Support Systems 46 (1), 411–419.

Temur, G.T., Balcilar, M., Bolat, B., 2014. A fuzzy expert system design for forecasting return quantity in reverse logistics network. Journal of Enterprise Information Management 27 (3), 316–328.

Thomassey, S., Happiette, M., 2007. A neural clustering and classification system for sales forecasting of new apparel items. Applied Soft Computing 7 (4), 1177–1187.

Turki, W., Mounir, B., 2014. A proposition of a decision support system for reverse logistics. In: 2014 International Conference on Advanced Logistics and Transport (ICALT). IEEE, pp. 120–125.

Vigneswaran, C., Ananthasubramanian, M., Anbumani, N., 2012. Neural network approach for optimizing the bioscouring performance of organic cotton fabric through aerodynamic system. Journal of Textile & Apparel Technology & Management 7 (3), 1–14.

Wong, W.K., Mok, P.Y., Leung, S.Y.S., 2013. 8–Optimizing Apparel Production Systems Using Genetic Algorithms. Optimizing Decision Making in the Apparel Supply Chain Using Artificial Intelligence, pp. 153–169.

Wong, W.K., Zeng, X.H., Au, W.M.R., 2009. A decision support tool for apparel coordination through integrating the knowledge-based attribute evaluation expert system and the T-S fuzzy neural network. Expert Systems with Applications 36 (2), 2377–2390.

Yao, Z., Wu, Y., et al., 2005. Demand uncertainty and manufacturer returns policies for style-good retailing competition. Production Planning and Control 16 (7), 691–702.

Zhai, C.J., Li, Y.J., 2011. Study on on-line retailer's return policy in B2C mode. Journal of Industrial Engineering & Engineering Management 25 (1), 62–68.

Vendor-managed inventory systems in the apparel industry

H. Chaudhry[1], *G. Hodge*[2]
[1]Lahore University of Management Sciences, Lahore, Pakistan; [2]North Carolina State University, Raleigh, NC, United States

11.1 Introduction

Globalization along with maturing markets have changed the ways business is conducted. The consumer-driven apparel retail industry especially faces a situation where increased competition is forcing retailers to introduce new products quickly, thus shortening product life cycles. The supply side, on the other hand, is constrained by the push to further reduce the prices by pushing production further away from consumption location to low-cost countries, thus increasing the product delivery times. Decreasing life cycles and increasing production times is resulting in increased costs to the retailers in the shape of supply chain mismatch costs. Supply chains that manage these costs efficiently tend to outsmart the competition.

The changing demand patterns leaves a supplier unable to respond in time due to lead time constraints. As a result, the customers start carrying excess inventory, while the suppliers, on the other hand, with fear of losing business, start carrying anticipatory stock; hence inventories build up at two locations within the supply chain without any correlation with the actual demand pattern (Choi and Sethi, 2010). Even this buildup of inventory does not serve a purpose because the supply chain is not demand driven, and hence the stock-out or over-stocking cost remains there at the customer/retailer end, while increased inventory costs continue. Apparel retailers are adopting different measures to manage these supply chain mismatch costs. Of these, one strategy that has gained much importance is quick response (QR), which is based on waiting for the actual demand to replenish within seasons. Within QR, there are multiple strategies, such as accurate consumer response (ACR), collaborative replenishment planning (CRP), and vendor-managed inventory (VMI), that have been highlighted in supply chain literature.

VMI gives a platform to coordinate the supply chain with an aim of increasing service level while reducing inventory costs. Vendor-managed inventory is a strategy in which the customer shares point-of-sale (POS) data with the vendor, and the responsibility of making replenishment decisions is handed over to the vendor (Chopra and Meindl, 2001; Simchi-Levi et al., 2003). Synchronized supply chains result in increased responsiveness and reduced inventory cost (Holweg et al., 2005). The VMI relationship between the TAL Apparel and JC Penny has symbolized the

Information Systems for the Fashion and Apparel Industry. http://dx.doi.org/10.1016/B978-0-08-100571-2.00011-7

success of technology-driven VMI relationship, where the vendor designs and then manages the shirt business and replenishes the stock directly from its production units (Hirsch, 2005).

The role of the vendor becomes increasingly important since the vendor not only has to manage the production as in the traditional purchasing model, but that vendor also has to assume the role of the buyer to coordinate the supply chain. However, empirical evidence on the role of the vendor, especially in terms of evolution along VMI implementation, is less conclusive. This study maps the evolution of a vendor in a VMI journey. This paper begins with looking at prior research in VMI followed by research design and a detailed account of a vendor's VMI implementation. The case description is followed by a discussion part along with conclusion and future research areas.

11.2 Vendor-managed inventory research

The research in the VMI area has been carried out using simulation/mathematical modeling techniques as well as using empirical studies. However, the majority of studies have used the first approach (Kauremaa et al., 2009). Disney and Towill (2003) have shown VMI to be better in terms of responding to abrupt demand changes as opposed to the traditional ordering mechanism. Many studies have shown decreased bullwhip effect (Disney et al., 2003), inventory reduction, as well as service level increase as a result of VMI implementation (Waller et al., 1999; Cachon, 2001). VMI gives more time to the vendors to plan their operations (Kaipia et al., 2002). Dong and Xu (2002) and Lee and Cho (2005) have looked at VMI profits, and they found higher buyer profits, while profits at the supplier's end varied. The modeling/simulation research has generally shown a positive impact of VMI adoption on supply chain efficiencies; however, there is less to be learned on the issues related to VMI implementation.

The empirical research has broadly looked at VMI implementation, application, enablers, and relationship classification. Tony and Zamalo (2005) have looked at VMI application, both upstream and downstream in the home appliances industry. They have highlighted high sales volume, the short distance between brand and the sales company, and technology as well as its knowledge as elements of VMI success. Vigtil (2007) explored the type of data needed by the supplier for VMI implementation using four case studies in automotive and industrial settings. According to Vigtil (2007), when the vendor chooses to stock, then information pertaining to the level of inventory and withdrawals are essential; whereas if the vendor chooses to order, then data related to incoming orders would be needed. Moreover, they found that electronic integration and data updates are not prerequisites.

Claassen et al. (2008) have looked at the enablers of successful VMI applications as well as performance outcomes from a buyer's perspective. They have explored small and large organizations within retail, construction, chemical, equipment, and electronics industries. The found that the nature of product impacts the strategy. Where strategic products for buyers are involved, they were highly involved, while for commodity products, the buyers wanted the suppliers to take charge. They also found strict rules implemented by the buyers for inventory limits. Inventory information in their respondents was shared daily, while forecasts were shared weekly or monthly. Claassen et al. (2008) found the effect of VMI on cost benefits to be the weakest. They have explained this as a lack of implementation of a complete VMI model. This fact has also been reported by Holweg et al. (2005). According to them, the buyer implements strict limits and does not give the vendor complete authority to drive the VMI model. As a result the supplier does not consider the total supply chain cost while making decisions. In addition, this behavior results in excess inventory in the system and hence diminishes the cost impact of VMI model.

Kauremaa et al. (2009) looked at five dyads from grocery, consumer goods, chemical, and paper industries. The study was aimed at developing a typology of VMI relations based on efficiency, material flow, and supply chain improvement goals. They found a basic VMI relationship with only replenishment responsibility transferred to the supplier, where the buyer is looking for operational benefits, and the supplier is looking for creating differentiation. From their data set, they found a cooperative VMI relationship where buyer and vendor work collaboratively toward achieving joint goals. They found such relations to be focusing on non-VMI areas as well while exploring possibilities for creating joint value. Kauremaa et al. (2009) also mentioned synchronized VMI relationships, which according to them are an extension of cooperative relations where decision-making is merged. They found a small share of VMI in supplier's total sales as well as a long production cycle as inhibiting factors of availing VMI benefits.

Elvander et al. (2007) have looked at different design dimension to categorize VMI systems. They have grouped design dimensions into four broader categories that include inventory control, information, decision-making, and integration. Within these, Elvander et al. (2007) have captured multiple alternatives from their data set and proposed a framework of dimensions based on the dominant choice. According to them, inventory is located both with the supplier and the customer, deliveries are made from the supplier's warehouse, while the supplier invoices the customer upon issuance of inventory. In terms of information-related dimensions, they found batch transactions from customer ERP systems with some evidence of online access to customer's ERP. In decision-making dimensions, they found continuous review and ordering practice, with the supplier making replenishment orders and shipment decisions. For their integration dimension, the findings were only applicable in the case of generic products delivered to many customers.

Holweg et al. (2005) have looked at collaboration classification based on collaboration in planning and inventory management dimensions. They have divided collaboration into three types. In type 1 planning collaboration, the customers share the information, and the vendor makes use of that information to do internal planning; however, the customer then places orders without any consultations. With type 2, vendor managed replenishment (VMR), the supplier has the information, but there are two decision points, and the information is not linked to the supplier production process. The reason cited is that the retailer is one of many customers serviced, so linking this with the production process that serves a large proportion of other customer can create problems. Type 3 is collaboration with one decision point.

Holmstrom (1998) studied a VMI system implementation between a vendor and a wholesaler. They have distributed the implementation into conceptualization, pilot, and full-scale implementation phases. According to Holmstorm (1998), VMI resulted in reduction in demand variability for the vendor, reduced stock, and better availability for the buyer, while the partnership gained from reduced lead time.

Past studies in the VMI area are predisposed toward investigating the impacts of VMI on buyers only (Kauremaa et al., 2009). Moreover, the major emphasis of these studies has been focused on capturing collaborative efforts in the grocery sector, but even in those dyads, a growing number of supplier companies have been skeptical of the collaboration results (Corsten and Kumar 2003). In addition, empirical studies on VMI implementation are still few compared to the simulation and modeling work done in this area (Kauremaa et al., 2009).

Multiple stages of VMI collaboration have been identified (Holweg et al., 2005; Claassen et al., 2008) with the true benefits of VMI being realized at higher stages. Similarly, Kauremaa et al. (2009) have identified existence of multiple levels of VMI relationships. However, there has not been any mention of what allows a vendor to graduate from one level of VMI model to the next, nor has it been detailed which resources and routines would be needed to move to a higher level of VMI model. In addition, in terms of benefits to the supplier, the literature has mixed views. This study is focused on answering the following research questions:

1. What are the prerequisites of graduating from a lower level VMI to a higher level VMI relationship model?
2. What are the benefits of VMI relationship to a vendor?

In addition to the these research questions, the findings would be useful for customers in learning ways they can help their vendors prepare for the next stages of VMI implementation.

11.3 Research design

The purpose of this study is to capture the evolution of VMI practices from a vendor's perspective. That requires an in-depth study of an organization that has progressed in implementing the VMI model. It is important to assess the sequence of events to generalize the findings based on the observations.

The objective of our case study is to identify patterns and variables that can be later tested in a broader setting to validate the findings. The number of cases to be studied depends on the phenomena being researched. Studying multiple cases gives an opportunity to compare and contrast practices among respondent organizations, while a single case study is more suited for a detailed analysis. Especially while covering an evolution, a single company, longitudinal study adds more value (Eisenhardt, 1989; Yin, 2003).

For an exploratory case study, the selection of organization plays an important role for the opportunity to study the desired phenomenon. Generally, purposive or convenience sampling is done based on the knowledge of the researcher regarding the industry. The organization selected for this study is one of the largest vertically integrated knitwear apparel manufacturers in Pakistan, with a history of working with leading international retailers. The company had been known to the researcher for over 10 years, and the researcher has written multiple case studies on this organization. The case company initiated the VMI model for a core underwear program for one of its major customer in 2005. Since then, the company has developed a VMI-based relationship with some additional customers and now manages a portfolio of products, including fashion items in additional to basic products, under the VMI program, thus providing an opportunity to study the evolution of a vendor on its VMI journey. For this study, interviews were conducted with the company's VMI project head and the VMI team members during the first half of 2015.

11.4 Case data

The case company is a vertically integrated knitwear garment manufacturer from Pakistan with in-house knitting, dyeing, cutting, and sewing facilities. The company had a state-of-the-art, home-grown ERP system running in the early 2000s that enabled the organization to track the history of every product to the machine and worker level. The development of this system started in the late 1990s with an initiative by the CEO, who had always been interested in managing the business using technology. The case company was known for its expertise in managing the undergarment commodity business and ran year-round programs for top US brands and retailers.

By 2005, the company had reached an annual sales level of approximately USD $100 million, out of which 50% was bought by a US-based mass retailer (Retailer A). Of this business, 50% was for an undergarment line that had 8–10 more styles for a year-round program. By that time, retailer A predominantly worked with the garment factories via importers. The importer would work with the retailers to finalize the programs and then placed quantities to garment factories. This arrangement saved the retailers from maintaining sourcing offices in vendor countries and helped them focus on the retail side of the business. However, a disconnect was felt at the retailer's end, especially from the point of view of product development. In 2005, Retailer A decided to move to the direct sourcing model for incremental business while keeping the importer managed model for the remainder. Retailer A approached the case company to discuss possibilities of directing business using a VMI model for the core undergarment program.

The case company had never worked on a VMI model before. The CEO took personal interest in exploring the VMI model as an opportunity for business expansion, but also for fear that the retailer might move this business to a competitor offering VMI services. A core team of two senior merchandizers and an IT manager was formed to initiate working on establishing capabilities to offer VMI services. Under this model the retailer, instead of placing individual purchase orders against which the company offered free-on-board terms, would place weekly replenishment orders to the company, and the company would replenish those quantities within a week from its US-based warehouse. In addition, the company had to establish electronic data interchange linkage with the retailer to get weekly POS data at the store level. Although the replenishment quantities were to be provided by the retailer, the store-level POS data was to be used to plan inventories in the supply chain.

The case company established a warehouse in the United States. In addition, another company was set up to make financial transactions with the retailer, since the VMI model required inventory ownership by the vendor until the warehouse with weekly transactions with the retailer was available. The company also needed a more advanced system because their existing IT system only offered visibility to the factory gate; whereas under the new system, the company was required to maintain visibility during transit as well as at the warehouse in the country of retail.

The company started conservatively with regard to maintaining inventory in the supply chain since it did not have the experience of working directly with POS data. To achieve a 100% fill rate, the company decided to load the warehouse with at least 4–6 weeks of inventory. In addition, the company also set up a warehouse in the country of origin to create a buffer against sale- and production-level variations.

After having spent almost a year under this model, supported by extra inventory in the system to maintain a 100% fill rate, the company started to monitor style-level data. The merchandise team that monitored the data realized that all styles within the program had different demand patterns. For some styles, demand was more or less evenly spread across the year, whereas for others, demand showed seasonal peaks.

The VMI team also started to look at the store-level data. Initially, they selected the top 100 stores in their product category and started looking at their weekly sales trends. They found that the stores did not have optimized inventories, especially at the SKU level. For example, some stores carried higher stock for size medium and lesser stock for size small, whereas based on the store-level demand data, medium was selling less, and the stores were stocking out in size small. Along with that, the VMI team started making their own forecasts while looking at the store level data. When the team compared their own forecasts and the orders they received from the retailer with the store level inventory data, they were surprised to see that store-level fill rate would have improved if their forecasts had been followed. They shared their forecasts and findings with Retailer A's team. Based on this feedback, the retailer started asking for regular inputs from the VMI team. According to the company's VMI team, they felt that the retailer buying team had started trusting their capabilities to manage the VMI business. Moreover, their role changed from reactive to more suggestive, and the relationship became more of mutual collaboration.

During that time, while the case company was maintaining a 100% fill rate from its warehouse in the country of retail, two of the Retailer A's vendors, one based out of South America and the other from North America, had product quality issues. During that time, as cotton prices shot up in the international market, their fabric vendors could not maintain the fabric quality, ultimately resulting in claims. On the other hand, the case company, because of a vertical setup and sales-level visibility, was able to book cotton for the program much earlier. In addition, they were able to achieve much higher production level efficiencies by dedicating resources to the program. The company's consistent quality and fill rate made the retailer divert some of their competitors' core underwear business to the case company.

The company until that time was known for its expertise in the undergarment commodity business. As the company graduated from VMR-level relationship to a more collaborative and suggestive role, it started to explore a polo shirt business line, as Retailer A also ran their polo programs on a VMI model. The polo business was different from the undergarment business, which ran throughout the year. Polo shirts market for 9-months, although there were some colors such as white, navy, red, and black that never ran out. The retailers had two to three additional colors that sold throughout the season. However, season end brought season ending inventory, which had to be sold out at marked-down prices.

The vendor's country of origin had a 4–5 week transit time to the country of retail. The vendor maintained 4–6 weeks of demand inventory in the warehouse, while the sewing work in process (WIP) was around 2 weeks. This meant that once the season started, based on the past year's data and current season's forecast, 13 week's production was already locked. For example, if a style had an initial projection of 100,000 pieces for the first month, the company would keep 100,000–150,000 pieces in the warehouse. To maintain the flow (to maintain a constant level of inventory in the warehouse) with 25,000,000 weekly replenishments, the warehouse must be provided with 25,000 pieces weekly from the home country. Four to five weeks of transit time meant another 100,000–125,000 pieces in transit. Moreover, to maintain a weekly outflow from the country of origin, sewing would be planned at 25,000 pieces per week. With 2 weeks of inventory in the sewing lines, meaning another 50,000 pieces, a style with a monthly sales projection of 100,000 would require approximately 300,000 pieces in the supply chain even before the start of the selling season. So the first response to initial sales trends would only reflect in the inventory in the supply chain beyond 3 month's forecasted demand. Although a lesser transit time favored the vendors with geographic proximity to the country of retail, they were also constrained one level further up in the supply chain in terms of fabric transit time. So, vendors with geographic proximity could respond to SKU-level variation within a color better than the case company, but they were equally constrained in the case of a color-level change because of fabric import from China. The case company could make SKU-level (ie, sizes within color) changes if it used one width for all sizes. Otherwise, the changes could only be made beyond 4 month's demand. But that still gave enough time to respond to demand changes for a 9-month-long program to respond to stock outs as well as to over stocking issues that resulted in season end inventories. However, programs shorter than 9 months, such as 6-month programs, did not give much time to respond to store-level demand variations.

The next level in the VMI journey was achieved when the case company started working on the polo program for a large mass retailer. Although the price was very competitive and did not leave much margin, the company took that as a challenge to further optimize their processes. Again, this was more of a VMR program with replenishments from the warehouse based on weekly forecasts made by the retailer's buyers.

Based on their experience and confidence, the company started to approach large customers with basic and semi-fashion products and started pitching their VMI services. The case company's performance landed them with the core underwear program of a warehouse club run by one of their customers. In this program the company was required to review weekly sales data from all club stores and then suggest weekly orders that would need to be replenished from the warehouse to the clubs. The club would enter the suggested quantities into their system and issue replenishment orders.

The case company's contracts with its customers, as opposed to the traditional purchase order–driven model, were not transactional. These were informal, revolving contracts with the needs for the next 2–3 months mutually revisited. The financial liability of forecasted production WIP was never formalized, though it always remained an implied liability on the retailer's part based upon mutual trust and understanding. The financial transaction took place in the country of retail. According to the VMI team, "There aren't many conflicts while you are running the VMI program for customers because you have crossed that threshold where the customer and vendor have a transaction order–based relationship. However, in the case of a VMI model, generally issues or discussion agendas relate to inventory buildup in the system, especially for the programs that see a sudden decline in sales. For year-round programs, inventory buildup is not much of a concern for the vendor or the retailer as that could be adjusted. But even in case of shorter duration programs, we support the customer by diverting WIP to other business lines, or the customer shifts those inventories to their other lines."

One of their customers, who got on board because of the VMI service capabilities, started with the replenishment from the country of retail model, where the company serviced replenishment orders from their warehouse in the country of retail. Later the customer demanded replenishment from the country of origin's warehouse. The attraction for the customer in this arrangement was a lower total cost of doing business; the product or transportation cost would not change, but the warehousing cost (lease plus maintenance) would be significantly lower in the country of origin. However, this model moved replenishment forecasting 5 weeks (transit time) away from the retail stores, and so inventories went up in the store. Later the retailer retracted from replenishment from country of origin and moved back to replenishment from the country of retail model.

By 2014, the case company had reached USD $100 million in business on VMI programs from different customers. A dedicated group of 25 personnel managed the accounts in terms of making liaison with the buyers, working with the data to make projections, and coordinating with the manufacturing division to make production and shipping schedules. In addition, the company maintained a separate IT team to support the VMI business. The company also had a separate team of developers and data entry personnel supporting the ERP operations company wide. The IT team for the VMI business regularly worked with the customers' IT department on updates. In addition, the IT team traveled to IT exhibitions to keep themselves abreast of the latest developments.

The IT team developed a smart phone application that gave a weekly update on the projections and inventories in the system and shared the data with their sales team in the country of retail. The sales team shared the data with the customers' buyers in their weekly meetings and shared the feedback with their team back home. Subsequently, the IT team developed a mobile application to give customers visibility into the inventory levels in the supply chain. Through their mobile devices the buyers could see inventory level at the warehouse, in transit, and in the manufacturing pipeline. The data also showed key performance indicators (KPIs) such as fill rates and annual sale trends as well as anticipated fill rate, sales history, projections, and forecasts.

Based on the company's experience in setting up VMI operations and then graduating to the next level of suggesting replenishment plans, some customers even asked the case company to help their other product vendors in the region in setting up VMI operations. The case company formed a separate organization comprised of their VMI supporting merchandizers and IT team. This team provided services to other vendors in the shape of setting up IT linkages and managing the financial and operational aspects of VMI. The team did the forecasting and liaison with the customer on behalf of other vendors and then forwarded the manufacturing and shipment flow to the vendors. The company managed VMI business for a sock and towel manufacturer.

According to the VMI manager at the case company, "The initial push toward the VMI model was more of a survival mode under the threat of losing the core business. However, our CEO's resolve and interest in using technology and systems to manage operations made the journey easier. However, this model of business involved significant upfront investment in the shape of setting up a company and warehouses in the country of retail operations. In addition, the investment needed in inventories that was required to support the model was a significant commitment from the company. However, our company's past performance and quality helped the customer to move from a traditional purchase order–based model to a VMI model."

According to the VMI team, "The first year was a learning phase, where we just wanted to ensure 100% fill rate even if we had to maintain some extra inventory in the system. However, as we gained confidence and got used to data management, we were able to reduce the inventories. Maintaining extra inventories also gave some confidence to the buyers who were skeptical of our data handling capability initially, and that's why at the start it was not a pure VMI model; rather, we were furnishing customer orders from our warehouse. The major reason was that the customer did not have trust in our capabilities. So it was more of a supervised VMR model as opposed to a partnership-based VMI relationship.

The next phase was when we started to experiment at our end. We started making forecasts, compared those with the customer orders, simulated the environments, and measured the impact under different scenarios. Once we had faith in our forecasting, we approached the customer and presented our analysis. Then we started processing data in parallel before the customers accepted our capabilities. Then the next phase was when we felt confident and started approaching different customers with our VMI services. However, for customers not going toward the VMI business, we felt it was either their lack of size or their outdated retail management systems. Some had outdated forecasting programs that could not incorporate the required variables. We were able to leverage the investment in technology and human resources. Next was moving on to the true VMI model where the customer had delegated the forecasting function to us. And then we started offering the consulting services to other vendors."

The VMI manager further stated, "Sometimes, organizational structure at the customer end did not let them make an optimal decision. The warehousing, inventory carrying, and financial costs are generally taken out of product stated cost in sourcing formula and are taken as corporate expenses, so the category buyers are not willing to take the risk because of longer forecast exposure. However, it was only a matter of starting up with just one program that would then would open up the horizons of speed, flexibility, and responsiveness, which is priceless to the buyers in retail world. The collaboration that was cultivated through use of combined intelligence, shared forecast, and read and react mechanisms developed strategic romance between vendor and retailer. On the other side, as far as benefits to vendors are concerned, it was never the price that makes VMI a lucrative business for a vendor; instead, it's the efficiencies that one derives from the model. The efficiencies come from better planning visibility, striking good contracts with our vendors, buying raw materials at better prices, and making shipping contracts for an entire year rather than on order-to-order basis. The returns are realized not in the first order, but rather the first few years are a period of investment."

According to the VMI head, "Because of our limitations of geographic distance from the country of retail, the minimum season we can manage under a VMI model has to be at least of 9 months because a 6-month cycle does not leave enough room to react to manage the sales variations. Nine months gives us some space to respond. However, if we were located next to the country of retail with 4–5 days of transit time and are able to reduce our lead time to 40 days, we can implement a VMI model even for short-season fashion products as well."

11.5 Discussion

The case company started the VMI implementation based on demand of the customer. So it was more of a push from the customer side rather than an initiative from the vendor side to begin with. The implementation started more from the marketing aspect of retaining a business by creating differentiation from the rest of the customer's vendors. This aspect has been mentioned in the literature, as some studies have reported VMI as a marketing tool (Vergin and Barr, 1999).

There are not many studies that have tried to explore the relationship of a vendor's exposure and expertise in information systems to successful implementation and evolution of VMI business. In this study, the case company's strong historical emphasis on systems helped them in transition from traditional to VMI systems. A dedicated team for the VMI business was set up. This has been cited in the previous literature to some extent, where Waller et al., (2004) noted that the responsibility for managing shared information lies with a VMI team.

In terms of requirements to establish VMI services, a significant financial commitment is needed from the vendor's side. This commitment is in terms of updating the internal system to create visibility not only within the organization but also to create chain-level visibility for managing the business. In addition, establishing data linkage with the customer is also needed. The vendor is required to set up the warehousing business in the country of retail. This requires investment in terms of infrastructure but also in terms of hiring human resources in the country of retail. More importantly, since vendors in the apparel business are manufacturing in low-cost countries, managing an entity in the country of retail can become a significant challenge from a management as well as financial perspective. This financial strength is seen by the retailer/customer as a prerequisite of starting a VMI model with a vendor as well as a commitment.

The case company started with additional inventories, especially during the first year, to show commitment and performance by ensuring a 100% fill rate from the warehouse. Achabal et al. (2000) have also mentioned that service level can be improved with an increase in inventory investment.

The setting up and running of the VMI model at the case company represents the first phase or level of relationship, which is more of one-way relationship considering the supplier sets up the warehouse and fills that up with inventory but has no say in deciding the amount and timing of replenishment (Holweg et al., 2005; Claassen et al., 2008; Kauremaa et al., 2009). Claassen et al. (2008) have also mentioned the importance of trust for venturing into a VMI relationship. However, based on the case data, there seem to be two levels of trust operating within the VMI environment. The first level of trust is one that lets the customer and vendor enter into a VMI relationship. In this, the customer is looking at the commitment from the vendor's side to support the relationship as well as the ability to invest in the VMI business through resources and technology. Even with this commitment and initial level of trust, the buyer is only willing to enter into the first level of a VMI relationship, where the sales data is shared with the vendor, but the vendor is not given the autonomy to make replenishment decisions. Rather, the customer dictates the replenishment quantities and time of replenishment. Based on the case data, it seems that the retailer lacked trust in the data handling capabilities of the vendor and relied on its buying team for decision-making. However, once the case vendor became familiar with the system and started developing its own forecast and compared the results of their forecast with the customer's forecasts, the customer showed a willingness to move to the next level of relationship (ie, where the vendor's forecast was also reviewed before making a decision). Although the relationship had moved on to the next level, the vendor was still not given autonomy to make independent decisions.

The literature has mostly explored and emphasized the importance of upstream data transfer (ie, from the retailer/customer to the vendor). The literature has also explored the means of upstream data transfer (Vigtil, 2007), though that study highlighted the importance of data transfer downstream at three stages. One was related to making visibility of inventories in the supply chain, focused more on making the customer feel safe in transition from the traditional system to a VMI model. Another was related to sharing of analysis by the vendor. This helped in establishing the credibility and the capability of handling data. And, a third was sharing advance data depicting a supply chain situation in a convenient way. This was more of a value-added service at a higher level of relationship.

The cost benefits to the supplier are not up front. Rather the up-front benefit is of locking in a customer in long-term business, and in reality, this "locking in" results in long-term gains. The initial 1–2 years are more of an investment phase (ie, investment in technology, resources, and more importantly excess inventory) during which the supplier learns the dynamics of VMI business. However, the actual benefits come from greater visibility into future business that lets the supplier plan and optimize their operations (Kapia et al., 2002). The case company also realized benefits in the shape of raw material buying, efficient utilization of resources, better planning, and better negotiation of rates with their partners, as mentioned by Cheung and Li (2002) and Disney and Towill (2003). However, as the relationship moves to the next level where the supplier makes the replenishment decisions, the entire chain benefits because inventory holding and replenishment is optimized.

The case company started with basic commodity type products for VMI business. From customer as well as vendor points of view, starting with a commodity product was beneficial as there is not any threat of end-of-season stock. Although the vendor started with excess inventory in the pipeline, that gave them enough time to understand the system before they could optimize the inventory. Once the vendor had proven its commitment to the program, the customer transferred a seasonal product on the VMI model. This product had a 9-month selling window. The long transit, as well as production time, did not let the vendor graduate further toward adopting the VMI model for shorter life cycle products. Kauremaa et al. (2009) have also mentioned long production time as one of the inhibitors of VMI benefits.

The case vendor also ran multiple types of VMI programs, such as replenishment from country of retail, as well as replenishment from country of origin. The advantage of replenishment from country of origin was a benefit in saving warehousing cost at the expense of moving forecasting process further upstream in the supply chain. This model could favor basic products with longer lead times with some extra buffer inventories at store level. The tradeoff is higher inventories at retail locations versus the cost of running a warehouse in the country of retail. The phases of relationships and nature of transactions, along with corresponding roles of the vendors and the buyers, are summarized in Fig. 11.1.

Phase	Nature	Vendor's capability requirement	Vendor's role	Buyer's role
Pre VMI	Traditional interaction with repetitive business	Manufacturing, process control, systems	Meeting cost, quality, delivery targets	Looking for commitment, no involvement of vendor in decision-making
Phase 1	Replenishment of basic products	Investment inventory HR technology infrastructure data visibility	Managing deliveries based on the schedule given by buyer, developing capabilities to handle data	Decision-making (quantity/del) still with buyer as buyer lacks trust in vendor's data handling capabilities
Phase 2	VMI extension to shorter life cycle products	Forecasting, supply chain synchronization	Replenishment decision-making	Supervision

Figure 11.1 Phases of VMI relationships.

11.6 Conclusion and future research

This study captured a detailed account of a vendor that started on a VMI journey and over the years graduated to a higher level of VMI relationship. The vendor's prior performance and commitment ensured a VMI program; however, subsequent performance and commitment in terms of investment in inventories and resources helped the vendor to understand the business model and gain the trust of the customer to graduate to the next level, or what Holweg et al. (2005) terms as movement from a VMR to VMI model. This requires the vendor to develop their human resources to understand the trends in POS data, past trends, and demand patterns of similar products or cannibalizing products plus the impact of promotional events to come up with the forecasts. Since this study is based on one case study, further research is needed to validate and generalize the areas in which a vendor must focus to graduate to higher level VMI relationships.

In terms of benefits, they are accrued more from better visibility of demand that leads to better planning and optimization of resources. In terms of products suitable for VMI business, longer life cycle products can be converted to a VMI model even with longer product lead times as well as transit times. However, products with shorter life cycles are supported by quicker turn around with shorter product lead times as well as transit times.

This study identified some issues related to buyer's behavior. The reluctance on the part of the buyers to let the vendor manage the replenishment decision-making needs to be further investigated, whether that results from lack of trust or whether that is driven by some other motive. Similarly, while choosing replenishment from country of origin versus country of retail, it needs to be researched whether there exists some benefit one way or the other. In addition, product characteristics and their relationship with the choice of replenishment model needs to be further ascertained. Finally, since this research was based on one case study, the results cannot be generalized to all situations. There is a need for a larger sample to validate the findings.

References

Achabal, D.D., McIntyre, S.H., Smith, S.A., Kalyanam, A., 2000. A decision support system for vendor managed inventory. Journal of Retailing 76 (4), 430–454.

Cachon, G.P., 2001. Stock wars: inventory competition in a two echelon supply chain with multiple retailers. Operations Research 49 (5), 658–674.

Cheung, K.L., Lee, H.L., 2002. The inventory benefit of shipment coordination and stock rebalancing in a supply chain. Management Science 48 (2), 300–306.

Choi, T.M., Sethi, S., 2010. Innovative quick response programs: a review. International Journal of Production Economic 127, 1–12.

Chopra, S., Meindl, P., 2001. Supply Chain Management: Strategy, Planning and Operation. Prentice-Hall, Upper Saddle River, NJ.

Claassen, M., Weele, A., Raaij, E., 2008. Performance outcomes and success factors of vendor managed inventory (VMI). Supply Chain Management: An International Journal 13 (6), 406–414.

Corsten, D., Kumar, N., 2003. Profits in the pie of the beholder. Harvard Business Review 81 (5). May 22–23.

Disney, S.M., Towill, D.R., 2003. The effect of vendor managed inventory (VMI) dynamics on the bullwhip effect in supply chains. International Journal of Production Economic 85, 199–215.

Disney, S.M., Potter, A.T., Gardner, B.M., 2003. The impact of vendor managed inventory on transport operations. Transportation Research Part E 39 (5), 363–380.

Dong, Y., Xu, K., 2002. A supply chain model of vendor managed inventory. Transportation Research Part E 43 (4), 75–95.

Eisenhardt, K.M., 1989. Building theories from case study research. Academy of Management Review 14 (4), 532–550.

Elvander, M.S., Sarpola, S., Mattsson, M., 2007. Framework for characterizing the design of VMI systems. International Journal of Physical Distribution & Logistics Management 37 (10), 782–798.

Hirsch, S., 2005. E-management – supplier becomes closer partner. International Trade Forum 3, 12–13.

Holmstrom, J., 1998. Business process innovation in the supply chain – a case study of implementing vendor managed inventory. European Journal of Purchasing & Supply Management 4, 127–131.

Holweg, M., Disney, S., Holmstorm, J., Smaros, J., 2005. Supply chain collaboration: making sense of the strategy continuum. European Management Journal 23 (2), 170–191.

Kaipia, R., Holmstorm, J., Tanskanen, K., 2002. VMI: what are you losing if you let your customer place order? Production Planning & Control 13 (1), 17–25.

Kauremaa, J., Smaros, J., Holmstorm, J., 2009. Patterns of vendor managed inventory: findings from a multiple case study. International Journal of Production and Operations Management 29 (11), 1109–1139.

Lee, C.C., Chu, W.H.J., 2005. Who should control inventory in a supply chain? European Journal of Operational Research 164 (1), 158–172.

Simchi-Levi, D., Kaminsky, P., Simchi-Levi, P., 2003. Designing and Managing the Supply Chain. McGraw-Hill, New York, NY.

Tony, A., Zamalo, E., 2005. From a traditional replenishment system ot vendor managed inventory: a case study from the household electrical appliances sector. International Journal of Production Economics 96, 63–79.

Vergin, R., Barr, K., 1999. Building competitiveness in grocery supply through continuous replenishment planning. Industrial Marketing Management 28, 145–153.

Vigtil, A., 2007. Information exchange in vendor managed inventory. International Journal of Physical Distribution & Logistics Management 37 (2), 137–147.

Waller, M., Nachtmann, H., Angulo, A., 2004. Supply chain information sharing in a vendor managed inventory partnership. Journal of Business Logistics 25 (1), 101–120.

Waller, M., Johnson, M.E., Davis, T., 1999. Vendor managed inventory in the retail supply chain. Journal of Business Logistics 20 (1), 183–203.

Yin, R.K., 2003. Case Study Research: Design and Methods, third ed. Sage Publications.

Enterprise resource planning systems for use in apparel supply chains

12

Brahmadeep, S. Thomassey
University Lille Nord of France, ENSAIT-GEMTEX, 2 allée Louise et Victor Champier, Roubaix, France

12.1 Introduction

One of the oldest industries that fulfills the basic needs of mankind is the apparel and fashion industry. It is labor intensive, raw material intensive, capital intensive, product intensive, and inventory intensive. New globalization policies have removed all economic barriers and have also created fierce competition with many challenges between industrialized nations. In a progressively information technology (IT)-driven global economic environment, it is indeed very important to promote and facilitate adoption of IT in the apparel industry to help this industry to remain in the national/international competitive market.

The application of IT has increased efficiency at all stages, starting from trend forecasting, designing, manufacturing, supply chain, and logistics to planning and replenishment at the retail stores. One innovation has been the enterprise resource planning (ERP) application, which is a tool that brings about effective coordination between the departments of an organization to direct the process by providing comparative information and analysis regarding trends and forecasts. It assists in effective management of the supply chain, just-in-time inventory, and information as well as enterprise logistics management, effective monitoring and control, accurate planning and scheduling of orders, better data predictions, quick response to queries, and online detailed information of orders [1].

ERP software is for running a business. ERP was coined as an extension of the concept of manufacturing resource planning software, which automated the process of keeping a manufacturing line supplied with materials to meet incoming orders. An ERP system is a suite of applications including financials, manufacturing, human resources, and other modules that together automate the back-office business administration functions of an enterprise. Leading ERP vendors include SAP, Oracle, People soft, and JD Edwards. ERP system refers to the integration and extension of a business's operational IT systems, with the end goals of making information flow within (and beyond) a company more immediate and dynamic; increasing the usefulness and shelf life of information; eliminating redundancy and automating routine processes; and making information system components more flexible. Departmental boundaries generally become softer, accessibility of data is increased for partner companies and customers, and the company's ability to respond to the marketplace is generally enhanced.

Information Systems for the Fashion and Apparel Industry. http://dx.doi.org/10.1016/B978-0-08-100571-2.00012-9

The ERP system is the latest high-end solution that IT has lent to business applications. The ERP system solutions seek to streamline and integrate operation processes and information flows in the company to synergize the resources of an organization, namely men, material, money, and machine, through information. Initially implementation of an ERP package had been possible only for very large multinational companies and infrastructure companies due to high cost involved. By 2015, many companies worldwide have chosen to implement an ERP, and it is expected in the near future that 60% of companies will be implementing some sort of ERP package since this will become necessary for gaining competitive advantage [2].

12.2 Enterprise resource planning systems in the apparel industry: review

An ERP system provides a complete technological solution to integrate and streamline the organization processes and ensure a smooth flow of information. It bridges the information gap across the organization and helps to integrate the resources of the business [3]. It also provides a solution to eliminate issues related to material management, productivity, customer service, cash flow, finance management, quality, inventory, delivery, etc. The ERP enables the organization to put systems such as a management information system (MIS), a decision support system (DSS), data management, and data mining in place and provides vital alerts such as early warnings when the processes deviate from the guidelines provided.

An ERP system can be defined as a software solution that addresses the enterprise needs taking the process view of the organization to meet the organizational goals, tightly integrating all functions of an enterprise It is an industry term for the broad set of activities supported by the multimodule application software that helps a manufacturer or other business manage all the parts of its business. An ERP system facilitates integration of company-wide information systems with the potential to go across companies.

12.2.1 Need for an ERP system

In the absence of an ERP system, a large industry may find itself with many software applications that cannot communicate or interface effectively with one another (see Fig. 12.1).

An ERP system helps business processes flow more smoothly and improves the efficiency of the fulfillment process. It leads to reduced inventory. Eventually, it decreases the overall business cost.

Literarily, an ERP system refers to enterprise resource planning software. In general, an ERP system is an industry term for the broad set of activities supported by multimodule application software that help a manufacturer or other business manage the important parts of its business, including product planning, parts purchasing, maintaining inventories, interacting with suppliers, providing customer service, and tracking

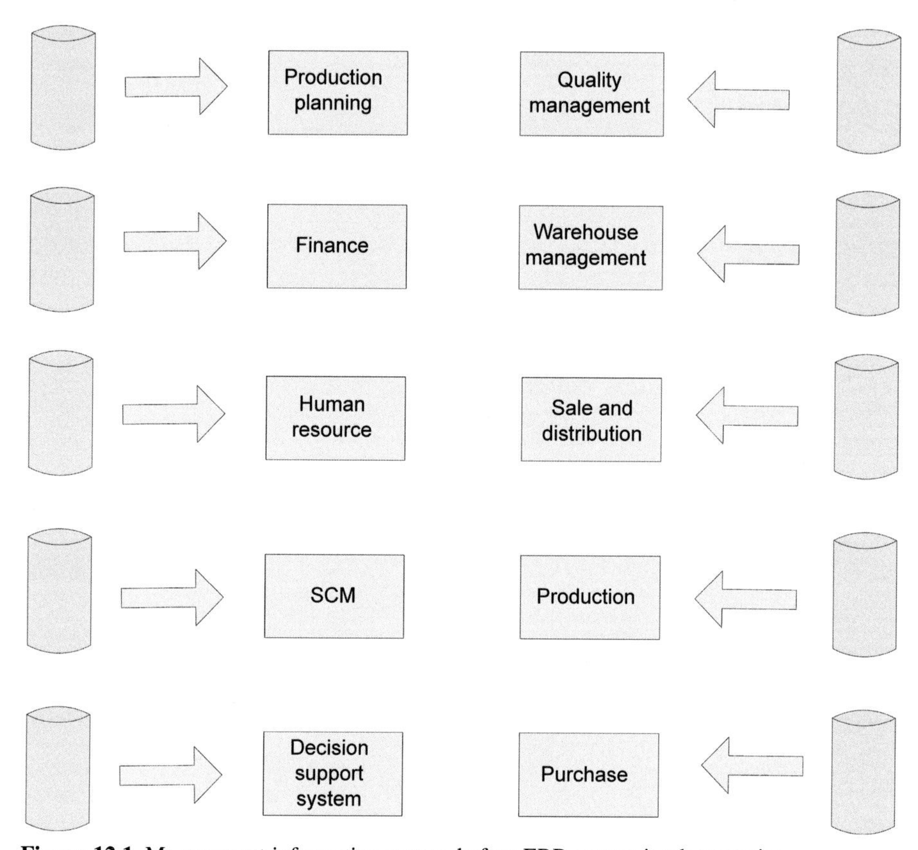

Figure 12.1 Management information system before ERP system implementation.

orders. Typically, an ERP system uses or is integrated with a relational database system. In practice, an ERP system does not live up to its acronym. It does not do much planning or resource planning.

Remember the word "E" for enterprise. This is the true ambition of the ERP system. It attempts to integrate all departments and functions across a company into a single computer system that can serve all those departments' particular needs. The integration streamlines internal business processes and improves productivity of a company.

To be concise, ERP system software offers the following benefits:

It integrates all aspects of the business processes including manufacturing, design, customer services, financial, sales, and distribution (see Fig. 12.2.) By integrating business processes and people anywhere in your company, you can enjoy more efficient workflow and improved productivity.

At its simplest level, the system provides a way to integrate all your business process. To get the most from the software, you have to get people inside your company to adopt the work methods outlined in the software. If the people do not agree with the method and the system has no flexibility to be customized, ERP system projects may fail. Therefore, you should choose ERP system software wisely.

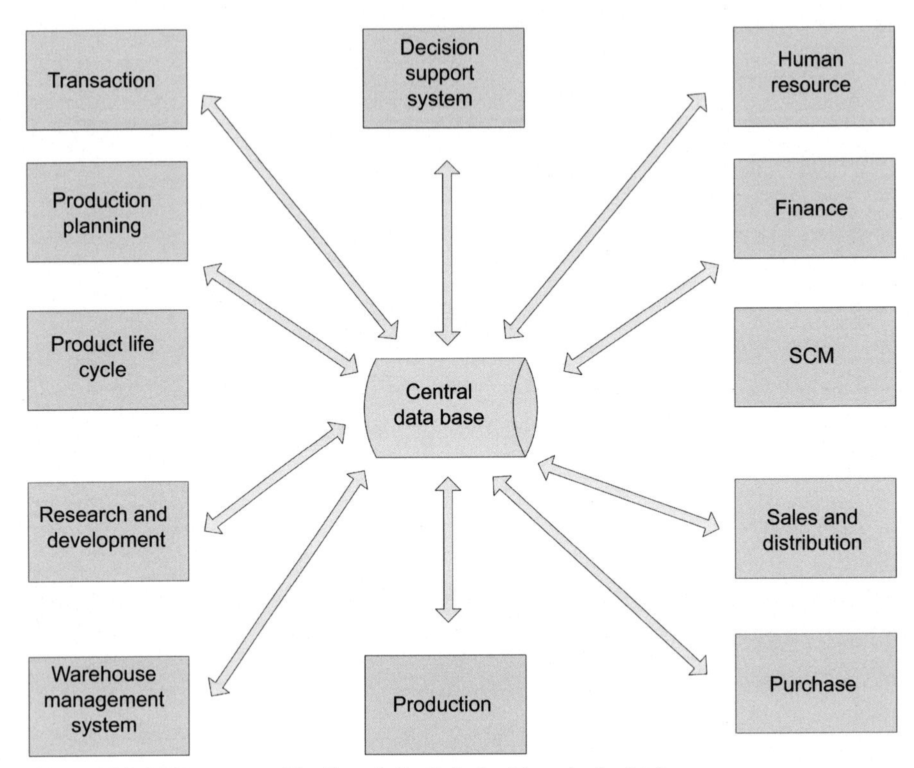

Figure 12.2 ERP system with all modules linked with a single database.

12.2.2 Benefits and efficiency of ERP systems

The textile industry generally concentrates on designing or manufacturing of clothes and also the task of distribution and use of the manufactured product. The textile industry has changed continuously in the past few years. There is a demand to plan out the resources and keep proper data.

An ERP system is software for business that can be customized according to the needs of the enterprise. The ERP system software is the latest high-end solution for performing the business efficiently. The software aims at keeping track of data and making work, processes, and information flow smoothly. An ERP system integrates all aspects of business processes including manufacturing, design, customer services, financial, sales, and distribution [4,5]. By integrating business processes and people anywhere in the company, they can enjoy more efficient workflow and improved productivity.

In 2015, retailers and consumers push for lower prices, better quality, and quicker delivery. An ERP system provides the right information to the right people at the right time anywhere in the world, enabling you to improve productivity, enhance decision-making, and promote communication between coworkers, customers, and vendors. It helps to reduce or eliminate duplicate work, and it automates operational tasks and provides easy access to information. ERP systems, therefore, can deliver significant time savings [6].

ERP systems have helped to increase the quality and efficiency of the manufacturing process. The manufacturing process experiences certain problems often because of miscommunication and lack of communication. It provides a solution by enhancing co-ordination by keeping an eye on the supply chain, warehouse, and logistics. It also helps greatly in the function of tracking the progress made in the manufacturing of the product. If any technical problem occurs, it can be tracked down easily. The customer can be answered easily with the statistics in hand, and his or her queries can be easily answered with the details of the status of the product. Long chains of communication are shortened, and the details can be shared through internet, thus avoiding any miscommunication. Manufacturing companies often find that multiple business units across the company make the same widget using different methods and computer systems. ERP systems standardize the manufacturing processes and improve quality [7–9].

An ERP system integrates the processes of the business across departments into a single information system, resulting in reduced operating cost. The problem of low inventory or reduced operating cost is ruled out. Whenever a resource is needed, it is available on time because everything has already been planned. The day-to-day management becomes easy because it keeps track of the warehouse also. Everything going into the data warehouse is recorded, thus planning for a specific day can be easily done. Due to the fact that every activity is recorded, the actual cost can be calculated easily. The unwanted ambiguity is ruled out, which makes the database user friendly. Since the product is being managed so well, its quality remains intact: what the world gets is a reliable product. Every small detail can be easily taken care of and smooth flow of activity takes place.

For both textiles and the garment sector, strategic planning is important, and an ERP system is designed to support this through resource planning. It facilitates report generation, which has to be updated every time a progress takes place. This report can be distributed among employees so that they also know which area has to be given more attention for the completion of the task. The textile and garment industry is ever changing, and thus it is important to know what the customers need and record it.

Advanced ERP systems software has the ability to be customized to the extent that screens can be remodeled, fields can be edited, and the architecture modified through progressive installation processes. In addition, these can be operated in a secured, Internet-based environment. These features provide flexibility and convenience in implementation and operation. Some multilingual ERP systems can perform automatic translation that enables almost every style detail to be viewed in several languages, such as English and Chinese. It improves the effectiveness of communication [10].

A few advantages and disadvantages of the ERP systems in the fashion industry are outlined in the following points:

- ERP systems connect the necessary software for accurate forecasting to be done; this allows inventory levels to be kept at maximum efficiency for the company to be more profitable,
- Integration among different functional areas to ensure proper communication, productivity, and efficiency,

- Design engineering (how to best make the product),
- Order tracking, from acceptance through fulfillment,
- The revenue cycle, from invoice through cash receipt,
- Managing interdependencies of complex processes bill of materials (BOMs),
- Tracking the three-way match between purchase orders (what was ordered), inventory receipts (what arrived), and costing (what the vendor invoiced),
- The accounting for all of these tasks: tracking the revenue, cost, and profit at a granular level.

ERP systems centralize the data in one place. Here are some benefits of an ERP system:

- Eliminates the problem of synchronizing changes between multiple systems consolidation of finance, marketing and sales, human resource, and manufacturing applications
- Permits control of business processes that cross functional boundaries
- Provides top-down view of the enterprise; real-time information is available to management anywhere, anytime to make proper decisions
- Reduces the risk of loss of sensitive data by consolidating multiple permissions and security models into a single structure
- Shorten production lead time and delivery time
- Facilitating business learning, empowering, and building common visions

Some security features are included within an ERP system to protect against both outsider crimes, such as industrial espionage, and insider crimes. A data tampering scenario, for example, might involve a disgruntled employee intentionally modifying prices to below the breakeven point in an attempt to interfere with the company's profit. ERP systems typically provide functionality for implementing internal controls to prevent actions of this kind. ERP system vendors are also moving toward better integration with other kinds of information security tools.

Problems with ERP systems are mainly due to the implementation issues and the inadequate investment in ongoing training for the involved IT personnel, including those implementing and testing changes, as well as a lack of corporate policy protecting the integrity of the data in the ERP systems and the ways in which it is used. Some of the major points on the disadvantages are mentioned in the following [11–15]:

- Customization of the system is limited.
- Reengineering of business processes to fit the industry standard prescribed by the ERP system may lead to a loss of competitive advantage.
- ERP systems can be very expensive.
- ERP systems are often seen as too rigid and too difficult to adapt to the specific workflow and business process of some companies; this is cited as one of the main causes of their failure.
- Many of the integrated links need high accuracy in other applications to work effectively; a company can achieve minimum standards, then over time, "dirty data" will reduce the reliability of some applications.
- Once a system is established, switching costs are very high for any one of the partners (reducing flexibility and strategic control at the corporate level).
- The blurring of company boundaries can cause problems in accountability, lines of responsibility, and employee morale.

- Resistance in sharing sensitive internal information between departments can reduce the effectiveness of the software.
- Some large organizations may have multiple departments with separate, independent resources, missions, chain of command, etc., and consolidation into a single enterprise may yield limited benefits.

12.2.3 SWOT analysis

The SWOT analysis (strength, weakness, opportunity, and threat) of the ERP systems in the fashion and apparel sector is detailed in the following subsections.

12.2.3.1 Strengths

Strength means the characteristics or factors that give an edge for a company over its competitors, the advantages over others in the market. The strength factors, depending on the organization, for ERP systems implementation are some of the factors listed in the matrix, such as cost, schedule, infrastructure, efficient and experienced manpower, and a long-term mission and vision. The organization that has a strong financial hand can go for any new change in the organization, and as an ERP system needs lots of money, cost can be one of the important strengths of the organization. The implementation is a long and time-consuming process, so it requires more time for better implementation in organization, and if the organization has the time, it will be more of a benefit to the ERP system. If the organization already has better infrastructure in hand, the implementation is easy, otherwise more time is needed to create a new setup for the new technology deployment. Human resources is a main strength, and if the organization has efficient and experienced manpower, thing are easily adopted and perfectly implement in most of the cases. Above all, a clear, advanced, long-term mission and vision in top-level management will always stand as a supporting poll for the ERP system development team as the parental tree for achieving the goals and objectives.

A few domain specific strengths are listed:

- Industry-specific functionality
- Dedicated support and implementation for professionals well versed in the textile and apparel industry
- Benchmarking abilities and world-class industry experience
- Minimum customizations
- Improve company performance with industry-specific optimization functions (fabric cut optimization, cost optimization, and optimize planning and scheduling)
- Integrated solution for all areas (fiber to garment) and all activities (sales to purchasing)

12.2.3.2 Weaknesses

Weaknesses are characteristics or factors that can be harmful if used against the firm by its competitors or disadvantages relative to the internal environment of the organization. During the ERP system implementation, the role and support of top-level management plays a vital role, and if decision-making is of a bureaucratic nature at the top level of management during the implementation, others in the organization may not have faith in management, adversely affecting the implementation. Sometimes not all in the organization may have as much technical savvy as the development team, yet if they will use the ERP system, they must understand the technology. If they failed in this effort, it will be a weakness for the organization. Sometimes, the people of the organization are not in position to change their traditional working environment nor to acquire the new technology in their working mode, or it can be dangerous for the organization to acquire the new technology. The organization hired the consultants for implementation and deployment of the ERP system, but after finishing the task, the consultants leave the organization. Here the problem arises in regard to the maintenance of the ERP system. If there is no group of people in the organization who have deep knowledge of ERP system technology, the consultation process must start again, increasing the cost of the process. Most of the organization fails to analyze the need of market and customer requirements, and by poor knowledge of the market, the ERP system implementation may fail.

A few domain specific weaknesses are listed:

- Too small of an investment in dedicated R&D, mostly customer-based modifications
- Too small of a customer base to provide support to help desk facilities
- No proper version control and no new releases
- Need to patch many solutions together
- Lack of integration
- Older technology, and no funds to invest in leading-edge technology

12.2.3.3 Opportunities

Opportunities are characteristics or favorable situations that can bring a competitive advantage in the market. The organization that is always looking for the new opportunities to achieve their objectives will never face failure in the market. Similarly, the organization that is planning for ERP system implementation needs to find opportunities in the market for the implemented technology adopted by the organization. While considering the opportunities in the market, the first element is the customer, and an ERP system will create the rapid and corrective responses to the request for proposals (RFPs) of customers. Further, it can create a consultant division in the market for the ERP system implementation for other organizations without an ERP system. With the help of rapid responses to the RFPs of customers, the ERP system creates the opportunities to make the customer satisfied at a certain level. With the help of an ERP system, we can have an idea about how and what data is flowing among the different processing activities in the organization, which keeps top-level management updated. The ERP system also gives a better understanding of the current organizational workflow and fund flow, which helps the organization to change its tactical decisions by creating new strategies for the improvements.

A few domain-specific opportunities are listed:

- Reduce need of customizations
- Change and optimize business organization

- Improve customer service and profitability
- Reduce inventory and WIP
- Minimize lead times
- Control costs
- React quickly to market changes
- Minimize investment and IT budget
- Fast and on-budget implementation

12.2.3.4 Threats

Threats are characteristics or unfavorable situations that can negatively affect the business or cause trouble for the organization. While considering the threats parameter in an ERP system implementation, the very first factor that comes to mind is security of data that travels from the different individuals, divisions, departments, offices, and plant, sometimes at the same location or different geographical locations, which needs even more security. One of the costliest threats for the ERP system implementation is the maintenance cost of the ERP system, which requires continuous monitoring, time, and people in the process. Sometimes, slow growth of the organization also has serious effects on the process, where the organizational growth is not up to the mark as decided by the top-level management during the implementation. Competition is the biggest threat for any organization, especially if the competitor is more advanced in technology. After completing the implementation, some people from the organization must have enough technical knowledge to solve the problems they may face in the future to reduce the dependency on the ERP system development consultant, who are outsiders to the organization, thereby reducing the cost of maintenance.

A few domain-specific threats are listed as follows:

- Company may close or sell off to others, losing focus
- Obsolete technology, no Internet access, etc.
- Supporting Hardware can become obsolete soon
- High turnover of employees may lead to loss of installation expertise
- High costs of upgrading to new modern software or hardware

Textile industry ERP system solutions may become a risk if selected poorly. The selection must not be based on the presentations and demonstration alone: all the aforementioned factors have to be well considered if success is to be achieved.

12.2.4 ERP systems in the apparel market: a brief summary

The following ERP systems are found on the market for use in the fashion and apparel industries:

12.2.4.1 Sync

Sync is a fully integrated ERP system software designed specifically for the apparel industry. This business management system manages the entire job of costing as well as the project management process, and it includes seven comprehensive modules [16].

Costing
- Costing software that allows calculating costs with accuracy

Purchasing
- Software to control the creation and processing of purchase orders

Stock control software
- Inventory control software to ensure effective control of stock levels

Mobile sales
- This apparel software module allows the creation of sales orders from remote locations

Task management
- Task management software makes it easy to keep track of tasks across multiple projects and allows job tracking

Financial integration
- Seamless integration into financial software (Pastel Partner, Pastel Evolution, QuickBooks, AccPac, and Microsoft Dynamics GP)

Reports
- Effective business management system generates critical reports at the click of a button

12.2.4.2 Ysoft

Apparel Industry Extensions (AIE) is an apparel system/solution built upon open-source technologies, primarily Compiere ERP system and customer relationship management (CRM) software. Compiere is the number one open-source ERP system software application on Sourceforge.net and features a comprehensive solution covering customer management and supply chain to accounting for small- to medium-sized enterprises in distribution and service industries. However, to reach wider audiences, Compiere is destined to be a generic ERP system application and, hence, may fall short of features/functions that are important to certain industries, for example, the apparel industry [17,18].

To address the specific needs of trading and sourcing business operations in the apparel industry, Compiere has been heavily customized and enhanced. AIE aims at helping those companies to manage their businesses more effectively with a robust, flexible, and economical system. Among the enhancements is the ability to handle products by style/color/size throughout the sales cycle.

The main features of Ysoft are listed next:

User-definable color and size codes
- This allows setting up unlimited number of colors and sizes for each style.

Garments data entry in grid window
- Style/color/size can be entered to the system in a grid window; this feature is available in both sales order and purchase order.

Assortment number
- Assortment number is used to group garments by the destinations within a customer's organization such that customers are not required to repackage their garments before dispatching them to their final destinations within their organizations; assortment number is instrumental to distribution order.

Assortment Details Report

- A nice looking Assortment Details Report is printed in style-grid format showing the quantities ordered for each combination of color and size; this report is intended to supplement the sales order printout, which contains quantity ordered per style without the detailed breakdown of colors and sizes.

Packing instructions

- Packing instructions, if available, can be entered in sales order (or purchase order) and even printed as part of the sales orders; packing instructions are used to indicate how garments should be packed for delivery, and if entered in sales order, are automatically copied to shipment and purchase orders to save duplicate data entry.

Merge garments of multi-style/color/size into a carton box

- This gives the highest flexibility in merging garments of different styles, colors, and sizes into a carton box; overflowing a carton box with more garments is permissible, and if remerging is necessary, it can be easily done with a click of a button.

Packing instructions printout

- An intuitive printout of packing instructions is developed to make the viewing of quantities of garments packed in different carton boxes easier; the printout is designed to cater to multistyle, color, and size.

Copy packing instructions from sales order to purchase order

- To save redundant data entry for packing instructions throughout the sales cycle, packing instructions that are entered in sales order are carried over to shipment and purchase order automatically.

Generate apparel products using style/color/size combination

- To make the creation of garment products easier and faster, a product generation window is provided for generating garment products using the specified style/color/size combination.

Order status tracking and email notification

- Sales orders and purchase orders can be tracked for progress. The system allows defining milestone templates for different type of orders. Each milestone carries an expected completion date, among others. If today's date is greater than the expected completion date of a milestone, the system will highlight the milestone to draw your attention when you are viewing the order status of the sales order. In addition, when you are away from the system, you can still get an email notification sent out by the alert. An order status information window presenting the status of all orders in tree structure by order and by style is provided for easy enquiry.

Exceptions alerts

- The system sends alerts by email for situations that you might like to act upon. The alert is user-definable, and hence, it can be used to alert basically anything.

Product catalog by style with style image

- A product catalog is specially designed to show/print some key information about a style along with its image/picture; information to be shown/printed is user-definable.

Product BOM definition report and copy

- It is a printout of all the BOM details of a product; this report is useful when customers are making a serious enquiry about a particular product.

BOM drop of sales orders into production order

- To facilitate the update of inventory of raw materials consumed by garment products in a sales order after shipment, a new feature similar to BOM drop was developed to make the copy of sales orders into production orders easier.

Generate consolidated material requirements plan for multiple orders
- This is one of the most important and powerful features of the system; it will explore the BOM definitions of the garments and compute and consolidate all raw materials that are required to manufacture all the garments in the order selected.

Other features include 3-month calendaring, auto-update year-end retained earnings brought forward, and auto-alert on prepayments and outstanding invoices.

12.2.4.3 VisualGEMS

VisualGEMS has been designed with the flexible working of the garment industry in mind. The functional breakup of the software has been organized to match with the distribution of garment organization into various departments. The facility of online referencing of master data and even related transaction data keeps the users free from the pressure of remembering codes. The integrated design of VisualGEMS allows one to establish and enhance interdepartment communication within the organization, which otherwise may become a major productivity bottleneck. Compatibility of information/data sharing needs between the various departments is built into the software and is the major pay-off of the integrated environment.

In spite of being a standard software package, VisualGEMS can be tailored to suit your specific requirements by using powerful template technology. User-definable templates of voucher and shipment documents can be flexibly used to meet the specific needs of the industry's document formats. VisualGEMS has been designed to meet the industry's present needs and to be flexible enough to accommodate new requirements [19].

The main features of VisualGEMS are listed next:

- Supports multicompany working
- Supports multiuser working with extensive control of user access permissions
- Supports multicurrency export sales
- Supports progressive implementation by a setup-based workflow
- Provides online master creation
- No codes to remember for items, parties, and accounts data
- Supports storage and printing of pictures/sketch of garment style
- Provides user with powerful decision-making tools like purchase control center for generation of purchase orders starting from a table of unfulfilled raw material requirements
- Maintains stocks at your stores as well as factories, processors, and job working units
- Allocates material to orders with goods received notes entry depending on EO delivery date, first come first serve, or weighted average basis
- Allows issue of raw material for an order from unallocated stock or from other orders
- Supports user-configurable reports
- Reports can be exported to MS Word, MS Excel, and text formats
- Extensive drill down from almost all reports up to voucher level for quick audit of data
- Supports multicurrency vouchers
- Financial accounting module supports creation of user-defined account books, which are very useful for multiple banks, cash, purchase, and sale for clear bifurcation of data and separate voucher numbering
- Covers extra needs of sales tax, TDS, and some special audit reports

- Purchase bills can be entered with item details
- Unique credit/expense registers for entering creditor's bills without item details
- Special multicolumn reports for cash/purchase/sales
- Consolidated outstanding report for getting picture of parties having sister concern
- Allows maintaining receivables/payables on bill-by-bill basis
- Add-ons for exporters for exhaustive tracking of pre-shipment/postshipment credits
- Transparent and easy migrations to new year with automatic data carry forward

VisualGEMS offers the following standard modules that can be configured to work in an integrated fashion:

- Export order management
- Purchase order management
- Import management
- Inventory management
- Production management
- Quota management
- Shipment management
- Financial accounting

12.2.4.4 e-Smartx

e-Smartx is a state-of-the-art, modern, cost-effective, and proven enterprise platform solution that meets the need of the apparel manufacturing industry. The effective support from installation to usage is the key success for both the user and the developer [20].

e-Smartx streamlines the total process cycle in the apparel manufacturing. Its workflow system provides superior business management capability for faster response to the customer demands. Internet-enabled, real-time MIS gives manufactures the capability to monitor the whole process from product development to exports across the globe. It integrates all the departments across multifactory locations, such as these:

- Merchandising
- Planning
- Purchase
- Production
- Inventory management
- Exports
- Finance
- Payroll

Advantages and benefits:

- Effective monitoring of overall operations of the company
- Single database across the globe
- Highly robust and real-time MIS
- Flexible and parameterized
- Greater efficiency and performance

- Continued support and constant technology migration
- Help reduce operation costs
- Supports strategic planning
- Internet access

12.2.4.5 ATOM

ATOM is a suite of application software designed for the textile industry. If you are a manufacturer of apparel, silk products, home furnishing, made-ups, or just fabric, then ATOM allows you to save costs, eliminate errors, and stay ahead of your competition [21].

Here are the key features of ATOM:

- ERP system designed exclusively for the textile industry
- Multicompany, multilanguage features
- Manage entire process chain, from knitting/weaving, cutting, and sewing in a single system
- Project and monitor sourcing and production
- Track raw materials availability automatically
- Monitor production orders in real time
- Visual dashboards for easy information access
- Reliable technology platform that supports a high volume of transactions
- Modular functions for need-based implementation

Key benefits:

- Increase operational efficiencies; maximize cash flow and ROI
- Reduce order-cycle times
- Always maintain inventories at proper levels
- Reduce wastage of fabric and accessories
- Accounting of chemicals and dyestuff in process house
- Make quick business decisions using real-time information
- Automate routine processes
- Manage in-house and job work production
- React to changing business needs swiftly
- Built by experts in fashion business
- Low cost of ownership

A basic comparison of the aforementioned apparel sector–specific ERP system software based on similarities and additional features is summarized in Table 12.1.

These are the **specialized ERP system software** for the apparel and fashion industry. Further, there are some other software that are used in the apparel and fashion industry, for example, **SAP, NOW,** and **Oracle**.

12.2.5 ERP system modules for the apparel sector

The previous section mentioned various ERP systems that are used by the fashion and apparel industries. All that are a part of the apparel industry have the following modules/components or extensions in the ERP system.

Table 12.1 Summary and comparison of the apparel-specific ERP systems mentioned in the literature

ERP system software	Differences: additional features	Similarities or common features
Sync	Mobile sales, financial integration (links seamlessly with financial software).	All basic modules required in apparel industry, for example, costing, purchase order management and tracking, production management, inventory and stock, packaging and shipment, and financial accounting.
Ysoft	Garment specification in good detail, parameters and specifications can easily be customized, very detailed order management system, 3-month calendaring system with auto-update and alert system, generate multiple requirements plan for multiple orders, exception alerts (user-defined).	
VisualGEMS	Tailor-made customizable platform, supports multitasking with extensive control of access permissions, supports progressive implementation by a setup-based workflow, online master creation, user friendly (no coded value) and better GUI, production planning and scheduling, resource allocation, payroll and human resource integrated, various add-ons and tools.	
e-Smartx	Faster response to customer demands, Internet-based real-time management information system, integrates all modules for online tracking, monitoring and control; flexible and parameterized, single database across the globe, greater efficiency and performance, continued support and constant technology migration.	
ATOM	Multicompany, multilanguage. Entire process supply chain, special modules for textile production, sourcing monitoring, visual dashboards for easy information access, decision-making in real time, low cost.	

Components/modules:

- Transaction
- Financials
- Distribution
- Human Resources
- Product lifecycle management
- Advanced applications
- CRM
- Supply chain management software
- Purchasing
- Manufacturing
- Distribution
- Warehouse management system
- Management portal/dashboard
- DSS

These modules can exist in a system or utilized in an ad-hoc fashion.

Manufacturing: engineering, bills of material, work orders, scheduling, capacity, workflow management, quality control, cost management, manufacturing process, manufacturing projects, manufacturing flow
Supply chain management: order to cash, inventory, order entry, purchasing, product configurator, supply chain planning, supplier scheduling, inspection of goods, claim processing, commission calculation
Financials: general ledger, cash management, accounts payable, accounts receivable, fixed assets
Project management: costing, billing, time and expense, performance units, activity management
Human resources: human resources, payroll, training, time and attendance, benefits
CRM: sales and marketing, commissions, service, customer contact, call-center support
Data services: various "self-service" interfaces for customers, suppliers, and/or employees
Access control: management of user privileges for various processes
DSS: forecasting, planning, and reducing overall cost by optimization

12.2.6 Implementation

Many organizations do not have sufficient internal skills to implement an ERP system project. This results in many organizations offering consulting services for implementation. Typically, a consulting team is responsible for the entire ERP system implementation, see Fig. 12.3.

ERP systems play an important role in the fashion and apparel industry. There are many cases of implementation globally. Also, there are many cases where the implementation is a huge challenge to the fashion companies [13,22].

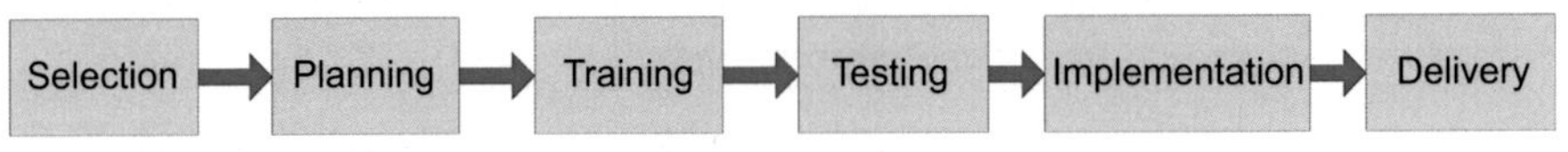

Figure 12.3 Basic ERP system implementation process.

For best business practices and successful business results, the ERP system application has to be completely implemented across all departments. The implementation of an ERP system will have a greater impact on the processes of the organization, functioning of the department, organization culture, employees, and the management [23].

ERP system implementation is a difficult task. Most of the implementation overshoots the implementation schedule, costs overrun budgets, under-delivers the business expectations, and delivers less business value [24]. It involves huge financial implications, a lot of human resource at different levels with different frequencies, and it requires lots of technical expertise and a dedicated approach, which makes it more difficult for smooth implementation. In general, ERP system implementation in an existing industry is a transformation process. Any transformation in the industry has to go through its rough patches of resistance, conflict, setbacks, bottle necks, and time delays. This also leads to extensive delay in implementation, which most of the time is treated as a failure to deliver in time [14,25].

The apparel industry works with stringent time frames and involves many technical processes. As the product life cycle becomes shorter with less lead time to deliver, the industry moves in to automation of its processes with state-of-the-art technology. Automating the processes of the industry through implementing an ERP system application will be one of the major initiatives taken by the organization to keep up the pace of the industry's growth.

Harvard Business Review [23] from September, 2011, states: "There were 1471 project implementations examined for comparing their budgets and estimated performance benefits with actual cost and result. Out of these, 27% of the projects had overrun the project cost even up to 200%. Panorama's study [26] reiterates that, 57% of implementations take longer than expected due to the unrealistic expectations regarding time frames."

12.2.6.1 Project planning

The implementation of an ERP system is an initiative taken by an individual or group of individuals or management in an organization. An organization has to identify its objective and need for implementing the ERP systems in line with the organization's goals. The ERP system implementation will not yield expected results if the objectives are not clearly defined. These objectives must be defined, keeping in mind parameters such as existing business needs, the processes, and future expansion plans. The organization must identify a team of technical experts who understand the business processes to define the requirements of the automation process. The expert team must study existing ERP system products available in the industry and the software used by peer groups.

12.2.6.2 Software selection

Product selection plays a key role in ERP system implementation. Successful implementation is determined by the product selection that determines the future course of business. The technical team plays an anchor role in software selection, and it must be transparent, critical, and analytical and should function with complete freedom of expression. The software should be flexible to meet the changes in business, scalable to grow with the company, less complex, user friendly, affordable to the organization, and compatible with the internal culture of the organization. The software should use the latest technology used in the industry. Software selection should not be influenced by an individual or a group, popularity or brand name and should not be biased.

12.2.6.3 Vendor selection

Vendor selection is also a key to successful project implementation. The technical team has to analyze the existence of the product in the market and the experience of the vendor by analyzing the number of successful installations, turnover, and experience in development and implementation. Technical expertise of the vendor's implementation team or the implementation partner is an important key for implementing an ERP system in an organization. The ERP system implementation can fail due to poor vendor selection and lack of after-sales support.

12.2.6.4 Customization

ERP system software are not ready-to-use products; these require minor changes to make them functional and to meet the complete requirements of the organization. The technical team in association with the experts of the external implementation team must determine the level of customization required, by keeping the parameters in mind such as these:

- Ready to use features of the software by understanding existing business processes of the organization.
- The technical team identifying the gaps that need to be bridged, so that the organization practice becomes akin to the ERP system environment; GAP analysis would determine the depth of restructuring required in an organization and provide necessary suggestions such as new reports, analysis, and better features; this process is inevitable even though it is expensive and time-consuming.
- Reengineering needs organizational restructuring and change in the processes that will help in enhancing productivity and eliminate processes that are not required; reengineering the processes must be taken to suit the requirements of the software and for the better process controls.
- The product must be user friendly; customization should not exceed 20% of the project scope.

Most of the implementation delays are caused by excessive time taken for customization. Misunderstanding the level of customization required or the clear-cut understanding of processes of the organization will result in unnecessary delay at the implementation stage.

12.2.6.5 Architectural design

Technical requirements of the organization determine the architectural requirements. Technical requirements include the number of modules required, which includes e-business and e-commerce applications. Architectural design is the backbone of the ERP system implementation. The implementation team should identify the type of deployment required, such as on premise or host the application at a data center or cloud (private or public). Based on this the organization should determine the number of user locations, data capturing points, access points, and report centers and determine the architectural requirements such as hardware, software, network, firewall, and other infrastructure requirements. The security levels should be clearly defined at the time of preparing the software requirement specification (SRS). Any ambiguity in architecture design will complicate and delay the implementation.

12.2.6.6 Cost of the software

A proper investment plan with clear-cut budget allocation is a key to the implementation success of ERP systems. Vendors never disclose the complete investment plan, which includes these points:

- Product pricing
- Hardware infrastructure
- Technology
- Training and implementation
- Maintenance

Improper fund flow will lead to a delay in implementation. A service level agreement must be signed with a complete plan of action and scope for deviations such as delay in implementations and accountability factors on both sides.

12.2.6.7 Data management

Capturing the right data at the right point is the key to scientific implementation, which includes elimination of duplicate and redundant data and capturing indirect data through data mining. The success of implementation is determined by the use of existing data stored in the organization in various formats using a data conversion method.

12.2.6.8 Preparing for the venture

The ERP system implementation team must have participation from all stake holders such as technical experts, top- and middle-level management, end users, and finance. The implementation may fail due to improper implementation methodology, poor budget allocation, fund flow, and nonavailability of infrastructure. Failure to draw blue prints, SRS for implementation, or defining the responsibility and accountability will lead to implementation failure. Lack of awareness among the employees about ERP systems and its advantages will lead to resistance to change.

Prior to the commencement of the implementation the organization must provide the right IT infrastructure prescribed by the implementation team. Nonavailability of infrastructure and technology may lead to unnecessary delays in project implementation.

The ERP system implementation may not yield expected results in time due to improper installation, configuration, customization, critical testing with real-time data, or training of trainers and end users. A conflict of interest between internal and external experts and resistance to change by employees will result in a delay of the implementation.

Failure to implement the ERP system application or delay in the process will affect the growth of the industry and result in a huge loss in terms of finance and time. In the long run, delay in implementation diverts the focus of the industry, thus challenging its basic existence. The ERP system implementation failure is not a threat to one organization; it is a threat to the apparel industry itself. It indirectly affects society and the growth of the nation. There is tremendous scope to study the difficulties faced by organizations in implementing the ERP system application in the apparel industry.

12.3 Case studies

12.3.1 Success

This section describes an ERP system implementation success story at a multibrand fashion retail chain in India. Pantaloons Fashion and Retail is the flagship retail company in India. It is one of the fashion retail pioneers catering to the entire Indian consumption space. Through multiple retail formats, the company has managed to connect a diverse and passionate community of Indian buyers, sellers, and businesses. This company has generated trust among the Indian customers through innovative offerings, quality products, and affordable prices that help customers achieve a better quality of life every day. Currently, the company serves customers via 103 stores across 49 cities in India. The company came into retail in 1997 with the launching of its fashion retail chain: Pantaloons in Kolkata [27].

More than 8 years after it entered into the retail business, Pantaloons realized that they needed an IT solution that would help to integrate their operations and help them to stay competitive in the rapidly growing Indian retail market. Pantaloons was regularly opening stores in the metro cities, and there was an urgent need for a reliable enterprise-wide application to help run its business effectively [28]. The company was looking for a solution that would bring all of its businesses and processes together. For Pantaloon retail before the implementation, most of its solutions were developed in-house. Retail Enterprise Management (REM), a distributed point of sales (POS)-based solution, was developed in-house and deployed at every store, which managed inventory promotions, sales, and customer profiles across all the outlets. With the expansion in business, the software was modified, enhanced, and stretched to accommodate the changing retail dynamics. Pantaloon had also piloted the use of RFID technology, and their experiment had been successful.

12.3.1.1 Need for ERP systems at Pantaloons

Pantaloon wished to improve their store operations and have a robust system to support their growth. An ERP system would help meet their requirements through the following:

- Improved financial tracking and reporting for all retail locations and business levels
- Deft handling of complexities in retail business requirements
- Enhanced decision-making by providing more granular, real-time information
- Support for the financial accounting needs of a rapidly expanding retail business
- Expediting the reconciliation and closing processes.

12.3.1.2 The solution

After a comprehensive evaluation of different options and software companies, the management at Pantaloon decided to go for SAP's retail solution. This was because SAP had a solution specifically directed at the retail sector that met much of Pantaloon's requirements. SAP being the market leader in the ERP system vendor space, it was believed that SAP was the best possible solution provider.

Some of the qualities of SAP retail solutions are that it supports product development, which includes ideation, trend analysis, and collaboration with partners in the supply chain; sourcing and procurement, which involves working with manufacturers to fulfill orders according to strategic merchandising plans and optimize cost, quality, and speed variables that must be weighted differently as business needs, buying plans, and market demand patterns change; managing the supply chain, which involves handling the logistics of moving finished goods from the source into stores and overseeing global trade and procurement requirements; selling goods across a variety of channels to customers, which requires marketing and brand management; managing mark-downs and capturing customer reactions, analyzing data, and using it to optimize the next phase of the design process.

12.3.1.3 The implementation

Three successive risks associated with the adoption of new technology were identified. The first risk was more of a technological one: whether the decision to opt for certain technology is correct or not. The other problem was this: the trained employees would start to leave the organization in search of better jobs armed with their new learning. So, there would be the case of trained employees leaving the organization creating a void very difficult to fill. The biggest of the three risks was planning the implementation and the danger of investment going awfully wrong. Initially, the entire focus was on validity of the solution and whether it met needs.

Several sessions were planned to educate and inform all the employees about the need for ERP systems and the benefits that would accrue to the company as a result of the implementation. The SAP implementation project was named and branded as SAARTHI. With everything in place, the next step was to form an implementation team. This team was comprised of experts from Pantaloons, consultants from SAP, the vendor, and Novasoft, the implementation partner. This team composition ensured that the team had expertise at all required levels, namely domain, process, project management, technical, and packaging [29]. Once this was done, a newsletter was started with the intention of keeping everyone informed about the project. The implementation was undertaken in three phases. In the first phase, the existing processes were documented and the gaps were identified. This was used to then arrive at new states of the solution.

In the second phase, the SAP platform was developed with the help of the template designed by Novasoft. SAP defined this template on the basis of their expertise in retail solutions and a careful evaluation of Pantaloons' needs. This technique is called accelerated SAP (ASAP), which is used to speed up the implementation. The last phase in this project was for stores to go live. This involved switching over to the new system and porting data from the legacy system to the new system. Before the SAP implementation, all the data was unorganized. This data had to be organized and ported to the new SAP application, and this was a major challenge. This was the most time-consuming activity during the implementation. Finally, the implementation project began and was completed in 6 months.

12.3.1.4 Postimplementation

In the period following the implementation, steps were taken to ensure the smooth functioning of the newly deployed system. In the 5 years after the completion of the implementation, several additional modules from SAP were implemented to keep the system up to date.

The key challenges after implementation were in managing the perceptions of people during the period of 6 months when the implementation was underway. People were not convinced about the need for an ERP system and did not see why it was necessary. Migrating unorganized data to an organized format was the next key challenge [30].

Pantaloons was not able to see immediate benefits from this implementation. They were already working on MAP (merchandise assortment planning), auto-replenishment, and purchase orders. They hoped to use these systems to optimize their inventory and cut it by about 2–4 weeks. The SAP ERP system is completely integrated with POS machines. Further, all the sales data generated in a day are uploaded into SAP overnight, automatically updating the inventory movement and financial changes. In the long run, they were able to experience improvements in several areas:

- Financial and strategic benefit
- Greater business insights through more accurate and timely financial information
- Lower costs via convergence of financial accounting and controlling
- Enhanced data quality
- Easier compliance with regulatory requirements via the SAP parallel accounting feature
- Operational benefits
- Real-time, continuous reconciliation of cost elements and expense accounts, freeing up personnel for more value-added activities
- Ability to close books 15–20% faster
- 5–10% reduction in accounts receivables

In 2011, Pantaloon Retail also won the Best run Business in Mobility Adoption from SAP. This reaffirmed that they had made good use of IT to make information available anytime, anywhere on any device [31].

Similar cases of the successful implementation of ERP systems in the fashion industry could be seen at Zara, H&M, Ikea, and Walmart [32].

12.3.2 Failures

This section describes an ERP system implementation failure story at a global sportswear retail chain. This case study is an analysis of the performance at Nike Inc. of the implementation of an ERP system to manage the supply chain, mainly on the issue of a new software failure that affected Nike adversely. The reasons behind the failure and the effects of the breakdown are discussed briefly.

Supply chain management (SCM) plays an important role in the successful running of any business. SCM has allowed companies to rethink their entire operation and restructure it so they can focus on its core competencies and outsource processes that are not within the core competencies of the company. But the opposite effect of the SCM also comes into view at times. There have been many cases in the past in which the SCM has proven to be a reason for poor financial performance of the company.

Nike had a good supply chain management wherein the company was in a practice of never manufacturing the products. Instead, the company used to outsource the manufacturing process directly to the suppliers. These suppliers were known as the contractors. Nike looked for a new system that was more suitable to the requirements that were at hand at that time. The company expected the new system to provide a streamline with its buyers and suppliers of raw materials and other requirements to lower the operating costs of the company. The new system was expected to be flexible enough to work in collaboration with different CRM applications employed to make order taking more effective and accurate to manage inventory effectively. One of the goals of Nike behind this was to reduce order-to-delivery time by nearly half.

It was in the year 1975 when the Nike introduced the Future Program. Under this program, Nike's global operations were broadly divided into five geographic regions. This was aimed toward obtaining better operations and effective systems. During this era, toward the end of the 1990s, the supply chain management seemed to be ineffective and inadequate. There were problems like ineffective forecasting and managing the changing trends. Further, in the year 2000, Nike launched the NSC project. This project aimed at implementing its ERP system software onto a single SAP platform. This project ultimately proved to be a big disaster for Nike's operations.

In the spring of 2001, Nike blamed i2 Technologies for a massive sales and earnings shortfall. Nike posted a profit of only $97 million that quarter—at least $48 million below forecast. Nike said the culprit was i2's demand forecasting and supply chain management systems. The supply chain software was supposed to reduce the amount of raw materials that Nike needed to produce its products. It was also supposed to help make sure Nike built more of the products customers wanted and fewer of the ones they did not. Instead, Nike was left with far too many of the wrong products and not nearly enough pairs of the right ones [33–35].

12.3.2.1 Failure

Nike introduced a new system to create a new supply chain system to effectively run the whole system from order taking to order delivery. The design of the new system was made to support the existing system for a streamlined communication between the company and its buyers and suppliers, lowering the overall operating costs. To make the new system more effective, Nike combined it with CRM applications and considered it suitable in managing inventory and other systems effectively. Nike went into contract with i2 for the installation of the main system, and for the supplements and supportive CRM applications it went to SAP AG and Siebel Systems Inc. The project was finalized in 5 years at an estimated cost of $400 million. The i2 software cost $40 million overall.

12.3.2.2 The reason

The total cost of i2's demand forecasting and supply chain management software was only about $40 million. The other $360 million was spent over 5 years on customer relationship and ERP system software from SAP. This whole project aimed at effective forecasting proved to be a big SCM failure for the company. This meant huge financial losses for Nike. The reasons for i2's poor software implementation were finally designated:

- Third-party integrator, inexperience of i2
- Customization
- Trying to forecast too far out ahead
- Pilot test
- Lack of training
- Inadequate information
- Problems in smooth integration
- Changing market conditions
- Review meetings

12.3.2.3 Reasons behind failure of new system

The first step that led to failure was the decision of Nike not to use the basic templates and required methodology that was developed by i2 specifically for the new system being provided. The software was capable of working separately from managing the demand systems of Nike, but Nike decided to customize it to make it work with its existing demand management system. So, the new system was customized, and changes were made in the basics to accommodate the essentials of the existing system. These changes in the new systems basics resulted in the slowing down of the new system applications, and ultimately, the users faced an increase in the wait time for loading. Another, nontechnical reason for failure was the employee turnover that occurred as a result of implementation of this new system. The major harm was caused by the resignation of the chief IT manager, who was involved in the decision-making to initiate the renovation of the supply chain, and all this happened before the system was installed properly.

Another main reason that led to failure was the avoidance of including a third-party integration in the implementation of the new system. The company also faced a lot of trouble in getting the system integrated with the running processes of the company, and the simultaneous implementation of SAP software was a lot of trouble. The problems got a genuine start when Nike started making use of the system in sending orders to its manufacturers in Asia when the system was immature. Because the system was not properly developed and used before maturity, the demand was improperly estimated. To its detriment, Nike overestimated demands that were in least demand and underestimated demands of those products that were in high demand at the time. These wrong estimates were compiled in company data form, and the system sent that flawed data to its manufacturers around the globe, mainly to Asia, which resulted in some manufacturers receiving double orders for the same product. At that time, officers of i2 were not of the view to start use of the software because of its immaturity. A few months later, it was found that Nike was manufacturing an exceptionally greater number of specific models and an extremely low number of others. Due to this, Nike put itself far away from a position to even meet demand of its retailers for some models that were in high demand.

But to their good luck the staff members of Nike and i2 tracked down the problematic element, and in a very short time, they had developed effective ways to get around those problems. Ways that were developed involved the changing of existing operational procedures or alternatively to take the huge step of writing the new software that would better meet the requirements at hand. Whatever the circumstances had been, however, with the passage of time, the problems were completely identified and rectified. But to Nike's bad luck, much time had passed and because of being late in eliminating the flaws, intensively serious problems of inventory handling were at hand.

The reasons for the failure were different as identified by different analysts; as one of the reason identified that both Nike and i2 made the same mistake of avoidance and not taking the precautionary measures that were critically appropriate with respect to the project. The new system was never put to a test before its implementation, so the glitches that would have been revealed much earlier weren't known. Another group of analysts stated that failure at Nike was mainly because it went into a new system implementation contract with i2, a company that had previously no experience in developing software systems for the apparel industry [33–35].

12.3.2.4 The result

This resulted in a huge profit loss for the company. It effected the Nike's reputation adversely. Massive sales and earnings shortfalls were observed. Thus Nike understood the importance of SCM and forecasting. Forecasting for Nike was extremely important, especially for determining quantities. An important feature that the company came to understand regarding its SCM was that it needed to pay attention to the "make-to-order" rather than the "make-to-stock" policy. Nike also understood that in their type of business, forecasting plays an important role in reducing the supply time. For all this to be obtained, it was important for the company to have the best forecasting. Nike further improved its supply chain management by installing a well-equipped and competent system. Its continuous interaction with the consumers on the web portal proved to be the best feature. Herein the consumers achieved a supreme level of customization of products. Nike also improved itself by proper analysis of the data and implementation. Feedback, as in any other industry, plays an important role in Nike today. Nike obtains continuous feedback from its suppliers and consumers.

Implementation is a complex procedure, and it must be tackled in the best way. Evaluating pros and cons of any project beforehand is also equally important. A pilot project can turn around the fate of any business if handled correctly.

12.4 Conclusion

ERP systems are playing a significant role in the field of fashion and the apparel industry. They give companies a clear advantage over others in the global competitive market. There are many benefits and advantages involving the use of ERP systems. This review highlighted the modules that are designed specifically for the apparel sector and suggest that ERP systems are going deeper in this domain. The SWOT analysis, implementation, and the case studies demonstrated both the positive and the negative sides of ERP systems. There are issues related to the implementation in some of the organizations that are to be sorted out by taking corrective measures. Finally, we believe that the integration of ERP systems in the fashion and apparel sectors will enable them to lead in the future.

References

[1] Sumner M. Enterprise resource planning. Prentice Hall; 2005.

[2] Leon A. Enterprise resource planning. Tata McGraw Hill; 1999.

[3] Sharma A. Risks in ERP implementation. CISA, CIA 17th Meeting of the INTOSAI Working Group on IT Audit, Tokyo, Japan, May 21–22, 2008.

[4] Davenport T. Mission critical – realizing the promise of enterprise systems. Boston (MA): Harvard Business School Publishing 2000.

[5] Soh C, Kien SS, Tay-Yap J. Cultural fits and misfits: is ERP a universal solution? Commun ACM 2000;43(4):47–51.

[6] Diwan P, Sharma S. ERP. PENTAGON Press; 2000.

[7] Umble EJ, Haft RR, Umble MM. Enterprise resource planning: implementation procedures and critical success factors. Eur J Oper Res 2003;146(2):241–57.

[8] Wang Y, Li H, Warfield J, Xu L. Knowledge management in the ERP era. Syst Res Behav Sci 2006;23(2):125–8.

[9] Laudon KC, Laudon JP. Management information systems. 12th ed. Prentice Hall; 2012.

[10] Sadagopan. ERP – a managerial perspective. Tata McGraw Hill; 1999.

[11] Hines T. From analogue to digital supply chains: implications for fashion marketing. In: Hines T, Bruce M, editors. Fashion marketing: contemporary issues. Oxford: Butterworth–Heinemann; 2001. p. 1–47.

[12] Yusuf Y, Gunasekaran A, Abthorpe MS. Enterprise information systems project implementation: a case study of ERP in Rolls-Royce. Int J Prod Econ 2004;87:251–66.
[13] Yu CS. Causes influencing the effectiveness of the post-implementation ERP system. Ind Manag Data Syst 2005;105(1):115–32.
[14] Themistocleous M, Irani Z, Love PED. Evaluating the integration of supply chain information systems: a case study. Eur J Oper Res 2004;159:393–405.
[15] Rothenberger MA, Srite M. An investigation of customization in ERP system implementations. IEEE Trans Eng Manag 2009;56(4):663–76.
[16] iSync Solution, http://www.isyncsolutions.com.
[17] Compiere ERP, http://www.compiere.com.
[18] YSOFT, https://www.ysoft.com.
[19] Visual Gems, http://www.visualgems.com.
[20] e-SmartX, http://www.indusmedia.in.
[21] Atom Fashion Solutions, http://www.lampsoftware.in.
[22] Hui PCL, Tse K, Choi TM, Liu N. Enterprise resource planning systems for the textiles and clothing industry. In: Cheng TCE, Choi TM, editors. Innovative quick response programs in logistics and supply chain management. Springer; 2010.
[23] Flyvbjerg B, Budzier A. Why your IT project may be riskier than you think. Harv Bus Rev September 2011;89(9):601–3.
[24] Miguel C, Thomaz W. How consultants can help organisations survive ERP frenzy. 1999. Paper submitted to Managerial Consultation Division (August 1999).
[25] Choi TM, Chow PS, Liu SC. Implementation of fashion ERP systems in China: case study of a fashion brand, review and future challenges. Int J Prod Econ 2013;146:70–81.
[26] ERP Report. Panorama consulting. 2010.
[27] Pantaloons. Pantaloons.com. 2015.
[28] Shah K. Pantaloon: ERP in retail. March 2007. Retrieved from Network Magazine India: http://www.networkmagazineindia.com/200703/casestudy02.shtm.
[29] Trivedi G. Pantaloons makes it a strategic function. 2010. Retrieved from CIO: http://www.cio.in/case-study/pantaloons-makes-itstrategic-function.
[30] Baburajan. Pantaloon retail: The problem is not with technology but gap in the perception and possibility. March 2007. Retrieved from: http://www.voicendata.com/voice data/news/165345/pantaloonretail-the-technology-gap-perception-possibility.
[31] SAP. SAP ACE Awards 2011. 2011. Retrieved from SAP: www.sap.com/india/campaigns/ace_awards/winners/index.epx.
[32] ERPLY Solutions. Erply.com. 2015.
[33] Farmer MA. I2-Nike fallout a cautionary tale. CNET News; 2001.
[34] BARRET L. Long strange trip: nike finally regains footing. Baseline Magazine; January 2003.
[35] Nike Profile, 2009. http://www.nike.com, World Supply chain disasters, http://www.scdigest.com.

Intelligent demand forecasting supported risk management systems for fast fashion inventory management

13

T.-M. Choi, S. Ren
The Hong Kong Polytechnic University, Kowloon, Hong Kong

13.1 Introduction and background

In the fashion apparel industry, fast fashion is a pertinent and well-developed business operations model (Choi, 2014). Under the fast fashion model, companies offer very trendy products with simple designs. The product life cycle of these fast fashion products is usually very short and the respective demand is very volatile (Choi et al., 2014). Renowned fast fashion brands such as H&M, Mango, and Zara are all well-established international leaders in fashion retailing.

In the literature, fast fashion operations have been rather widely examined over recent years. To be specific, by modeling the core fast fashion business elements, Caro and Gallien (2007) employ the dynamic programming approach to identify the optimal retail assortment plan for fast fashion companies. Later, Caro and Gallien (2010) construct a comprehensive fast fashion retail network model and analytically examine the optimal inventory planning problem with multiple items. They investigate the problem with reference to Zara's industrial practices. They also propose the optimal inventory allocation rule in the respective fast fashion retail network. Cachon and Swinney (2011) pioneer an important study on fast fashion supply chain systems with the considerations of forward-looking strategic consumers. The authors incorporate the core fast fashion operation elements into the analytical model. They reveal how offering short lead times with an improved product design can help fast fashion companies gain a competitive edge in the presence of strategic consumers. Li et al. (2014) investigate the channel coordination issue for a fast fashion supply chain. The authors analytically construct the newsvendor problem-based inventory model and conduct a mean-variance analysis. They focus on exploring how the buyback policy can be used to achieve supply chain coordination with multiple risk-averse retailers. Counterintuitively, they reveal analytically that a simple returns policy may be used to successfully coordinate the fast fashion supply chain system in the presence of multiple retailers under some conditions. Most recently, Hill and Lee (2015) empirically explore the sustainable brand extension strategies of fast

Information Systems for the Fashion and Apparel Industry. http://dx.doi.org/10.1016/B978-0-08-100571-2.00013-0

fashion retailers. Other related studies on fast fashion business models and operations can be found in Choi (2014) and Caro and Martınez-de-Albeniz (2015).

To support fast fashion business models, two functional areas are critical and fundamental. These are: (1) demand forecasting and (2) inventory management. As a matter of fact, owing to the features of fast fashion products, demand forecasting is indeed very challenging. This point is easily reflected by the fact that there are basically no, or very few, historical data available to assist demand prediction for fast fashion products (because the product life cycles are very short and most products are new products). As a consequence, even the advanced artificial intelligence (AI) methods, the simple statistical models, or the more sophisticated hybrid models may not function well in forecasting demand for fast fashion products (see Liu et al. (2013) for a review on the fashion sales forecasting models commonly explored in the literature; see Thomassey (2014) for a review on some more issues around sales forecasting in the apparel industry). Notice that demand forecasting is highly related to inventory management as fast fashion companies will plan their product assortment and optimal inventory policy by making reference to the product demand prediction. The accuracy of demand forecasting would hence directly affect the performance of inventory management. With the understanding on the fact that there is no perfect demand forecasting, and fast fashion demand forecasting tends to be inaccurate, the level of risk associated with demand forecasting driven inventory planning for fast fashion companies is naturally high. As a result, there is a genuine need to develop proper risk management systems for fast fashion inventory management.

Owing to the challenges brought by demand forecasting for fast fashion operations, this chapter examines how we can develop intelligent demand forecasting supported risk management systems to enhance fast fashion inventory management. Related future research directions are also discussed. We believe that the findings are useful for both practitioners and academicians.

The rest of this paper is organized as follows. Section 13.2 explores the demand forecasting supported inventory control problems. Section 13.3 reviews the inventory control models with risk considerations. Section 13.4 discusses the intelligent fast fashion demand forecasting supported risk minimization inventory control model. Section 13.5 concludes the chapter with a discussion on future research.

13.2 Demand forecasting supported inventory control

As we discussed in Section 13.1, demand forecasting for fast fashion companies is a challenging task. In fact, from the real-world operations perspective, fast fashion companies have to conduct demand forecasting for their products within a very short period of time or nearly in "real time," for a lot of stock-keeping units (SKUs), and in the presence of very limited historical data. To have an accurate forecast almost becomes a "mission impossible"! Thus, it is necessary to have efficient computerized information systems to provide support. Nowadays, in many fashion retailers, radio frequency identification (RFID) technology (Gaukler, 2011) has been implemented (eg, Marks and Spencer, and Zara) for product checking, consumer behavior tracing, and inventory planning. If properly implemented in fast fashion, the RFID technology can help fast fashion companies to quickly get the demand information at retail sales floors and update it into the point-of-sales (POS) system. This is a quick way of having the inputs on consumer demand. In addition, if the fast fashion companies are also equipped with the ERP systems (Choi et al., 2013), the information received from the RFID system and stored in the POS system would be synchronized quickly with the enterprise database in the central server. This directly means that the fast fashion companies will be able to get the latest information almost in real time. With the popular intelligent demand planning modules available in most commercial ERP systems, the fast fashion companies will then be able to come up with the demand forecast in a timely manner. Notice that recent research has revealed that the panel data-based forecasting models can be very versatile and useful for conducting fashion demand forecasting (Ren et al., 2015), compared to the sophisticated AI models and the simple one-dimensional time-series statistical models. Thus, fast fashion companies may consider incorporating the panel data-based demand forecasting models into their ERP system's intelligent demand planning module.

After conducting demand forecasting by using the intelligent demand planning module, fast fashion companies will employ the forecasted demand time series to plan its inventory. In the literature, there are only very limited studies explicitly connecting demand forecasting and the respective inventory planning and control. We review some important works as follows. First, Gardner (1990) examines the impacts brought about by forecasting on inventory planning in a distribution network. He finds that the specific choice of forecasting is influential and affects the corresponding inventory investment significantly. Spedding and Chan (2000) investigate the selection of forecasting models for inventory control in a batch manufacturing setting. They argue that the traditional forecasting models are not capable of fulfilling the requirements of forecasting for short-life products. They thus propose a Bayesian dynamic linear model for conducting time series demand forecasting. Gardner and Diaz-Saiz (2002) report a case study on forecasting and inventory planning of a distributor of "product parts." They find that the right classification of the seasonal time series and the use of proper decomposition procedure can help enhance forecasting results and inventory planning significantly. Aviv (2003) explores the supply chain inventory management problem with the ARIMA-based time series demand forecasting model. He also examines the benefits of various types of information-sharing agreements. Yelland (2009) conducts an empirical evaluation of a Bayesian-based demand forecasting model for inventory control of items with low-count time series. More recently, Chung et al. (2012) investigate the new-release movies demand forecasting at Blockbuster. They use real data and reveal the financial impact brought about by the forecasting model on Blockbuster's retail operations. The above papers have explored the forecasting models and inventory control. They can hence provide the foundation and literature support for the development of the intelligent risk management systems with demand forecasting driven inventory control.

13.3 Inventory models with risk considerations

Risk is a term commonly used in the real world and it is an important issue for companies. In fact, in operations, risk is associated with the control problems when there are uncertain outcomes and some outcomes are unfavorable. In fast fashion inventory control, risk management is definitely one critical issue on senior management's agenda.

Traditionally, most prior studies in the stochastic inventory management literature assume the companies (and their decision makers) are risk neutral and focus solely on exploring the optimal inventory policy with a goal of either minimizing the expected inventory cost or maximizing the expected operational profit. However, it is known that the operational objectives of most companies, including fast fashion companies, are not solely on maximizing the expected profit. Instead, they are more concerned with the issues such as target profitability (Shi et al., 2011), value at risk (Tapiero, 2005), and return on investment on each assortment buying plan (Choi, 2007). As a result, it is important to consider these risk-related objectives in planning the optimal inventory decision. In the literature, some studies have examined the optimal inventory control for the single-item newsvendor problem by using the risk-averse utility function approach (Anvari, 1987; Chung, 1990; Eeckhoudt et al., 1995). However, since the utility function approach suffers a major drawback of the tremendous difficulty of assessing its analytical closed-form expression, it is not being applied in real-world decision making. As a result, some studies employ the implementable and intuitive mean-variance (MV) models for studying inventory control with risk minimization. Established by Nobel laureate Harry Markowitz in the 1950s (Markowitz, 1959), the MV formulation has been widely applied for exploring financial portfolio investment problems with risk consideration. Under Markowitz' MV framework, an optimal decision is one which optimizes the investment problem with the consideration of both payoff and risk in which payoff is measured by the expected return (ie, "mean") and risk is quantified by the variance of return (ie, "variance"). In the inventory control literature, Lau (1980) is the first piece of research which specifically explores the newsvendor inventory problem with a mean-standard-deviation optimization objective function. After that, a number of studies are devoted to applying the MV framework to analyze supply chain inventory management-related problems. These include Lau and Lau (1999), Agrawal and Seshadri (2000), Choi et al. (2008), Choi (2011), and Li et al. (2014). Note that all of the above reviewed risk-related inventory models only examine the situation when there is one product item under inventory control and demand is assumed to be stationary and demand distribution is given.

For multi-item inventory control, the constrained multi-item newsvendor problem, termed the newsstand problem, is first examined in Lau and Lau (1996) and then further explored in Lau and Lau (1997). Lau and Lau argue in their above works that demand in real-world situations would seldom fit exactly the commonly assumed distributions such as the normal distribution. They hence propose a novel three-step demand distribution modeling procedure for estimating the statistical demand distribution for modeling the multi-item "newsstand problem." Later on, based on the discrete choice analysis model (Akiva and Lerman, 1985) and stochastic optimization framework, Vaagen and Wallace (2008) investigate the apparel assortment planning problem when there are two states of the world on the popularity of the apparel products. They note that fashion products differ by style and color, with correlated demands among items. Thus, they propose a stochastic optimization model to help develop the optimal product portfolio with hedging between expected profit and downside risk under the mean-variance (MV) framework (Choi and Chiu, 2012). From their analysis, they prove how mis-specifying the statistical demand distributions can lead to inferior results with an improper hedging.

13.4 An intelligent fast fashion demand forecasting supported risk minimization inventory control model

After reviewing the literature on demand forecasting-based inventory control and risk-related inventory management, we propose a new optimization model for developing the intelligent fast fashion forecasting-based risk minimization inventory control system. Here, "intelligent" refers to the fact that the system is smart and it can use the available information and automatically provide the optimal inventory assortment plan with risk minimization.

To be specific, from the intelligent demand planning module of the fast fashion company's ERP system, we will obtain the predicted future demand time series, and we describe it as follows:

$d_{t,i}$ = demand for SKU (ie, item with specific color, size, style, etc.) i, at time t based on the demand forecasting result.

Following the MV framework and the related studies, we consider a multi-item inventory planning problem with hedging among items. We aim at minimizing the semivariance of the product portfolio payoff with a bound on the return on investment. Since purchasing budget is also a very important factor in inventory planning for fast fashion, a constraint is added on it. Following the approach adopted by Vaagen and Wallace (2008), we consider (1) there are two states of the world which stand for "good market condition" and "bad market condition" respectively, (2) the maximum level of variety I (number of style–color combinations, which is time-independent [as it is predetermined for the whole season]) is strategically defined, and flexible to adapt both states of the world S. The following optimization model can be constructed to yield the optimal portfolio for time period t.

$$\min \quad SV_t(\pi_t) = \sum_{\text{Scenario } S} p_t^s \left(\Delta\pi_{t,s}^-\right)^2$$

subject to:

$$R_t(\pi_t) \geq \alpha_t$$

$$\sum c_{t,i}x_{t,i} \leq C_t$$

$$\Delta\pi^{-}_{t,s} \geq \overline{\pi}_t - \pi^{s}_t \quad \forall s \in S$$

$$y^{s}_{t,i} \leq d^{s}_{t,i} \quad \forall i \in I, \ s \in S$$

$$y^{s}_{t,i} \leq x_{t,i} \quad \forall i \in I, \ s \in S$$

$$x_{t,i} \geq 0 \quad \forall i \in I$$

$$y^{s}_{t,i} \geq 0 \quad \forall i \in I, \ s \in S$$

$$\Delta\pi^{-}_{t,s} \geq 0 \quad \forall s \in S$$

where:

$d^{s}_{t,i}$ = demand for SKU i, under scenario s at time t, based on the forecasting result.
$x_{t,i}$ = stocking quantity of SKU i at time t.
$y^{s}_{t,i}$ = sale for SKU i scenario s at time t.
π^{s}_t = profit on scenario s at time $t = \sum_{\text{SKU } i \in I} \left\{ -c_{t,i}x_{t,i} + \left[v_{t,i}y^{s}_{t,i} + g_{t,i}\left(x_{t,i} - y^{s}_{t,i}\right)\right]\right\} \quad \forall s \in S.$
$\Delta\pi^{-}_{t,s}$ = negative deviation from expected profit for scenario s at time t.
p^{s}_t = probability of scenario s at time t.
$v_{t,i}$ = unit selling price of SKU i at time t.
$c_{t,i}$ = unit purchasing cost SKU i at time t.
$a_{t,i}$ = unit required space for SKU i at time t.
$g_{t,i}$ = unit salvage value of SKU i at time t.
π_t = portfolio payoff at time t.
$\overline{\pi}_t$ = expected profit over all scenarios $= \sum_{\text{Scen } s} p^{s}\pi^{s}$.
$R(\overline{\pi}_t)$ = return on investment (ROI) $= \overline{\pi}_t / \Sigma c_{t,i}x_{t,i}$.
α_t = lower bound on the ROI of the whole assortment portfolio.
C_t = upper bound on the purchasing budget.

Note that the above proposed stochastic optimization problem is just one example for the intelligent fast fashion demand forecasting-based risk minimization inventory control system. Companies can adjust the constraints and the objective of the problem to fit their own operational goals and situations, which also means creating a customized stochastic optimization problem.

13.5 Concluding remarks and future research

In this chapter, we have discussed and examined how fast fashion companies can develop intelligent demand forecasting supported risk management systems for their inventory management. We have reviewed some representative literature on areas including fast fashion operations, demand forecasting supported inventory control, and inventory models with risk considerations. The related computerized information systems are also proposed.

For future research, on one hand, in terms of intelligent forecasting methods for inventory control, one most promising direction is to examine further how different intelligent forecasting methods can be further combined to develop new and more robust methods (Choi et al., 2011; Liu et al., 2013). This extension potentially can yield even better forecasting performance for fast fashion companies and hence reduce the level of risk associated with the respective inventory control systems. As pointed out by Yesil et al. (2012), it is critically important to consider the specific needs of each individual fast fashion company before we can develop the right demand forecasting model to achieve the optimal forecasting result. As a consequence, there is a fruitful area for future research on whether there is a generalizable forecasting framework for conducting forecasting across multiple companies in fast fashion with the goal of reducing risk in inventory management. On the other hand, it is important to examine the real-world implementation and conduct the cost-and-benefit analysis of the specific forecasting models for inventory control systems with risk considerations. The proposed optimization model in Section 13.4 provides an example for the respective studies. This direction of exploration is crucial because even though higher forecasting accuracy is always preferable to a lower one, how much "more" business value the fast fashion companies would gain with the consideration of risk by having this "higher forecasting accuracy" is largely unknown. Thus, it is usually hard for fast fashion companies to decide if it is wise to invest on the implementation of a better demand forecasting system. In addition, how to incorporate a better demand forecasting model into the existing business intelligent operations management system in the fast fashion companies for making scientifically sound decisions on production, inventory, and distribution with risk analysis is another critical issue which deserves future research. Lastly, in a multiagent supply chain environment with high demand uncertainty and information asymmetry, what the agents can do to coordinate the supply chain in the presence of the intelligent demand forecasting tool and the related risk minimization inventory control model is a very challenging area for further investigations.

Acknowledgment

This paper is partially supported by the internal research funding provided by The Hong Kong Polytechnic University (account number: PolyU 152020/15E).

References

Agrawal, B., Seshadri, S., 2000. Risk intermediation in supply chains. IIE Transactions 32, 819–831.

Akiva, B., Lerman, S.R., 1985. Discrete Choice Analysis: Theory and Application to Predict Travel Demand. The MIT Press, Cambridge, MA.

Anvari, M., 1987. Optimality criteria and risk in inventory models: the case of the newsboy problem. Journal of the Operational Research Society 38, 625–632.

Aviv, Y., 2003. A time-series framework for supply-chain inventory management. Operations Research 51 (2), 210–227.

Cachon, G., Swinney, R., 2011. The value of fast fashion: quick response, enhanced design, and strategic consumer behavior. Management Science 57 (4), 778–795.

Caro, F., Gallien, J., 2007. Dynamic assortment with demand learning for seasonal consumer goods. Management Science 53 (2), 276–292.

Caro, F., Gallien, J., 2010. Inventory management of a fast-fashion retail network. Operations Research 58 (2), 257–273.

Caro, F., Martınez-de-Albeniz, V., 2015. Fast fashion: business model overview and research opportunities. In: Agrawal, Smith (Eds.), Retail Supply Chain Management, pp. 237–264.

Choi, T.M., 2007. Pre-season stocking and pricing decisions for fashion retailers with multiple information updating. International Journal of Production Economics 106, 146–170.

Choi, T.M., 2011. Coordination and risk analysis of VMI supply chains with RFID technology. IEEE Transactions on Industrial Informatics 7 (3), 497–504.

Choi, T.M. (Ed.), 2014. Fast Fashion Systems: Theories and Applications. CRC Press.

Choi, T.M., Chiu, C.H., 2012. Risk Analysis in Stochastic Supply Chains: A Mean-Risk Approach. In: International Series in Operations Research and Management Science. Springer.

Choi, T.M., Chow, P.S., Liu, S.C., 2013. Implementation of fashion ERP systems in China: case study of a fashion brand, review and future challenges. International Journal of Production Economics 146, 70–81.

Choi, T.M., Li, D., Yan, H., Chiu, C.H., 2008. Channel coordination in fashion supply chains with agents having mean-variance objectives. Omega 36, 565–576.

Choi, T.M., Hui, C.L., Liu, N., Ng, S.F., Yu, Y., 2014. Fast fashion sales forecasting with limited data and time. Decision Support Systems 59, 84–92.

Choi, T.M., Yu, Y., Au, K.F., 2011. A hybrid SARIMA wavelet transform method for sales forecasting. Decision Support Systems 51 (1), 130–140.

Chung, C., Niu, S.C., Sriskandarajah, C., 2012. A sales forecast model for short-life-cycle products: new releases at Blockbuster. Production and Operations Management 21 (5), 851–873.

Chung, K., 1990. Risk in inventory models: the case of the newsboy problem - optimality conditions. Journal of the Operational Research Society 41, 173–176.

Eeckhoudt, L., Gollier, C., Schlesinger, H., 1995. The risk-averse (and prudent) newsboy. Management Science 41, 786–794.

Gardner, E.S., 1990. Evaluating forecast performance in an inventory control system. Management Science 36 (4), 490–499.

Gardner, E.S., Diaz-Saiz, J., 2002. Seasonal adjustment of inventory demand series: a case study. International Journal of Forecasting 18, 117–123.

Gaukler, G.M., 2011. Item-level RFID in a retail supply chain with stock-out-based substitution. IEEE Transactions on Industrial Informatics 7 (2), 362–370.

Hill, J., Lee, H.H., 2015. Sustainable brand extensions of fast fashion retailers. Journal of Fashion Marketing and Management 19 (2), 205–222.

Lau, H.S., 1980. The newsboy problem under alternative optimization objectives. Journal of the Operational Research Society 31, 525–535.

Lau, H.S., Lau, A.H.L., 1996. The newsstand problem: a capacitated multiple-product single-period inventory problem. European Journal of Operational Research 94, 29–42.

Lau, H.S., Lau, A.H.L., 1997. Some results on implementing a multi-item multi-constraint single-period inventory model. International Journal of Production Economics 48, 121–128.
Lau, H.S., Lau, A.H.L., 1999. Manufacturer's pricing strategy and returns policy for a single-period commodity. European Journal of Operational Research 116, 291–304.
Li, J., Choi, T.M., Cheng, T.C.E., 2014. Mean-variance analysis of fast fashion supply chains with returns policy. IEEE Transactions on Systems, Man, and Cybernetics: Systems 44 (4), 422–434.
Liu, N., Ren, S., Choi, T.M., Hui, C.L., Ng, S.F., 2013. Sales forecasting for fashion retailing service industry: a review. Mathematical Problems in Engineering 2013 (9 pages).
Markowitz, H.M., 1959. Portfolio Selection: Efficient Diversification of Investment. John Wiley & Sons, New York.
Ren, S., Choi, T.M., Liu, N., 2015. Fashion sales forecasting with a panel data-based particle-filter model. IEEE Transactions on Systems, Man, and Cybernetics – Systems 45 (3), 411–421.
Shi, C.V., Yang, S., Xia, Y., Zhao, X., 2011. Inventory competition for newsvendors under the objective of profit satisficing. European Journal of Operational Research 215, 367–373.
Spedding, T.A., Chan, K.K., 2000. Forecasting demand and inventory management using Bayesian time series. Integrated Manufacturing Systems 11 (5), 331–339.
Tapiero, C.S., 2005. Value at risk and inventory control. European Journal of Operational Research 163, 769–775.
Thomassey, S., 2014. Sales forecasting in apparel and fashion industry: a review. In: Choi, et al. (Eds.), Intelligent Fashion Forecasting Systems: Models and Applications, pp. 9–27.
Vaagen, H., Wallace, S.W., 2008. Product variety arising from hedging in the fashion supply chains. International Journal of Production Economics 114, 431–455.
Yelland, P.M., 2009. Bayesian forecasting for low-count time series using state-space models: an empirical evaluation for inventory management. International Journal of Production Economics 118, 95–103.
Yesil, E., Kaya, M., Siradag, S., 2012. Fuzzy forecast combiner design for fast fashion demand forecasting. In: International Symposium on Innovations in Intelligent Systems and Applications, pp. 1–5.

Index

'*Note:* Page numbers followed by "f" indicate figures, "t" indicate tables.'

Printed in the United States
By Bookmasters